Advances in Tissue Banking - Volume 6

ADVANCES IN TISSUE BANKING

Advances in Tissue Banking - Volume 6

ADVANCES IN TISSUE BANKING

editor

Glyn O Phillips

Research Transfer Ltd, UK

World Scientific

EW JERSEY • LONDON • SINGAPORE • BEIJING • SHANGHAI • HONG KONG • TAIPEI • CHENNAI • TOKYO

Published by

World Scientific Publishing Co. Pte. Ltd.
5 Toh Tuck Link, Singapore 596224
USA office: 27 Warren Street, Suite 401-402, Hackensack, NJ 07601
UK office: 57 Shelton Street, Covent Garden, London WC2H 9HE

British Library Cataloguing-in-Publication Data
A catalogue record for this book is available from the British Library.

Advances in Tissue Banking — Vol. 6
ADVANCES IN TISSUE BANKING

ISBN 978-981-238-088-3 (hardcover)
ISBN 978-981-277-697-6 (ebook for institutions)
ISBN 978-981-4487-99-3 (ebook for individuals)

For any available supplementary material, please visit
https://www.worldscientific.com/worldscibooks/10.1142/5039#t=suppl

ADVANCES IN TISSUE BANKING

International Advisory Board

LIST OF CONTRIBUTORS

ELBA AGUSTÍ
Department of Cardiovascular Surgery and
Cardiovascular Tissue Bank, Hospital Clinic
University of Barcelona, Villarroel 170, 08036 Barcelona, Spain

JOSE A. ANDRADES
Department of Cell Biology and Genetics, Faculty of Sciences
University of Málaga, Campus Universitario de Teatinos
29071-Málaga, Spain

MARIAN ALTARRIBA
MIR of Ophtalmology, Hospital Virgen del Camino
Irrunlarrea, 4, 31008 Pamplona
Navarra, Spain

S.Y. BATRAKOV
Central Institute of Traumatology and Orthopaedics (CITO)
Priorova ul. 10, Moscow, Russia

J. BECERRA
Department of Cell Biology and Genetics, Faculty of Sciences
University of Málaga, Campus Universitario de Teatinos
29071-Málaga, Spain

J. BELLÓN
Departament of Morphological Sciences and Surgery
Faculty of Medicine, University of Alcalá
Ctra N-II Km 33,6, 28871-Alcalá de Henares
Madrid, Spain

JOLANTA BOCHENSKA
BMT Unit, K. Dluski Hospital/Institute of Immunology and Experimental Therapy
R. Weigl, 53-114 Wroclaw, Poland

C. ZAVANELLA BOTTA
Department of Cardiac Surgery
Complejo Hospitalario Juan Canalejo
15006 La Coruña, Spain

JULIA BUJÁN
Department of Morphological Sciences and Surgery
Faculty of Medicine, University of Alcalá
Ctra N-II Km 33,6, 28871-Alcalá de Henares
Madrid, Spain

M. RODRÍGUEZ CABARCOS
Cryobiology Unit
Complejo Hospitalario Juan Canalejo
15006 La Coruña, Spain

F. CARBONELL-UBEROS
Tissue Bank and Immunohaematolgy Service
Transfusion Centre of Valencia, Avenida del Cid
65-A, 46014 Valencia, Spain

RAMÓN CARTAÑÁ
Department of Cardiovascular Surgery
and Cardiovascular Tissue Bank, Hospital Clinic
University of Barcelona, Villarroel 170
08036 Barcelona, Spain

MIGUEL A. CASTRO
Department of Cardiovascular Surgery
and Cardiovascular Tissue Bank, Hospital Clinic
University of Barcelona, Villarroel 170
08036 Barcelona, Spain

M. GONZÁLEZ CUESTA
Pathology Department
Complejo Hospitalario Juan Canalejo
15006 La Coruña, Spain

ANDREI A. CZITROM
University of Texas Southwestern Medical Center
Advanced Surgical Institute
Medical City Dallas Hospital
7777 Forest Lane, Suite C-707
Dallas, Texas, 75230 USA

M.J. DALMAU
Service of Cardiac Surgery
University General Hospital of Valencia
Avda. Tres Cruces, s/n, 46014 Valencia, Spain

S. PÉRTEGA DÍAZ
Statistics Department
Complejo Hospitalario Juan Canalejo
15006 La Coruña, Spain

DOROTA DLUBEK
BMT Unit, K. Dluski Hospital/Institute of Immunology
and Experimental Therapy, R. Weigl
53-114 Wroclaw, Poland

LUC DUINSLAEGER
Military Hospital Queen Astrid,
Bruynstreet, 2B-1120, Neder-over-Heembek
Brussels, Belgium

JAROSLAW DYBKO
BMT Unit, K. Dluski Hospital/Institute of Immunology
and Experimental Therapy, R. Weigl
53-114 Wroclaw, Poland

CARMEN DE FELIPE
National Transplant Organisation (ONT)
c/o Sinesio Delgado 8
E-28029 Madrid, Spain

MARIA J. FELIX
Catalan Transplant Organisation (OCATT)
Mejía Lequerica, 1
Pavelló Hèlios 3, la planta
08028 Barcelona, Spain

P. FILGUEIRA FERNÁNDEZ
Pathology Department
Complejo Hospitalario Juan Canalejo
15006 La Coruña, Spain

C. RODRÍGUEZ FERNÁNDEZ
Cryobiology Unit
Complejo Hospitalario Juan Canalejo
15006 La Coruña, Spain

FÁTIMA RODRÍGUEZ FERRO
MIR of Ophthalmology, Hospital Virgen del Camino
Irrunlarrea, 4, 31008 Pamplona
Navarra, Spain

P. FILGUEIRA
Pathology Department
Complejo Hospitalario Juan Canalejo
15006 La Coruña, Spain

IVÁN GARCÍA
Department of Cardiovascular Surgery
and Cardiovascular Tissue Bank, Hospital Clinic
University of Barcelona, Villarroel 170
08036 Barcelona, Spain

N. DOMENECH GARCÍA
Unit of Experimental Surgery
Complejo Hospitalario Juan Canalejo
15006 La Coruña, Spain

TIIT HAVIKO
Clinic of Traumatology and Orthopaedics
Tartu University Clinics, L. Puusepa 8
51014 Tartu, Estonia

F. HORNERO
Service of Cardiac Surgery
University General Hospital of Valencia
Avda. Tres Cruces, s/n, 46014 Valencia, Spain

XUN HU
Tissue Transformation Technologies
175 May Street, Suit 600
Edison, New Jersey 08837, USA

J. SÁNCHEZ IBÁÑEZ
Transplant Co-ordination Office
Complejo Hospitalario Juan Canalejo, As Xubias, 84
15006 La Coruña, Spain

SAID I. ISMAILOV
Department of Endocrine Surgery
Institute of Endocrinology
56, H. Abdullaev str.
Toshkent 700125, Uzbekistan

MORITOSHI ITOMAN
Department of Orthopaedic Surgery
Kitasato University School of Medicine
1-15-1 Kitasato, Sagamihara
Kanagawa 228-8555, Japan

TOSHIHIRO IZUMI
Department of Orthopaedic Surgery
Kitasato University School of Medicine
1-15-1 Kitasato, Sagamihara
Kanagawa 228-8555, Japan

HITOSHI ITO
Takasaki Radiation Chemistry Research Establishment
1233 Watanuki, Takasaki
Gunma, Japan

MIGUEL JOSA
Department of Cardiovascular Surgery
and Cardiovascular Tissue Bank, Hospital Clinic
University of Barcelona, Villarroel 170
08036 Barcelona, Spain

I.A. KASYMOV
Central Institute of Traumatology and Orthopaedics (CITO)
Priorova ul. 10, Moscow, Russia

A. KOMENDER
Department of Transplantology
Central Clinical Hospital
Medical University, ul. Chalubinskiego 5
02-004 Warsaw, Poland

J. KOMENDER
Department of Transplantology
Central Tissue Bank, Medical University
Centre of Biostructure Research
ul. Chalubinskiego 5
02-400 Warsaw, Poland

PEETER LAASIK
Elva Hospital, Elva, Estonia

ANNA LABA
BMT Unit, K. Dluski Hospital/Institute of Immunology
and Experimental Therapy, R. Weigl
53-114 Wroclaw, Poland

ANDRZEJ LANGE
BMT Unit, K. Dluski Hospital/Institute of Immunology
and Experimental Therapy, R. Weigl
53-114 Wroclaw, Poland

M.V. LEKISHVILI
Central Institute of Traumatology and Orthopaedics (CITO)
Priorova ul. 10, Moscow, Russia

J.L. LEQUERICA
Biomedicine Institute of Valencia
Spanish Council for Scientific Research
Jaime Roig 11, 46010 Valencia, Spain

H. MALCZEWSKA
Department of Histology & Embryology
Central Clinical Hospital
Medical University, ul. Chalubinskiego 5
02-400 Warsaw, Poland

R.O. FERNÁNDEZ MALLO
Cryobiology Unit
Complejo Hospitalario Juan Canalejo
15006 La Coruña, Spain

AMELIA MARAVÍ-AZNAR
DUE-ICU Nurses, San Miguel Clinic
Pamplona, Navarra, Spain

ENRIQUE MARAVÍ-AZNAR
MIR of Otorhinolaryngology
Pamplona, Navarra, Spain

ENRIQUE MARAVÍ-POMA
Transplant Co-ordination and DUE-Nurses
Hospital Virgen del Camino
Irrunlarrea, 4, 31008 Pamplona
Navarra, Spain

WOJCIECH MARCZYNSKI
Institute of Traumatology, Orthopaedics and Neurosurgery
Central Clinical Hospital Military School of Medicine
128 Szaserow str., 00-909 Warsaw, Poland

G. MATHEU
Pathology Department, Complejo Hospitalario Juan Canalejo
15006 La Coruña, Spain

SHINICHI MATSUMOTO
Department of Surgery, Division of Transplantation
University of Washington Medical Center
Seattle, WA 98195, USA

CARLOS-A. MESTRES
Department of Cardiovascular Surgery
and Cardiovascular Tissue Bank, Hospital Clinic
University of Barcelona, Villarroel 170
08036 Barcelona, Spain

H. MINEHARA
Department of Orthopaedic Surgery
Kitasato School of Medicine
Kitasato 1-15-1, Sagamihara
Kanagawa, 228-8555, Japan

VINCENTE MIRABET
Tissue Bank and Immunohaematology Service
Transfusion Centre of Valencia
Avenida del Cid, 65-A
46014 Valencia, Spain

S.P. MIRONOV
Central Institute of Traumatology and Orthopaedics (CITO)
Priorova ul. 10, Moscow, Russia

J.A. MONTERO
Service of Cardiac Surgery
University General Hospital of Valencia
Avda. Tres Cruces, s/n
46014 Valencia, Spain

CARMEN ARENAZ MONTERO
Transplant Co-ordination and DUE-Nurses
Hospital Virgen del Camino, Irrunlarrea, 4
31008 Pamplona, Navarra, Spain

AGUSTINA MARTÍN MONTERO
Transplant Co-ordination and DUE-Nurses
Hospital Virgen del Camino, Irrunlarrea, 4
31008 Pamplona, Navarra, Spain

JAIME MULET
Department of Cardiovascular Surgery
and Cardiovascular Tissue Bank, Hospital Clinic
University of Barcelona, Villarroel 170
08036 Barcelona, Spain

T. NARUSE
Department of Orthopaedic Surgery
Kitasato School of Medicine
Kitasato 1-15-1, Sagamihara
Kanagawa, 228-8555, Japan

L.B. NUGMANOVA
Department of Endocrine Surgery
Institute of Endocrinology
56, H. Abdullaev str.
Toshkent 700125, Uzbekistan

C. ANDIÓN NÚÑEZ
Cryobiology Unit
Complejo Hospitalario Juan Canalejo
15006 La Coruña, Spain

ALINA OSTROWSKA
Tissue Transformation Technologies
175 May Street, Suit 600
Edison, New Jersey 08837, USA

TOMASZ PACUSZKO
BMT Unit, K. Dluski Hospital/Institute of Immunology
and Experimental Therapy, R. Weigl
53-114 Wroclaw, Poland

S. PÉRTEGA
Statistics Department
Complejo Hospitalario Juan Canalejo
15006 La Coruña, Spain

J.F. PERTUSA
Department of Animal Biology
University of Valencia
Burjassot, Spain

S. PIQUER
Biomedicine Institute of Valencia
Spanish Council for Scientific Research
Jaime Roig 11
46010 Valencia, Spain

JOSÉ L. POMAR
Department of Cardiovascular Surgery
and Cardiovascular Tissue Bank, Hospital Clinic
University of Barcelona, Villarroel 170
08036 Barcelona, Spain

M.M. RASHITOV
Department of Endocrine Surgery
Institute of Endocrinology
56, H. Abdullaev str.
Toshkent 700125, Uzbekistan

A.H. REDDI
Center for Tissue Regeneration and Repair
Department of Orthopaedic Surgery
University of California Davis, Medical Center
Research Building 1, Room 2000
4635 Second Ave, Sacramento
CA 95817, USA

E. RENDAL
Cryobiology Unit, Complejo Hospitalario Juan Canalejo
15006 La Coruña, Spain

VICENTE RIAMBAU
Department of Cardiovascular Surgery
and Cardiovascular Tissue Bank, Hospital Clinic
University of Barcelona, Villarroel 170
08036 Barcelona, Spain

JOSEFINA RIPOLL
Transplant Co-ordination and DUE-Nurses
Hospital Virgen del Camino, Irrunlarrea, 4
31008 Pamplona, Navarra, Spain

M. RODRÍGUEZ
Cryobiology Unit, Complejo Hospitalario Juan Canalejo
15006 La Coruña, Spain

T. DÍAZ ROMÁN
Unit of Experimental Surgery
Complejo Hospitalario Juan Canalejo
15006 La Coruña, Spain

A.S. SAMKOV
Central Institute of Traumatology and Orthopaedics (CITO)
Priorova ul. 10, Moscow, Russia

J. SÁNCHEZ
Transplant Co-ordination Office
Complejo Hospitalario Juan Canalejo
As Xubias, 84
15006 La Coruña, Spain

JACINTO SÁNCHEZ-IBÁÑEZ
Human Tissue Transplant Programme
Juan Canalejo Hospital, As Xubias, 84
15006 La Coruña, Spain

E. SANZ
Biomedicine Institute of Valencia
Spanish Council for Scientific Research
Jaime Roig 11, 46010 Valencia, Spain

R. SEGURA
Department of Vascular Surgery
Complejo Hospitalario Juan Canalejo
15006 La Coruña, Spain

D. S´LADOWSKI
Department of Transplantology
Central Tissue Bank
Medical University
ul. Chalubinskiego 5
02-004 Warsow, Poland

R. BRIAN STEVENS
University of Washington Medical Center
Department of Surgery, Division of Transplantation
Seattle, WA 98195, USA

D. MICHAEL STRONG
Puget Sound Blood Center and Program
921 Terry Avenue
Seattle, WA 98104, USA

N. TAKAHIRA
Department of Orthopaedic Surgery
Kitasato School of Medicine
Kitasato 1-15-1, Sagamihara
Kanagawa, 228-8555, Japan

RAFAEL TEIJEIRA
Pathology Service of the Navarra Legal Medicine Institute
of the Government of Navarra
Pamplona, Navarra, Spain

THE VIRGEN DEL CAMINO HOSPITAL TISSUE TRANSPLANT WORKING GROUP
Transplant Co-ordination and DUE-Nurses
Hospital Virgen del Camino, Irrunlarrea, 4
31008 Pamplona, Navarra, Spain

TOSHIYUKI TOMINAGA
Department of Orthopaedic Surgery
Kitasato University School of Medicine
1-15-1 Kitasato, Sagamihara
Kanagawa 228-8555, Japan

ELZBIETA URBANOWSKA
Department of Haematology, Oncology, and Internal Medicine
The Medical University of Warsaw
ul. Banacha 1a, 02-097 Warsaw, Poland

P. VACA
Tissue Bank and Immunohaematolgy Service
Transfusion Centre of Valencia, Avenida del Cid
65-A, 46014 Valencia, Spain

M.G. VASILIEV
Central Institute of Traumatology and Orthopaedics (CITO)
Priorova ul. 10, Moscow, Russia

M[a] ESTHER RENDAL VÁZQUEZ
Cryobiology Unit, Carretera del Pasaje s/n
Hospital Teresa Herrera
15006 La Coruña, Spain

WIESLAW WIKTOR-JEDRZEJCZAK
Department of Haematology, Oncology, and Internal Medicine
The Medical University of Warsaw
ul. Banacha 1a, 02-097 Warsaw, Poland

BARBARA WYSOCZANSKA
BMT Unit, K. Dluski Hospital/Institute of Immunology
and Experimental Therapy, R. Weigl
53-114 Wroclaw, Poland

MITSUO YAMAUCHI
CB# 7455, Dental Research Center
University of North Carolina, Chapel Hill
NC 27599-7455, USA

PREFACE

The pace of tissue banking and transplantation continuous to accelerate and this volume of Advances in Tissue Banking series again reflects the growth of this inter-disciplinary subject. Our objective is to provide a review of these developments within the covers of one volume, without the need to plough through countless specialised journals for an overview of the particular subject. Of necessity, there has been the need to be selective in order to emphasise the most rapidly changing areas.

Regulatory control of tissue graft production and quality cannot any longer be ignored. The Food and Drug Administration (FDA) in the USA is now moving to replace the self-regulation peer review system which has hitherto operated. The controls being introduced will inevitably lead to closure of many of the smaller unregulated tissue banks. Europe is trying to come to grips with cross-national regulation and Japan is entering the field for the first time. Detailed comparisons are given here and the Spanish tissue donation model described outlines the need for detailed planning to include all sectors of the health system.

Musculosleletal tissue graft applications continue to underpin the bread an butter operations of all multi-tissue banks. However, this volume describes the extensive and long experience in this area in Russia, Poland, Estonia and Japan, not previously documented in Western literature. Whereas old and trusted methods continue, albeit with important technical innovation, the scientific introduction of bone morphogenetic proteins (BMPs), support more effective osteoinductive bone healing and combined with tissue engineering in cartilage and other tissues offer a new dimension to bone reconstruction. This section will interest all orthopaedic surgeons because of the mastery of the contributions.

Cell processing and preservation have not previously been included in these volumes, but since this area is being grafted on to the tissue bankers' portfolio, we now describe new developments in this field also. Stem cells and islet cell processing and preservation now offer the armoury of the cell biologist to the transplantologist. This section illustrates that the tissue banker must become a scientist as well as a technologist.

The vascular area has to some extent become a stand-alone situation, and the term "homograft" is still sentimentally retained, as if to emphasise the distinctiveness of this field. The principles of processing and preservation for other viable and/or preserved grafts are now extended to this areas also, as clearly demonstrated by this section.

The 27 contributions by leading specialists in their subject make for rich cocktail. For the diversity and imagination introduced into this volume, I must thank my two Guest Editors: Dr. Jacinto Sánchez-Ibáñez and Professor Anna Dziedzic-Goclowska. May I thank them for the uncomplaining way the have put up with my badgering. I hope that they will now feel that the end justifies the means!

Finally, may I thank those who have reported so positively on the previous volume, for their suggestions and constructive comments. Please continue to contact us with your ideas for contributions, either by yourself, your colleagues or even to appraise us of a subject you consider should be covered. Finally, may I thank my colleagues at World Scientific, particularly Yunisa, for their careful work in guiding this volume through the press.

Glyn O. Phillips
Editor-in-Chief

CONTENTS

SECTION I:

REGULATORY AND DONATION SYSTEMS

Advances in Tissue Banking Vol. 6

1 OBSERVATIONS ON THE EVOLVING REGULATORY PROGRAMME OF TISSUE BANKING BY THE FOOD AND DRUG ADMINISTRATION IN THE USA

ANDREI A. CZITROM
University of Texas Southwestern Medical Center
Advanced Surgical Institute
Medical City Dallas Hospital
Dallas, Texas, USA

1. Introduction

The Food and Drug Administration (FDA), an agency of the Department of Health and Human Services of the US Federal Government, has recently proposed a comprehensive set of regulations aimed at manufacturers of human cells, tissues, and cellular and tissue-based products ("HCT/P's"). The effort of the FDA in this direction began in 1997 when the regulatory plan was announced in two documents entitled "A Proposed Approach to the Regulation of Cellular and Tissue-Based Products" and "Reinventing the Regulation of Human Tissue (Department of Health and Human Services, 1997a,b). As a background to these proposals, it is important to know that the FDA had previously issued interim regulations (21 CFR 1270) effective December 14, 1993 based on Section 361(a) of the Public Health Service (PHS) Act, designed to prevent the spread of communicable diseases from contaminated persons, animals and articles (Department of Health and Human Services, 1993). On July 29, 1997 the FDA issued final regulations (21 CFR 1270) that require facilities engaged in recovery, screening, testing, processing, storage or distribution of human tissue to ensure that medical screening and infectious disease testing have been performed and that records documenting this be available for inspection by the FDA. These regulations contain provisions for the retention, recall or destruction of human tissue for which documentation is not available (Department of Health and Human Services, 1997c).

Since 1997 the FDA has published three proposed rules to implement the approach. The first proposal, "Establishment Registration and Listing for Manufacturers of Human Cellular and Tissue-Based Products" was published on May 14, 1998 (Department of Health and Human Services, 1998). This proposal became a final rule, "Human Cells, Tissues, and Cellular and Tissue-Based Products; Establishment Registration and Listing", published on January 19, 2001 (Docket No. 97N-484R) (Department of Health and Human Services, 2001b). The rule was implemented effective April 4, 2001 and requires that all HCT/P establishments register with the agency and list their products.

The second proposed rule, "Suitability Determination for Donors of Human Cellular and Tissue-Based Products" was published on September 30, 1999 (Docket No. 97-484S) (Department of Health and Human Services, 1999). It represents FDA's plan to take over the wide and comprehensive regulation of donor screening and testing for risk factors and clinical evidence of communicable disease agents and disease in the cell and tissue banking industry.

The third proposed rule, "Current Good Tissue Practice for Manufacturers of Human Cellular and Tissue-Based Products; Inspection and Enforcement" published on January 8, 2000 (Docket No. 97-484P) deals with establishing a programme of "Current Good Tissue Practice" (CGTP) requirements and inspection/enforcement provisions (Department of Health and Human Services, 2001a). All the FDA rules and proposals can be accessed by those interested in details at ***http://www.fda.gov/*** link "Federal Register" where they can be sorted by the given docket numbers.

It is evident that both the professionals engaged in cell and tissue banking and clinicians who use the banked cells or tissues will be affected by FDA regulation. One would like to pose the question as to why this type of government regulation is going forward. One would also want to examine the FDA proposals in light of professional standards that are already in place. Finally, one would like to have a look at the likely impact that such government control would have on providers and users of cells or tissues. As a clinical surgeon and scientist who spent most of a carrier close to the tissue banking and transplantation disciplines, I choose to look at this subject through the lens of a professional association that unites cell and tissue bankers and transplant clinicians, the American Association of Tissue Banks (AATB). Currently, the AATB represents 69 AATB-accredited tissue banks and 1200 individual professionals engaged in the procurement, processing, storage and distribution of human tissues. The AATB has been setting standards for tissue banking in the US on an ongoing basis since its formation in 1976 (Kaprisin and Woll, 2001). It is therefore more in the position than any other group or individual to shape and mold the evolving regulatory programme. The following observations will summarise a number of current issues that are debated while the government regulatory intent is taking shape.

2. Intent of the Proposed FDA Regulation

The proposed FDA regulation has arisen as a result of the agency's intent to provide oversight over a wide range of human cellular and tissue-based products that are intended for implantation, transplantation, infusion, or transfer, now and in the future. Clearly, the agency is acting with the intent to protect public health in the face of rapidly evolving technological innovations. The proposal intends to put in place comprehensive rules to regulate HCT/P's under section 361 of the PHS Act (42 U.S.C.264) rather than under section 351 of the PHS Act applicable to drugs, medical devices and biological products.

The urgency of establishing FDA regulation of human cellular and tissue-based products at this time is justified by the agency's concerns regarding compliance with industry standards by establishments across the country involved in the transplantation of reproductive tissues. Such tissues are donor oocytes or embryos used in assisted reproductive technology (ART) facilities and associated laboratories, and donor sperm distributed by sperm banks. It is unknown how many of the producers of reproductive tissues follow industry standards. The FDA acknowledges that a number of these facilities are in line with AATB or American Society of Reproductive Medicine (ASRM) standards but many others are not. Consequently, this particular industry poses increased risk to the population for disease transmission.

The risk of Human Immunodeficiency Virus (HIV) transmission by reproductive tissues should decrease with FDA oversight as it already has in the case of Therapeutic Donor Insemination (TDI) since HIV testing was recommended by the Center for Disease Control (CDC) in 1985 (Center for Disease Control, 1985). The benefits of FDA oversight are going to be most obvious for the risk of Hepatitis B (HBV) and Hepatitis C (HCV) transmission by reproductive cells. Given yearly prevalence rates of 27.6% for HBV and 9.8% for HCV in oocyte and sperm donors, the risk of transmission is very real. The population at risk each year is estimated to include 1600 to 4700 women undergoing *In Vitro* Fertilisation (IVF) with donor eggs, and 1300 newborns delivered as a result of that procedure, and 34 200 to 70 000 women receiving TDI, and 8800 newborns delivered as a result of that therapy. While these figures are alarming, even more concern is generated by the lack of uniform compliance of reproductive cell facilities with industry standards (Department of Health and Human Services, 1999). While the ASRM has published guidelines for donor screening, it does not provide oversight or accreditation of facilities involved in infertility treatment. Thus, it is clear that the intent of the FDA is to close loopholes and deficiencies rather than interfere with industry standards already in place.

The FDA acknowledges the current industry standards set by the professional associations representing manufacturers of non-reproductive tissue. These establishments include eye banks, tissue banks, marrow and peripheral blood stem cell donor centers, and cord blood banks. Professional associations such as the AATB, Eye Bank Association of America (EBAA), American Association of Blood Banks (AABB) and the Foundation for the Accreditation of Hematopoietic Cell Therapy (FAHCT) have been setting standards and accreditation procedures for producers of non-reproductive tissues for years. Consequently, the FDA estimates that 100% of these entities are already in compliance with industry standards and thus will be impacted minimally by the proposed regulation (Department of Health and Human Services, 1999).

3. Suitability Determination — Proposed Rule

This will be discussed with focus on new concepts introduced by the FDA, which require critical input from tissue bankers and clinicians as the regulatory process evolves.

3.1. Purpose and scope

Under "Purpose and Scope" (Proposal 1271.1), the FDA proposes to distinguish between two groups of human cellular and tissue-based products. The first group will be products regulated solely under the authority of section 361 of the PHS Act ("361 products"). The second will be products regulated under both section 351 of the PHS Act as drugs, medical devices and/or biological products and section 361 of the PHS Act. The FDA proposes that "361 products" be "minimally manipulated", intended for only "homologous use", be not combined or modified by addition of a drug or device and must not have systemic effects. Thus it is clear that the regulation of "361 products" will be less burdensome and more streamlined than that of drugs, medical devices and/or biological products. However, the criteria proposed for defining these products are undergoing evolving re-evaluation.

3.2. Definitions

In "Definitions" (Proposal 1271.3) the FDA holds that "minimal manipulation" for structural tissue is processing that does not alter the original relevant characteristics of the tissue relating to its utility for reconstruction, repair or replacement. For cells or nonstructural tissues, "minimal manipulation" means processing that does not alter the relevant biological characteristics of cells and tissues. The FDA defines "homologous use" for structural tissue-based products as use of the tissue for the same basic function that it fulfills in its native state, in a location where such structural function occurs normally. For cellular and nonstructural tissue-based products "homologous use" is defined as use of the cells or tissues to perform the same functions that they performed in the donor. It comes immediately to mind of a tissue banker or a clinical surgeon that these definitions are problematic. We shall discuss this issue in more detail when analysing the AATB position.

3.3. General requirements

Under "General Requirements" the FDA includes proposals on: determination of donor suitability (1271.50), records of donor suitability (1271.55), quarantine pending determination of donor suitability (1271.60), and quarantine and disposition of human cellular or tissue-based products from an unsuitable donor (1271.65). Of note here are three *special exceptions* where the regulation would not bar the implantation, transplantation, infusion or transfer of unsuitable tissue: 1) the product is for family related allogeneic use, 2) the product contains reproductive tissue from a directed donor, and 3) there is a documented urgent medical need for the product. The FDA will require that such unsuitable products be labelled with a "BIOHAZARD" legend. These exceptions will not be valid for products coming from donors who are recipients of xenotransplants or have been in contact with recipients of xenotranplants (to prevent the transmission of unknown animal pathogens, including retroviruses). This latter concern and the exceptional circumstances listed above have no direct AATB Standard equivalent although some of these issues are addressed under "exceptional release" rules of the AATB (American Association of Tissue Banks, 2001a).

3.4. Donor screening

"Donor Screening" (Proposal 1271.75) deals with the review of the donor's medical history and the evaluation of risk factors for communicable disease. The FDA proposal consists of two parts related to donor screening: 1) General requirements, and 2) Specific communicable disease screening requirements.

It is stressed under the *general requirements* that a donor medical history means a documented dialog with the donor, if alive, or a documented interview with the next of kin or other individual knowledgeable about the donor, if not alive. This issue is particularly important to prevent transmission of transmissible bovine encephalopathies (TSE) including Creutzfeldt-Jakob disease (CJD). According to the FDA, behavioural or cognitive changes of a cell or tissue donor would uncover a possible indication of TSE when all other screening and testing procedures are negative.

The *specific requirements* stipulate screening of all donors for HIV, HBV, HCV, and TSE, including CJD. A special screening concern are donors of reproductive cells or tissues when those cells or tissues are recovered through methods that could lead to the transmission of sexually transmitted or genitourinary diseases. Another concern of FDA is to exclude donors who have received a xenotransplant or have been in close contact with a xenotransplant recipient. Of note here is that this rule would affect only recipients of live xenotransplants and not those recipients who are transplanted with nonliving biological products such as porcine heart valves or porcine insulin.

3.5. Donor testing

"Donor Testing" (Proposal 1271.80 and 1271.85) includes general and specific requirements to assure reducing the risk of communicable disease transmission by defining how to test donors of cells or tissues intended for transplantation or implantation. *General requirements* state that a specimen from the mother of a foetal or neonatal donor would be acceptable for testing. It is proposed to require that the donor specimen be collected at the time of recovery of cells or tissues from the donor or within 48 hours after recovery. This is different from the current AATB standard which says 7 days prior to or after donation.

The FDA proposal defines exceptions to this requirement in living donors and allows collection up to 7 days prior to recovery if: 1) recovery is risky to the donor, 2) transplantation is necessary before results of testing at recovery are known, and 3) extensive processing is necessary before results of testing at recovery are known. There is no corresponding AATB standard for these exceptions (American Association of Tissue Banks, 2001a).

It is proposed that testing should be performed with FDA-licensed, approved or cleared donor screening tests. A proviso is made for *Chlamydia trachomatis* and *Neisseria gonorrhea* for which there are no FDA-licensed screening tests available and thus tests labelled for the detection of these organisms may be used. The proposed rule holds that a donor whose specimen tests are repeatedly reactive must be determined unsuitable. Exceptions to this are repeatedly reactive tests to Cytomegalovirus (CMV) and non-*Treponemal* screening tests for syphilis.

In the former case, product information sent with the cells or tissue will be evaluated by the physician who will than make a decision about the use of the product. In the latter case, a negative specific *Treponemal* confirmatory test is needed to override the previous false positive result. The general requirements also address the plasma dilution problem after transfusion of blood, colloids or crystalloids. A specimen taken after blood loss but before transfusions or infusions is suitable. In certain instances, an algorithm may be used to calculate dilution (Department of Health and Human Services, 1999).

The *specific requirements* identify the tests to be done on all donors of viable and non-viable cells or tissue as follows: HIV type 1 (FDA-licensed screening test for anti-HIV-1), HIV type 2 (FDA-licensed screening test for anti-HIV-2), HBV (FDA-licensed screening test for hepatitis B surface antigen "HbsAg"), HCV (FDA-licensed screening test for anti-HCV), and *Treponema pallidum* (FDA-cleared test for syphilis). In addition to these tests, donors of viable, leukocyte-rich cells or tissues (stem cells and reproductive cells or tissues) will have to be tested for evidence of infection with Human T-lymphotropic virus (HTLV) types I and II (FDA-licensed screening test for anti-HTLV I/II) and CMV (FDA-cleared test for anti-CMV).

Donors of reproductive cells and tissue will also have to be tested for infection due to genitourinary disease agents *Chlamidia trachomatis* and *Neisseria gonorrhea,* unless it can be shown that procurement ensures freedom from contamination with organisms that may be present in the genitourinary tract. The proposed rule would require retesting of donors of reproductive cells or tissues that can be reliably stored at least 6 months after the date of donation, thus establishing a 6 months minimum quarantine on these cells and tissues.

Retesting for HBV is most appropriately done with the HBV core antibody screening test rather than the surface antigen test used for the original screening. For all other banked tissues and cells from living donors, the FDA recommends but does not propose to require that, where appropriate and feasible, all donors be retested 6 months after donation and that a 6-month quarantine period be used.

For donors of *dura mater,* the proposed rule advocates an assessment and testing protocol that includes a full brain autopsy with examination of the dura mater by a qualified neuropathologist, testing for protease-resistant prion protein (investigational at present), donor medical history interview relevant to TSE risk, prevention of cross-contamination, and use of a chemical protocol to reduce infectivity according to previously published guidelines (Department of Health and Human Services, 1999). For *corneal tissue,* such testing for TSE would be desirable but is not feasible because of the need to transplant corneas within days of procurement. The AATB has no corresponding standard for TSE prevention (American Association of Tissue Banks, 2001a).

3.6. Exceptions

Under "Exceptions" (Proposal 1271.90) the proposed rule addresses situations where donor suitability determination would not be required and defines certain labelling requirements. Two situations are described were a determination of donor suitability would not be required: 1) in the case where autologous cells and tissues are removed and stored for later use in the same patient, and 2) in cases where reproductive cells or tissues are donated by a sexually active partner of the recipient.

The FDA still recommends testing and screening as a safety measure in these circumstances and, while not required, recommends that they be done as part of CGTP standards or as felt appropriate by the attending physician. The labelling requirements proposed are: 1) in exceptions from determination of donor suitability requirement, the label "NOT EVALUATED FOR INFECTIOUS SUBSTANCES" is to be used, 2) if any screening or testing in the case of "exceptions" is performed and indicates the presence of a disease agent the product is to be labelled with the "BIOHAZARD" legend, and 3) autologous banked cells and tissues are to be labelled "FOR AUTOLOGOUS USE ONLY" to prevent inadvertent allogeneic administration.

3.7. Drug and device amendments

"Drug and Device Amendments" contains the proposal to amend existing regulations 210 and 211, with respect to human cellular and tissue-based products regulated as drugs and/or biological products, and regulation 820, with respect to human cellular and tissue-based products regulated as devices, to incorporate CGTP standards and comply with the donor suitability procedures now being proposed.

4. Impact of Proposed Suitability Determination Rule

The FDA has prepared an "Initial Regulatory Flexibility Analysis" to look at the economic impact of the proposed rule on the substantial number of small entities that likely would be affected. These entities are providers of non-reproductive tissue (eye tissue, conventional tissue, stem cell) and reproductive tissue (ART facilities and sperm banks). The donor suitability cost analysis determined a total one-time cost of $252 800 to $1 264 000, a recurring cost of $723 000 to $1 340 000, and a total annualised cost of $759 000 to $1 520 000. The annualised cost per facility analysis showed a very small increase ranging from 0.004 to 0.6% of annual revenues and therefore not significant (Department of Health and Human Services, 1999).

The FDA argues that the proposed regulation would enhance both public health and public confidence in the safety and utility of transplanted cells and tissues, while imposing a minimum burden on the affected industry sectors. The FDA proposed rule is subject to review by the Office of Management and Budget under the Paperwork Reduction Act of 1995. The estimated annual reporting burden has been reported by FDA to be a total of 263 007.5 hours and the estimated annual record-keeping burden to be 25 200.5 hours. The FDA believes to be in compliance with the Paperwork Reduction Act of 1995 as the additional requirements for collection of information are seen as minimal. The FDA judges that the proposed rule has no significant impact on the human environment and thus has neither performed an environmental assessment nor provided an environmental impact statement.

5. Current Good Tissue Practice — Proposed Rule

This proposal is presented under multiple sections containing "Standard Operating Procedures" (SOP) and "Record-Keeping" provisions, and a "Regulation" section (Department of Health and Human Services, 2001a). This latter section is the core of the proposed rule delineating the FDA's concept of good tissue practice.

Subpart D of the "Regulations" deals with general rules of CGTP, exemptions and alternatives, establishment and maintenance of a quality programme, organisation and personnel, procedures, facilities, environmental control and monitoring, equipment, supplies and reagents, process controls, process changes, process validation, labelling controls, storage, receipt and distribution, records, tracking and complaint file.

Subpart E of the regulations addresses additional requirements for establishments described in 1271.10 and regulated solely under section 361 of the PHS Act and the regulations under this proposed part. It addresses applicability, reporting, labelling and claims.

Subpart F of the regulations describes inspection and enforcement of establishments under 1271.10. It covers applicability, inspections, human cellular and tissue-based products offered for import, orders of retention, recall, destruction, and cessation of manufacturing.

6. AATB Position and Response

The AATB is the most prominent professional association of tissue bankers and transplant physicians and surgeons in the US. With the exception of ocular tissue, the AATB member banks provide most of the commonly used structural tissues for clinical use in the US. As an individual surgeon member of the AATB and as a practicing orthopedic oncology surgeon, the author has had a close contact with the AATB over many years. Consequently, it is with great interest that he has been following the evolution of the governmental regulatory programme spearheaded by the FDA. The existing high professional AATB standards have already been in place for many years and are reviewed and re-adjusted on an ongoing basis (Kaprisin and Woll, 2001). It is well established on a national and international level that the AATB standards are the key to the accreditation of tissue banks in the US and are guiding standards for the world. It is implicit therefore that "Donor Suitability Determination" and "Good Tissue Practice" rules have been advocated and required for accreditation of tissue banks by the AATB in the US for many years.

One might now ask if the new FDA proposals differ significantly from the existing professional/accreditation AATB standard. The answer is no, the differences are small and frequently related to simple terminology. A complete comparison of FDA-proposed Donor Suitability Requirements and CGTP regulations with AATB Standards for Tissue Banking has been compiled and published by the AATB (American Association of Tissue Banks, 2001a).

This comparison shows that one can find an AATB standard already in place for most of the FDA-proposed suitability and good tissue practice regulations. While this makes it relatively easy for AATB professionals and tissue banks to satisfy eventual government law, it behooves them to be proactive in shaping the final rulemaking and stay vigilant to recognise and correct faulty concepts or wrong terminology.

Nevertheless, the reader is hoped to get a feeling on how proposed government regulations are shaped and molded by professionals who are going to be affected by the very same regulations within the framework of a democratic society.

7. Donor Suitability Issues

The AATB supports strongly mandatory screening and testing to prevent the transmission of communicable diseases. However, in response to the proposed "Suitability Determination Rule" by the FDA, the AATB has come forward critically with regard to the FDA's means for implementation, terminology related to criteria used to classify cell and tissue products, and jurisdictional determination (American Association of Tissue Banks, 1999, 2000). The AATB objects to the definition of "Donor Medical History Interview" (1271.3(o)) as including only face-to-face dialogues and argues that other forms of communication such as written or telephone exchanges must be included.

With regard to the identification of additional "Relevant Communicable Disease Agents or Disease Means" (1271.3(y)), the AATB has asked the FDA to explain the procedures it will use to identify such and asks the FDA to consult with and use the resources of the AATB. The requirement in the proposed rule that the donor blood test specimen be collected "At the Time of Recovery or within

48 Hours" (1271.80(b)) concerns the AATB because it would exclude pre-mortem samples. The AATB argues for permitting pre-mortem testing of cadaver donors because such donors are generally hospitalised and undergo comprehensive documentation. Thus, the exposure of such donors to disease agents is limited.

The FDA has already proposed to permit testing of living donors up to seven days prior to recovery while the potential exposure of such donors to disease is greater because they are not hospitalised. Post-mortem samples are often affected by haemolysis and haemodilution and this problem gives validity for drawing samples pre-mortem. The requirement by FDA that testing be performed "Using Only FDA Licensed, Cleared or Approved Products" (1271.80(c)) is too broad and the AATB asks that the FDA clarify in what circumstances tissue establishments can use tests that are not FDA-licensed. The AATB has also asked that not only laboratories certified under the "Clinical Laboratory Improvement Amendments of 1998" (CLIA), but also laboratories certified by "State Departments of Health and Human Services" and some "Foreign Laboratories" be permitted to test provided that their requirements are equivalent or more stringent than the FDA's requirements.

The AATB has also brought up the important point that clinical laboratories certified under CLIA should not require registration with the FDA, because the proposed rule provides already that such laboratories should be used. Facilities whose only role in tissue processing is "testing" are excluded from registration and listing requirements because they are under contract to a registered establishment (per FDA proposed rule). Facilities engaged in "screening" of prospective tissue donors should be deemed "establishments" requiring registration and listing because they are not necessarily governed by CLIA.

The AATB has raised specific objections to FDA's criteria and procedure for jurisdictional determinations related to two definitions contained in the "Suitability Determination" proposal: "minimal manipulation" and "homologous use". In addition, the AATB has the questioned FDA's criteria and procedure related to the "Tissue Reference Group" (TRG).

The definition by the FDA of "minimal manipulation" is processing that does not alter the tissue's original relevant utility for reconstruction, repair or replacement (for structural tissue) or does not change the biological characteristics of cells or tissues. The definition by the FDA for "homologous use" is the use of cellular or tissue-based products for replacement or supplementation when used for the same basic function they perform in the donor and in a location where such function normally occurs (for structural tissue products) (Department of Health and Human Services, 1999).

The AATB has put forward the view that these definitions offer imperfect and uncertain guidance for determining which tissues should be regulated as devices or drugs. While these definitions are debated, the AATB has been invited to give input on the "classification of bone dowels as medical devices". It has become evident that the TRG representing individual product sponsors has been trying to issue "recommendations" for regulation by the FDA affecting an entire class of tissue-based products. Since such action coming from the TRG could have broad effects on the tissue industry, the AATB has recommended that: 1) TRG Meetings be announced in the Federal Register, 2) TRG meetings be open to the public, and 3) TRG standard operating procedures should direct the Executive Secretary of the Group to publish the group's findings and the basis for its decisions and that the standard operating procedures should require the Group to explain jurisdictional determinations (American Association of Tissue Banks, 1999).

The AATB finds the "minimal manipulation" definition of the FDA very problematic (American Association of Tissue Banks, 2000). As defined, this could lead to the widespread classification and regulation of many currently available bone grafts as medical devices and thus disqualification from less stringent regulation as Section 361 products. The proposed criterion of "minimal manipulation" is very difficult to apply in a consistent and unbiased manner and it could be the source of legal challenges. Bone is processed and machined to shape in a variety of ways in a wide range from deep frozen to freeze-dried to demineralised forms. Since there is no evidence that such processing presents risks to patients and there is no "approved" processing technology, it is the view of the AATB that the imposition of regulatory controls by the definition of "minimal manipulation" is not warranted.

The AATB has come forward with specific and detailed criticism of the definition "homologous use" (American Association of Tissue Banks, 2000). The criteria of the definition are based on the misconception that holds that, if a tissue is transplanted for the same use and in the same anatomical site from which it was recovered, its use is more basic and less risky. This misperception ignores the standards of surgical practice and tissue banking. Surgeons select tissues for qualities and characteristics that often do not take into account the "original basic function" of the tissue or its "anatomical site" but rather have the qualities that are best to treat a specific problem in a patient. Bone grafts originate from a variety of donor sites and are used routinely by orthopaedic surgeons in hip replacement, trauma and oncology. Bone grafts intended for use in spinal fusions are very common applications of grafting in both orthopaedic surgery and neurosurgery.

In these cases, the "homologous use" criterion would disqualify the bone graft because it is used in a different location and for a different function (fusion by bone instead of maintaining the joint between the vertebrae which is normally filled with cartilage). A similar confusion occurs in the case of a fusion of any joint in the body. Based on these considerations, the AATB has proposed that bone implants be considered and recognised as "tissues" (regulated as Section 361 tissues) when used for the same basic characteristics (not functions) that they have inherently, regardless of the anatomic site from which they were recovered or the site in which they are implanted.

8. Good Tissue Practice Issues

The AATB has provided detailed comments on the FDA's proposed rule establishing CGTP requirements and inspection/enforcement provisions (Department of Health and Human Services, 2001a). The response of the AATB (American Association of Tissue Banks, 2001b) is organised in three parts: 1) a background reaffirming the Association's consistent support for balanced government regulation aimed at insuring the safe and successful clinical use of human tissues provided for transplantation, 2) reservations concerning some of the proposed CGTP regulations either because (a) "they impose requirements on the tissue community that are disproportionate to the level of risk associated with conventional tissues (Section 361)" or (b) "they are based upon provisions of AATB standards that are not intended to be mandatory", 3) reservations concerning the lawfulness of some of the provisions of the proposed rule.

The AATB endorses those provisions of the FDA's proposed CGPT rule that are specifically and directly designed to address the risk of disease transmission to prospective recipients from infected tissues. Provisions of the proposed CGPT rule found burdensome and unnecessary are requirements "to validate all software changes by tissue banks", a requirement that FDA has not established for drugs or medical devices.

The authority FDA claims, in this regard, exceeds the agency's powers with respect to drugs and medical devices under the Food, Drug and Cosmetic (FD&C) Act. This very fact contradicts the FDA's own expressed intention to subject "361 products" to regulatory requirements that are more modest than those applicable to products regulated as drugs or devices under the FD&C Act.

The AATB has noted that some provisions of the proposed CGTP rule have been borrowed from the AATB Standards for Tissue Banking (American Association of Tissue Banks, 2001a). While using AATB standards eases compliance with final regulations, it is inappropriate for the FDA to convert all the AATB standards into binding legal regulations without considering carefully the extent to which these requirements are essential to protect public health.

The AATB has voiced comments on specific provisions of the FDA proposal related to CGTP. In the following analysis, the AATB position is summarised according to the order in which the relevant provisions appear in the FDA proposal:

- Section 1271.3: the definition of "distribution" should indicate that it does not include intra-company transfers of human cellular or tissue-based products.

- Section 1271.150(a): the CGTP requirements include "the function and integrity" of the products. The AATB wishes this part to be deleted because it is too ambiguous.
- Section 1271.150(b): it is implied that establishments that engage other facilities for a step in the manufacturing process should be "responsible" for compliance of the FDA requirements. The AATB wishes this wording to be changed to indicate that establishments "must have a system in place" designed to ensure that the work is performed in compliance with FDA rules.
- Section 1271.155: dealing with exemptions, the FDA wording is deemed by the AATB to be inconsistent. The AATB recommends that the FDA address oral requests for exemptions through oral replies and that such exemptions have an immediate effect without the need for "written confirmation".
- Section 1271.160(a): the FDA requires that establishments maintain a "quality programme". The AATB thinks that these "quality programmes" should be commensurate with the manufacturing steps performed and the types of tissues involved. The AATB requests that the FDA include in the final regulation language that distinguishes between "quality programmes" with particular reference to products regulated under Section 361 as opposed to Section 351 of the FD&C Act applicable to drugs and medical devices.
- Section 1271.160(d): discusses audits and FDA inspection. The AATB requests that FDA clarify that internal audits are not subject to FDA inspection and refer to the previous general policy of not allowing access of inspectors to company audit records for other categories of medical products.
- Section 1271.160(e): requires "validation of computer software" and "validation of software changes". The AATB has serious concerns about the term "validation" because it is vague and it is not clear whether it corresponds to existing AATB standards. The AATB would like to see "validation" to be tailored to the type of processing used for a particular tissue. Thus, in current AATB standards, validation is required for shipping containers intended for tissues that must be maintained at other

than ambient temperature, while the standards require only verification or confirmation for other aspects of tissue processing.

- Section 1271.180: states that "any deviation from a procedure shall be authorised in advance by a responsible person, recorded, and justified". The AATB position on this issue is that deviation cannot always be authorised in advance because of unforeseen circumstances. Technical staff is trained to make decisions and document deviations. The AATB proposes that deviations be recorded at the time of occurrence and that the specific deviation be approved by a responsible person prior to the release of the tissues affected by the deviation.
- Section 1271.190: deals with provisions that try to define standards for the construction and maintenance of facilities that manufacture human cellular and tissue-based products. The AATB finds these provisions too broad and recommends that the language be modified to specifically define requirements to prevent the transmission of communicable disease agents and to ensure orderly handling without mix-ups.
- Section 1271.200(a): describes requirements for equipment and states that: "it should not have any adverse effects on the products". The AATB wants the reference to "adverse effects" deleted because, absent a demonstration of effect on the risk of contamination with communicable disease agents, equipment selection and placement is not an appropriate focus of regulatory concern for tissues regulated under Section 361.
- Section 1271.200(e): deals with requirements for records of maintenance for individual pieces of equipment and the need to have maintenance records available "at each piece of equipment". The AATB finds these provisions extremely burdensome for the tissue community and unnecessary to protect public health. Items that are subject to control according to lot numbers and simple or disposable items should be exempt from record keeping requirement. Records should not have to be available "at" but rather "for" each of the pieces of equipment as there is no public health justification to keep records in close physical proximity to equipment.

- Section 1271.220: the FDA puts forward provisions for ensuring that human cellular or tissue-based products are not contaminated and "maintain their function and integrity", furthermore that processing material be removed or limited to an amount that "does not adversely affect the product's "function and integrity". The AATB requests that references to "function and integrity" be deleted from this section and that instead the provisions should simply state that the products are not contaminated and will not transmit disease. The AATB also argues that tissue establishments should be allowed to determine whether residues elicit adverse reactions and label products regarding the presence of residues while procedures are in place for the removal of potentially toxic processing materials.
- Section 1271.220(d): deals with rules that prohibit pooling of cells or tissues from two or more donors. The AATB has proposed clarification to the term "pooling" which refers to commingling biological material from more than one donor in a single immediate container.
- Section 1271.230(a): states that in cases where the results of a process cannot be "fully" verified by subsequent inspection and tests, the process shall be validated and approved according to established procedures. The AATB requests the deletion of the term "fully" from this provision, as it is too broad and subject to inconsistent application. Once a process has been validated, and changes are required that do not increase the risk of disease transmission, a written justification should be sufficient.
- Section 1271.230(b): states: "a claim for sterility or viral inactivation shall be based on a validated process". The AATB recommends that the FDA recognise that claims for sterility for bone and soft tissue grafts may be based on determination by verification, not validation, if validation is not feasible.
- Section 1271.230: deals with validation for reduction of transmission of TSE while preserving the clinical utility of grafts. The AATB disagrees with this provision because no safe levels of TSE are known presently and it would be inappropriate to imply that processing can produce safe tissue.

- Section 1271.260(b): deals with storage temperatures and durations stating that these must ensure "product function and integrity" to prevent "product deterioration". The AATB has problems with the wording used by the FDA because concepts such as these are undefined and beyond the FDA's legal authority to ensure prevention of transmission of communicable diseases.
- Section 1271.270(e): describes rules for the retention of records for 10 years. The AATB recommends simplification of the language of this provision and has put forward the proposal that tissue establishments that cease to exist use their best efforts to maintain records rather than being compelled to do so.
- Section 1271.290: describes "product tracking" and the assignment of "identification codes". The AATB proposes to use the term "tracing" in the place of "tracking" to avoid confusion with medical device tracking regulation. The AATB would like the provisions of this section modified so that it recognises current practices of the industry and limitations that tissue establishments have in obtaining tracing information. The AATB requests that the FDA clarify that a single identification code may be used for an entire lot of structural tissue of the same type from the same donor even if distributed in more than one immediate container.
- Section 1271.320: deals with complaint procedures and their review, evaluation and documentation. The AATB requests that this section define that the complaint requirements apply only to tissues that are released for distribution. The AATB objects to the references to "tissue function and integrity" related to the definition of "complaint" in this section.
- Section 1271.420: describes provisions applicable to the importation of human cellular and tissue-based products and states that the importer must notify the director of the district of the FDA having jurisdiction over the port of entry through which the product is imported or his designate. The AATB has requested a small modification of the wording of this section to clarify that these requirements apply solely to tissues intended for "clinical use" and not to products intended for research.

In the preamble to the whole set of CGTP rules, the FDA requests "consultation from the States on preemption issues raised by the proposed CGTP rule". The AATB wants that the FDA clearly state in the final rule that its provisions "preempt state tissue regulations".

The AATB indicates clearly that some of the labelling provisions of the CGTP proposal exceed FDA's statutory authority because their relationship to the prevention of disease transmission is too attenuated. The provisions that aim to assure "function and integrity" are, according to the AATB, beyond the agency's authority under Section 361(a) of the PHS Act. Furthermore, the AATB alleges that certain of the proposed investigative and enforcement provisions of the FDA are invalid either because they cannot be justified by the disease transmission purpose of Section 361(a), or because they represent unlawful attempts to claim powers that Congress has only selectively conferred on the FDA in the FD&C Act.

Section 361 of the PHS Act is a 1944 re-codification of two quarantine laws passed in 1890 and 1893, respectively. The first law passed in 1890 authorised the President to direct the Secretary of the Treasury to effectuate rulemaking to prevent the spread of four specified infectious diseases (cholera, yellow fever, smallpox or plague), once introduced into the US. The second law passed in 1893 enabled the Secretary of Treasury to make rules and regulations that are necessary to prevent the introduction of the four specified diseases into the US.

Congress enacted the PHS Act in 1944 because of increasing air travel and the nation's involvement in World War II. This legislation reorganised and codified several scattered laws dealing with public health, including the 1890 and 1893 laws. The legislative history of the 1944 PHS Act, according to the AATB, demonstrates that the intention of Section 361(a) is to authorize federal action in the context of quarantine measures and their enforcement.

Congress enacted the precursor to the modern FD&C Act in 1906 giving the federal government limited enforcement powers with respect to food and drug articles. Thereafter, when Congress intended for the federal government to exercise greater authority to enforce laws related to safety, efficacy or labelling of food or therapeutic products, it conferred new authority specifically and narrowly without giving the federal government omnibus authority.

However, another provision in the 1944 statute, Section 351, empowers the federal government specifically to exercise authority to enforce the labelling and packaging of biological products. Section 361 contains no such directive and therefore, the AATB postulates that Congress did not intend for Section 361 to empower the federal government to impose product-labelling requirements (American Association of Tissue Banks, 2001b).

The FDA purports to rely on Section 361(a) of the PHS Act to support the enforcement provisions of the proposed CGTP rule. Since that statute was enacted before the FDA existed, several of the authorities claimed by the FDA in the CGTP proposal have no counterpart in the FDA's enabling statute, the FD&C Act. The AATB argues that Congress has been selective and often reluctant in granting the FDA investigative and enforcement authorities. Therefore, the interpretation of Section 361(a) as permitting the FDA to assert powers of investigation and enforcement, according to the AATB, is faulty and not compatible with the general strategy of Congress to grant such powers only for specific categories of products in specific circumstances.

The proposed CGTP rule would give the FDA unlimited authority to examine records with the use of photographs and videotapes. The AATB argues that historically, the FDA's efforts toward obtaining inspection authority over the years have been repeatedly curtailed by Congress. Thus, efforts by FDA to assert authority to inspect shipping records in the 1940s was rebuffed by the 1952 Cardiff decision of the Supreme Court (US Supreme Court, 1952).

Initially, trials by the FDA to win inspection authority under the FD&C Act failed in 1961 but eventually in 1962 legislation was enacted which gave the FDA explicit authority to inspect manufacturing facilities for prescription drugs but not for over-the counter (OTC) drugs or other products regulated under the FD&C

Act (US Congress, 1962a). Similarly, initial efforts by the FDA to obtain inspection authority for medical devices were initially rejected by Congress in 1974 and 1975 (US Congress, 1974, 1977).

Eventually, in 1976 Congress enacted the Medical Device Amendment legislation, which gave the FDA some very limited inspection authority for restricted medical devices, specifically denying power to inspect financial, sales, pricing, personnel or research data (US Congress, 1976a,b). Over the last decade, during the Bush and later the Clinton administrations, the FDA tried but did not succeed in winning additional investigative and enforcement authority. The AATB believes, based upon knowledge of congressional intent over the years, that the FDA's request for broad, unrestricted authority via Section 361(a) of the PHS Act is unjustified and must only be considered if directly related to the disease prevention goal of Section 361(a).

The proposed CGTP rule also includes provisions related to submit "adverse reaction" and "product deviation" reports. Specifically, these reports are not limited to the "transmission of communicable diseases" but also include "product contamination" and "product function and integrity".

Historically, Congress has given the FDA mandatory reporting authority only in respect to certain products and always carefully tailored in form of amendments to the FD&C Act. Under the Drug Amendments of 1962, drug manufacturers were required to report data "relating to clinical experience" and data "otherwise obtained", a response to the thalidomide scare (US Congress, 1962b). The Safe Medical Devices Act of 1990 provided for the reporting of most serious adverse events regarding medical devices, a response to the fatal strut fractures in the replacement heart valves manufactured by Shiley in the late 1980s (US Senate, 1990). The AATB argues that it is inappropriate and illegitimate for the FDA to require tissue establishments to submit reports concerning all adverse reactions and product deviations. The reporting requirements must be linked to the disease prevention goal of Section 361(a) and not to issues of "product deviation" or "function and integrity".

The CGTP proposal in section 1271.440 would give the FDA officials authority to issue administrative orders requiring tissue establishments to cease their operations when found non-compliant with the rules. Although a provision is made for the facilities to be able to request a hearing within 5 days after issuance of the order, there is no provision for opportunities for a challenge of the question of lacking compliance, and the order is of potentially infinite duration. In the view of the AATB, this administrative injunction provision is unconstitutional because it does not meet the most minimal standards of Due Process. AATB members have a constitutional right to practice their occupation and the deprivation of this right without giving the opportunity to present the case that the government action is unwarranted constitutes, in the view of the AATB, violation of the Due Process Clause (US Supreme Court, 1973). The AATB foresees a need for a legal challenge of this provision if the FDA does not change it.

The AATB has put forward a proposal for a co-operative mechanism whereby the FDA could use existing resources of professional accrediting organisations such as the AATB for the co-ordination of inspection activities. This is in line with a recommendation by the Office of the Inspector General at the Department of Health and Human Services. The co-operation would include the joint training of inspectors, recognition of AATB accreditation as fulfillment of CGTP requirements, extension of inspection cycles for accredited tissue banks and other joint resources aimed at saving time and expense for the FDA (American Association of Tissue Banks, 2001b).

The AATB argues against the proposed 180 days implementation requirement post publication of the final rule and is in favor of a phased implementation over 1 year to ensure an orderly transition.

10. Conclusion

The continuing dialog between the government represented by the FDA and the professionals of cell and tissue banking represented by the AATB is a great source of information reflecting directly the state-of-the-art of current knowledge in the field of cell and tissue banking. The analysis of the evolving regulatory programme proposal of cell and tissue banking by the FDA, as it is continuingly molded and shaped by critical input by the AATB, allows one to understand the complexity of problems associated with the science and clinical practice of cell and tissue banking. This very analysis is also the natural history and evolution of a much needed government regulation of cell and tissue banking in the US. The purpose of such regulation at this particular point in time is to assure the continuing safety of recipients of cells and tissue-based products against the transmission of communicable diseases.

While the FDA is working to implement this rulemaking under Section 361(a) of the PHS Act, it has proposed criteria such as "homologous use" and "minimal manipulation", and provisions such as "product function and integrity", that are not related or aimed at disease transmission risk and therefore beyond the authority granted to FDA under Section 361(a) by its statutory provision. Continuing critical interchange between government and professional associations will shape the future regulation towards implementation to achieve the ultimate safety goal to the advantage of our patients.

11. References

AMERICAN ASSOCIATION OF TISSUE BANKS (1999). AATB responds to FDA's rule: Suitability determination for donors of human cellular and tissue-based products, *AATB Information Alert*, December 29, **9**(4), 1–8.

AMERICAN ASSOCIATION OF TISSUE BANKS (2000). AATB addresses FDA public meeting on human bone allograft, *AATB Information Alert*, September 5, **10**(3), 1–4.

AMERICAN ASSOCIATION OF TISSUE BANKS (2001a). Government affairs committee reviews proposed good tissue practice regulations, *AATB Information Alert*, February 19, **11**(4), 3–25.

AMERICAN ASSOCIATION OF TISSUE BANKS (2001b). Association responds to FDA regarding proposed good tissue practice regulations, *AATB Information Alert*, May 8, **11**(5), 2–34.

CENTER FOR DISEASE CONTROL (1985). Provisional public health service inter-agency recommendations for screening donated blood and plasma for antibody to the virus causing acquired immunodeficiency syndrome, *Morbidity and Mortality Weekly Report,* January 11, **34**(1), 1–5.

DEPARTMENT OF HEALTH AND HUMAN SERVICES, FOOD and DRUG ADMINISTRATION (1993). Human tissue intended for transplantation. Interim rule, *Federal Register,* December 14, **58**, 65514–65521 (21 CFR 1270).

DEPARTMENT OF HEALTH AND HUMAN SERVICES, FOOD and DRUG ADMINISTRATION (1997a). Proposed approach to regulation of cellular and tissue-based products; availability and public meetings, *Federal Register,* March 4, **62**(42), 9721–9722.

DEPARTMENT OF HEALTH AND HUMAN SERVICES (1997b). *HHS News,* February 28.

DEPARTMENT OF HEALTH AND HUMAN SERVICES, FOOD and DRUG ADMINISTRATION (1997c). Human tissue intended for transplantation, *Federal Register,* July 29, **62**(145), 40429–40447 (21 CFR 1270).

DEPARTMENT OF HEALTH AND HUMAN SERVICES, FOOD and DRUG ADMINISTRATION (1998). Establishment, registration and listing for manufacturers of human cellular and tissue based products. Proposed Rule, *Federal Register,* May 14, **3**(93), 26744–26755.

DEPARTMENT OF HEALTH AND HUMAN SERVICES, FOOD and DRUG ADMINISTRATION (1999). Suitability determination for donors of human cellular and tissue-based products. Proposed rule, *Federal Register,* September 30, **64**(189), 52696–52723.

DEPARTMENT OF HEALTH AND HUMAN SERVICES, FOOD and DRUG ADMINISTRATION (2001a). Current good tissue practice for manufacturers of human cellular and tissue-based products; inspection and enforcement. Proposed rule, *Federal Register,* January 8, **66**(5), 1507–1559.

DEPARTMENT OF HEALTH AND HUMAN SERVICES, FOOD and DRUG ADMINISTRATION (2001b). Human cells, tissues, and cellular and tissue-based products; establishment, registration and listing. Final rule, *Federal Register*, January 19, **66**(13), 5447–5469.

KAPRISIN, D. and WOLL, J. (eds.) (2001). *Standards for Tissue Banking*. American Association of Tissue Banks, McLean, Virginia.

US CONGRESS (1962a). *Public Law* No. 87-781, #201, 76 Stat. 780, 792.

US CONGRESS (1962b). *Public Law* No. 87-781, 76 Stat. 780, 782.

US CONGRESS (1974). *93rd Congress*. Calendars of the US House of Representatives and History of Legislation, 137, 232.

US CONGRESS (1976a). *House of Representatives Report* No. 94-853, at 46–47.

US CONGRESS (1076b). *Public Law* No. 94-295, #6, 90 Stat.539, 581.

US CONGRESS (1977). *94th Congress*. Calendars of the US House of Representatives and History of Legislation, 136, 152.

US SENATE (1990). *Senate Report* No. 101-513, at 23.

US SUPREME COURT (1952). *US versus Cardiff*, No. 73 S. Ct. 189, 344 US 174.

US SUPREME COURT (1973). Thomas S. Gibson *et al.*, Appellants v. L.M. Berryhill *et al.* No. 71-653, May 7, 411 US 564, 571.

Advances in Tissue Banking Vol. 6

2 HUMAN TISSUES, MEDICAL DEVICES AND MEDICINAL PRODUCTS: CURRENT SITUATION IN THE EUROPEAN COMMUNITY

MARIA J. FELIX

Catalan Transplant Organisation (OCATT)
Barcelona, Spain

CARMEN DE FELIPE

National Transplant Organisation (ONT)
Madrid, Spain

1. Introduction

At a recent meeting held in Porto between June 14–16, 2000, in the framework of the Portuguese presidency of the European Union, on "Therapeutic utilisation of Human Organs and Tissues", the participants from the member States came out in favour of drafting a Community Directive on the Utilisation of Human Tissues, and the Board of Health of Ministers of the European Union was informed.

This news brings up one of the most important request in the health area related to the utilisation of human tissues, since it will make it possible to establish a legal harmonisation in European Union countries.

This article analyses the background of this situation which has made it possible that substances with identical origin and destination to be treated differently depending on country-specific legislation.

2. Background

The use of human-origin substances as single option or therapeutical alternative, either in the form of transplant or implant, is an uncontested fact from which more than 400 000 patients a year all over the world can benefit. The use of these components with therapeutical aims ranges, include from organ transplant to the transplantation or implantation of tissues and products derived from both of them.

The different parts of the human body may be used with therapeutical aims, for research and for *in vitro* diagnostic techniques. There is also another group of human-origin substances (collagen, fibrin, hormones, immunoglobulins, etc.) which are regarded as medicinal products or devices depending on their action mechanism.

The use of some of these components, such as organs for transplant, for therapeutical purposes aims is well known and clinically and socially accepted. In some cases, organs constitute the only therapeutic alternative for the recipient of the transplant, and in others cases they are the best therapeutical option to improve quality of life. The results obtained and media coverage have provided this therapeutical option a great social and scientific prestige. It must be remembered that the organ transplant system also has transplant organisations in place to protect and monitor related activities.

These organisations, which co-ordinate with each other, look after transparency and the co-ordination and follow-up of all processes involved in this activity. The immediacy between getting the organs and the actual transplant, the absence of manipulation and the impossibility of long-term storage facilitate the traceability of procedures and the coordination of the processes.

On the other hand, the therapeutic use of tissues, cells and derived products of organs and tissues, which enjoys less exposure in the social scenario, is subject to multiple manipulations and in most cases to a delayed use due to this possibility of long-term preservation. Besides these differences, the growing incorporation of biotechnological innovations in some human-origin derived products applied with clinical finality, leads the consideration of human tissues and derived products or organs and tissues as medicinal products or medical devices to be questioned. As the type of human-origin derived products used for clinical purposes increases, the legislation becomes more ambiguous, even in Spain, where there is a specific Royal Decree (RD) regulating the utilisation of human tissues.

3. Human Tissue — Regulatory aspects in the European Community

Spanish Law defines Human Tissue as "All parts of the human body, including surgical residue and cells. It also includes products incorporating human-origin tissues or cells and their derived products" (RD 411/96).

The clinical finality of tissues used may be authologous or allogenic and may come from a living or a dead donor. Clinical methods of use vary according to the type of tissue. Sometimes they are applied as a dressing to favour the regeneration of damaged tissue, sometimes they are given by infusion, and sometimes they are implanted or transplanted by means of surgery of varying degrees.

Processing and preservation procedures are more or less complex depending on the tissue type, although their common finality is to guarantee tissue viability and to preserve their function for clinical use. Moreover, all of them share a common characteristic: to guarantee safety of clinical use. To achieve this target by means of the conduct of as many serological and bacteriological tests as may be necessary, based on a meticulous screening of donors geared towards eliminating the possibility of the transmission of disease to the extent that scientific knowledge will permit.

At this moment, the regulatory situation in the different countries of the European Union is variable, and some countries do not even have regulatory provisions in place. Notwithstanding, there are recommendations issued by the World Health Organisation, the Council of Europe and different European Scientific Societies have their Good Practices in related activities.

The rules aimed at regulating organ and tissue transplant in Europe are:

- Resolution 29(78) of the Committee of Ministers of the Member States of the Council of Europe on the harmonisation of the legislation of Member States on the obtaining, implantation and transplantation of human substances.
- Recommendation 5(79) of the Committee of Ministers of the Member States of the Council of Europe on the international transport and exchange of human substances.
- Resolution of the European Parliament A3-0074/93 on the prohibition of the trade of organs for transplantation.

Directive 95/46 CE is applicable to the protection of safety and free movement of personal data.

These rules have always been interpreted as pertaining to human organs, although the reference to "human-origin substances" affords identical treatment to human origin organs, tissues and cells.

The only documents in the European scenario that specifically provide for human tissues and derived products are:

- Recommendation No. R (94) by the Committee of Members of the Member States of the Council of Europe on human tissue banks
- Document: Ethical Aspects of Human Tissue Banking issued by the European Group on Ethics in Science and New Technologies of the Council of Europe (EGESNT. CE), July 1998.

Both documents constitute, in the first case, a recommendation to European Union member countries and, in the second case, the opinion drawn up by a group of experts of the European Union. Therefore, compliance with the opinions and recommendations thus issued is not mandatory, although they should be taken into account when the European Union decides to address the regulation of this issue in compliance with article 152 of the

Treaty Establishing the European Community, by virtue of which a high level of protection of human health is guaranteed (point 4 section a), measures that establish high levels of quality and safety of human origin organs and substances, as well as of blood and blood-derived products; these measures will not prevent any Member State from maintaining or introducing stricter protection measures.

4. Medical Device and Medicinal Product — Legal Definition

The Community Directive 93/42/CEE, and in Spain the Royal Decree 414/1996 of March 1, 1996, define the **medical device** as any instrument, device, equipment, material or other article used alone or in combination, including computing programmes used for them to work properly, designed by the manufacturer for use in human beings or with the following aims:

- Diagnosis, prevention, control, treatment or alleviation of a disease
- Diagnosis, control, treatment, alleviation or compensation of a lesion or a deficiency
- Investigation, substitution or modification of the anatomy or of a deficiency
- Regulation of conception.

And which does not carry out the main action that is targeted inside or on the surface of the human body by pharmacological, immunological or metabolic means, but to whose function such media may contribute.

The Community Directive 65/65/CEE and in Spain Law 25/1990, of December 20 on medicinal products defines **medicinal product** as any substance and its associations or combinations designed to treat or prevent diseases in human beings or animals. Any substance or combination of substances which are given to human beings or animals to perform a medical diagnosis or restore, correct or modify physiological functions in human beings or animals are regarded as medicinal products.

The following criteria are established to decide which regimen should be applied, i.e. the medicinal product or medical device regime:

- The product's form of presentation
- The method used to obtain the main action.

This latter criterion, based on the mechanism of action, is crucial in the classification of the product. Typically, the function of medical devices is obtained by physical media (mechanical action, physical barrier, substitution or support to organ or body functions) and the action of medicinal products is attained generally by pharmacological, immunological or metabolic mechanisms. The mechanism of action of a product can be deduced by the labelling and claims of the manufacture and the scientific data that provide for its mechanism of action. While the manufacturer's claims are important, it is not possible to regard a product as pertaining to one or the other category contrary to current scientific data, and manufacturers are obliged to scientifically justify their proposal.

Pharmacological action is taken to mean the interaction of the molecules of the product in question and the cell component, of the receiver, which gives rise to a direct response or by blocking the response of another agent. While it is not a totally reliable criterion, the presence of a dose-response correlation is indicative of pharmacological effect.

Immunological mechanism is taken to mean the action involved in a specific immune reaction obtained by the stimulation and/or mobilisation of cells and/or products.

Metabolic mechanism is taken to mean the action that involves an alteration, including the halting, beginning or change in the rate of a normal chemical process that participates in or is available for the normal functioning of the body. The fact that a product is metabolised does not imply that its mechanism of action is metabolic. A medical device may be assisted in its function by pharmacological, immunological or metabolic mechanisms, although as soon as these mechanisms are no longer auxiliary with regard to the purpose of the product, the product will be regarded as a medicinal product.

The determination of the border between medical device, active implantable medical device and medicinal product has long been the subject of great debate in the European Union. Indeed, some products are assessed by both organisations.

5. Current Situation and Consequences

The absence of specific regulation in the European Union for human tissues and derived products has made it possible for legislation developed in some countries to be specific and strict, in keeping with the existing legislation on organ transplants, observing ethical aspects, health safety, distribution outside the market circuits (therefore with prohibition of commercialisation) with specific authorisation to obtainment centres and banks and implant centres as well as specific authorisation for the importation and exportation of human tissues, while other countries have enacted legislation classifying them as medical devices, and some other countries have no legislation whatsoever in place.

This lack of legislative harmonisation means that there are two circuits for the distribution of tissues/cells/derived products depending on the countries, with differential aspects that are of vital importance (affecting procedures of donation, obtainment and studies carried out on the donor and the tissue, traceability, serum libraries , preliminary studies and follow-up of the recipient, as well as the type of centres were they can be used) and of enormous social and health import, as can be seen in Table 1.

This paradox affords these MD "free movement" with CE stamp, whereas tissues, depending on the country, require a specific entry permit, explaining beforehand and on case-by-case basis the studies performed and the results obtained, which must satisfy the requirements for tissues in the target country. Moreover, in recent years EU standards have been approved or documents or reports have seen the light of day, related, directly or indirectly, to the therapeutic application of human tissues, such as:

(1) The approval of Community Directive 98/44/CE of the Europe an Parliament and of the Council of July 6, 1998 on the legal protection of biotechnological inventions.

Table 1. Comparison of legislation between tissues and medical devices.

	Tissues	Medical devices
Recruitment of donors	Exhaustive	Unknown
Register of donors	Yes	No
Serol. and bacteriol screening	Rigorous	Not specified
Repetition of test/autopsy	Recommended	Not specified
Donor serum library	Yes	No
Register of receivers	Yes	No
Study of the receiver	Mandatory	Not specified
Traceability	Mandatory	Not specified
Processing	Tissue banks	The trade
Quality control	Health area	The trade
Follow-up tissue/recipient	Yes	No
Distribution	Health circuits	Market circuits
Commercialisation	Prohibited	Accepted

This Directive defines as patentable any new inventions that entail inventive activity or which may have an industrial application, even if the object is a product made of or containing biological matter (a matter containing self-reproducible genetic information or reproducible in a biological system). Biological matter isolated from its natural environment or produced by means of a technical procedure can also be an invention. The human body, or the simple discovery of one of its elements, included the partial sequence of a gene, are not deemed patentable inventions. On the other hand, an isolated element of the human body (cell cultures) or that which is otherwise obtained by means of a technical procedure, including the total or partial sequence of a gene, may be a patentable invention. The objective of this Directive is to guarantee the free movement of biotechnological products patented by means of a harmonisation of the legal provisions of the Member states in order to clarify the legal framework of these products.

(2) The drafting of the document "Point to Consider on Human Somatic Cell Therapy" by the European Agency for the Evaluation of Medicinal Products.

This document defines cell therapy as the administration of preparations of duly adapted somatic cells for transfer to an individual. Its objective is the reconstruction of the functions of cells and tissues and entails the administration of modified antologous or allogenic cells for the repopulation of a specific organ or tissue of the receivers. These products may be an expanded and/or purified population of mother cells or more differentiated cells that perform a specifically defined physiological function, or may be cells manipulated, for example, in combination with non-cell matrices for the production of artificial engineered tissues. The document is a guide for the characterisation and quality control of human somatic cells, barring transfusable blood products (platelets or red blood cells) and for the control of the cell culture procedures used in industrial production to guarantee product and process.

(3) The Steering Committee on Bioethics (CDBI) of the Council of Europe in the draft additional protocol to the Convention on Human Rights and Biomedicine (CDHB) pertaining to the transplantation of human-origin organs and tissues dated July 25, establishes the criteria to be taken into consideration in the extraction and transplantation of organs and tissues.

This project establishes the ethical principles of health safety that should be observed in organ and tissue donation and transplant procedures. However, the explanatory report that accompanies the document states that "the term transplant is interpreted in the traditional sense and does not include the utilisation of human origin tissues under the form of medical devices or medicinal products. Although it does say that the principles contained in the protocol, mainly those pertaining to safety, traceability, information and consent in such form of use should be applied under the 'mutatis mutandis' form". Or in other words, tissues/cells eventually regarded as medical devices or medicinal products will not be subject to the provisions of article 20 on the Prohibition of profits or to those of article 21 on the Prohibition of the traffic of organs and tissues since tissues/cells used as medical device or medicinal product may be bought or sold.

The finalistic concept of utilisation makes it possible to establish different ethical treatments for substances of identical origin as it recognises that human-origin derived products whose form of administration is not traditional implantation may be regarded as medicinal products or medical devices susceptible to different protection and under a different legal framework, and even acknowledges the added value which the use of such elements affords the device in itself.

6. Considerations on Ethical Aspects and Health Safety

There are certain vital aspects that should be addressed in undertaking the process of European-wide legislative regulation. These aspects are closely linked to the origin and destination of human-origin derived products, regardless of whether they are regarded as tissues/medicinal products or medical devices and whether they are implanted, transplanted, taken or applied.

(a) Donors should be duly screened to prevent the possibility of transmission of diseases. These donors should be identified and studied.

(b) The altruism of the donation, profiteering must be absent from any interventions carried out from source to target, which does not mean that there are no associated costs or expenses, although any costs invoiced (extraction, evaluation, preservation, processing, transport and transplantation or implantation) should be only those generated by the activity.

(c) The serological and bacteriological screening of donor and tissue, destination of organs, tissues and substances of human origin must be rigorous with a view to preventing transmission of diseases. Tissue processing/preservation centres should be reasonably expected to quarantine the tissues and wait for the results of complementary studies such as autopsies, or, in the case of live donations, the repetition of certain analytical parameters after 6 months depending on the analytical methods used initially. The possibility of carrying out analytical determinations in some of the receivers of transplanted organs or the use of more specific and more sensitive tests for tissue screening should also be considered.

(d) Processing and preservation methods guaranteeing, insofar as may be possible from the scientific standpoint, tissue viability according to the clinical use to which they are to be put. Following the application of these intermediate procedures of handling and conservation, controls targeting the verification of the absence of contamination and tissue and cell viability depending on their destination (finished product control) are required.

(e) Traceability, which must be guaranteed in all human-origin derived products, be they regarded as tissues/products/medicinal products, guaranteeing confidentiality at all times.

(f) The maintenance of a serum library, both for its scientific aspect as well as for the possibility of applying controls and health monitoring.

(g) Pre-study of the tissue recipient, including points as important as (1) the information to be provided on requesting informed consent, (2) serological screening, and (3) post-implantation and transplantation follow-up.

(h) Establish the minimum conditions of compliance by centres or institutions where each procedure is carried out. These centres must be specifically authorised by the Competent Authority to carry out each procedure.

(i) The ethical aspects that apply to all human origin substances for use in human beings should not be based on the finalistic concept of utilisation, but on origin and destination. All human-origin derived products for human use should enjoy the same consideration regardless of how they are eventually classified. This criterion would make it possible to protect and respect the dignity of the donor and the safety of the recipient, complying, moreover, with one of the principles defended by the Convention on Human Rights and Biomedicine "*the protection of dignity, identity and integrity of people*" (Article 1).

7. Conclusions

Taking the above considerations into account, the planned Community Directive specific or human tissues recently agreed to may make it possible to regulate and harmonise a hitherto rather ambiguous situation.

In the making of the future Directive, which shall probably be very laborious, the considerations geared towards guaranteeing aspects related to health safety and the ethical treatment afforded to all donated substances should be addressed from an all-embracing standpoint.

Furthermore, it will also be necessary to establish the criteria that will determine when a human-origin substance is to be regarded as human tissue, medical device or medicinal product, taking into account the fact that they are of the same origin and destination as those classified as human tissues, and therefore enjoying identical safety measures and ethical considerations.

7.1. Health safety

This future CD (Community Directive) must relate to quality standards or Good Manufacturing Practices (GMP) established for each and every one of the procedures to which each type of tissue is to be submitted. These GMP, drawn up by scientific societies comprising expert professionals in the subjects, must be updated periodically to attain a "dynamic safety guarantee" that will make it possible to bring in the latest scientific knowledge.

7.2. Ethical aspects

As far as ethical aspects are concerned, it would suffice for the draft additional protocol to the CDHB, or more specifically its explanatory report, to address human-origin derived products on the basis of origin and destination, avoiding any exclusion of substances from donation. The suppression of the finalist concept of utilisation would afford ethical protection to all human-origin derived products that are to be used in human beings regardless of whether the clinical use is a transplant, implant, transfer, administration or application.

Across-the-board legal and technical criteria established by a CD, plus the ethical protection that would be granted by the additional protocol to the CDHB based on origin and destination, would accomplish two pivotal objectives:

- Respect and protect the rights and freedoms of donors and recipients of organs, tissues, cells and substances of human origin.
- Provide maximum guarantee for health safety of all human-origin substances to be used in human beings.

Advances in Tissue Banking Vol. 6

3 LEGAL SITUATION OF TISSUE TRANSPLANTATION IN JAPAN

M. ITOMAN, T. IZUMI, N. TAKAHIRA, T. NARUSE, T. TOMINAGA and H. MINEHARA

Department of Orthopaedic Surgery
Kitasato School of Medicine, Kitasato 1-15-1
Sagamihara, Kanagawa, 228-8555, Japan

1. Introduction

In Japan, there was a long period of time that people did not feel comfortable about the idea of clinical organ and tissue transplants. The problem was due more to an emotional attachment from relatives over the decreased, and the lack of a social consensus in general, than it was a medical or biological problem. Due to this, there was a delay in all fields of clinical transplantation. Even today with approved law for organ transplantation, there are people who express a negative opinion about organ transplantation. There is also the issue of the unpopular "donor card", which gives the donor's consent. Although there were no laws at the time, doctors have been performing the transplantation of tissue long before any laws were set on the issue. This report explains the transplantation of organs and tissues in Japan, especially from the bone transplant point of view.

2. Historical Background

In 1907, Mr. Tomita's report was the very first clinical bone allograft report from Japan, before reports such as xenograft, autograft, etc. can be found. The first organised bone bank was developed in Department of Orthopaedic Surgery of Kyushu University in 1953. It was the Frozen Bone Bank. It began with a freezer (–18°C) that Prof. Amako had purchased at department store. They used bones and amputated limbs that had been removed during surgery and retrieved the bones from living donors and kept them frozen until they could be used for grafting. There were no laws about transplantation at that time. Retrieving organs and/or tissues from living or deceased donors was illegal.

The Ministry of Health and Welfare (MHW) was very concerned about the importance of the issue. MHW issued a Ministerial Ordinance # 304 in 1954, to give permission to retrieve tissues from deceased donors and the clinical use of these tissues. According to the Ordinance # 304, it was not illegal to retrieve some tissues, such as bones from deceased donors, which will be used for transplantation. The purpose of the transplantation must be used as a treatment of a serious condition of the patients. This must first be determined by medical judgement with ethical consideration. It was added, as a condition that "it is necessary to obtain a written agreement from the deceased's family, next of kin." At this point, clinical bone transplantation, as well as the cornea retrieved from agreed donors, had already begun.

After this period, various institutions were using different methods, such as freeze-dried bone, decalcified bone and AAA bone, but none of these were what would be called a bone bank. These methods did not become popular, due to their unpractical methods of processing, unlike the method of freezing, as in Kyushu University Bone Bank, which has spread nation-wide. The Japanese Orthopaedic Association (JOA) has taken a nation wide survey that shows that institutional bone banks provide frozen preservation using deep freezers. The 1995 nation-wide survey also showed that the number of bone banks has grown to 125. Therefore, we can say that the demand of the bone allograft has been growing (Table 1).

Table 1. Bone banking in Japan (according to the nationwide survey in 1990 and 1995 by the JOA).

Methods of process-preservation	Number of institutions	
	1985–1989	1990–1994
Freezing	93	125
Freeze-drying	3	14
Liquid nitrogen	3	3
Defatting	2	2

Note: in 50 institutions, imported allografts were used (freeze-dried bone and fascia).

All the bone banks, other than Aichi Bone Bank and Kitasato University Hospital Bone Bank (KUBB), only preserve femoral heads that are retrieved when they have total hip replacement surgery for femoral fracture or osteoarthritis of the hip point. From the survey, 50 institutions showed that they import bones from countries such as the USA. JOA has been working on a bone allograft survey for the last five years.

In 1991, JOA issued the first guidelines for orthopaedic transplantation, due to the benefit of bone allografting. These guidelines show how to conduct bone allograft safely with ethical and medical consideration. In 1992, there was a case in the USA, where patients were infected with transmitted HIV from organs and bone allografts. The American Association of Tissue Banks (AATB) had emphasised sterilisation of tissues and inactivation of viruses. The JOA renewed their guidelines to include the sterilisation of tissues and the method to inactivate viruses.

Since 1971, when KUBB started, a deep freezer (−80°C) had been used. In the beginning, the same method as in Kyushu University was used to preserve the bones, which were retrieved from a living donor, in a sterilised airtight container. Due to the increase of activity at the hospital, the demand of bone allografts has increased year by year. In 1978, a discussion with law experts about the legal rights of retrieving bones from a cadaver determined that retrieving tissues, such as bones, is legally recognised for the purpose of grafting from a cardiac dead cadaver. Since 1954, MHW had made sure to have the donation agreement signed by the next of kin. Since 1978, with "The Agreement of Bone Retrieval", the retrieval of bones from the patients who have died in Kitasato University Hospital has begun. This was the very first retrieval of cadaver bones in Japan. In 1994, "Ethical Rules for Bone Transplantation" and a "Technical Manual" were submitted to Kitasato University Medical School and the Kitasato University Hospital's ethical committee and have been approved. In accordance with the sterilisation, preservation and processing methods of the technical manual, the retrieval of bones, ligaments, and tendons is done from living and cadaveric donors. The technical manual for transplantationis thoroughly adhered to.

In 1992, four University Hospitals in Aichi-prefecture and few other institutions, gathered and founded the Aichi Bone Bank, the very first regional bone bank in Japan. Mainly, they are retrieving bones at the University of Aichi Health and Allied Medical. The processing and preservation is done at Hachiya Orthopaedic Hospital, the head office of Aichi Bone Bank. The bones are then distributed to member institutions.

In 1958, MHW completed the Law for Cornea Transplantation, which says that the cornea must be retrieved from a cadaver. As soon as this law was concluded, Eye Banks were built and spread throughout Tokyo, Hyogo, and other locations, with the support of the Lions Club. Also, at the same time, kidney transplantation had been started with donors of living relatives and became very popular in a very short time. In 1978, MHW issued the Law for Kidney Transplantation, since there were over 500 cases of living kidney donors and the need of law to regulate the procedure was requested by the Japanese Transplantation Society (JTS). This law made it legal to retrieve kidneys from cardiac dead cadavers and illegal to retrieve kidneys from a beating heart cadaver. Therefore, it was required for a doctor to stop the patients heart, which was kept running by respirator or intensive care, and to show the patients family that ECG was flat. Clearly, transplanted kidneys that had blood flow stopped before the retrieval did not show as

Table 2. Organ transplantation in Japan (accoding to registration of the JTS).

Organs transplanted	Donor		
	Cadaver (brain death)	Living relatives	Total (brain death)
Heart	6 (6)		6 6(6)
Liver	6 (6)	41	47 (6)
Kidney	158 (12)	550	708 (12)

good results as compared to the ones from living donors. Due to the reasons stated above, kidney retrieval from living donors is still used (Table 2).

In 1968, Professor Wada performed the first heart transplantation at Sapporo Medical College, retrieving a heart from a 17-year-old male, who had drowned, and transplanting the heart to a male in his 20's, who had a very critical heart condition. At that time there was no clear diagnostic criteria for brain death, and no consensus on organ transplantation within Japanese society. This heart transplantation was so sudden that everyone paid attention to the event. There was a lot of opposition towards Prof. Wada's performance, e.g. he was sued as a murder by people who were against the organ transplantation.

At this time, JTS was at the beginning of a serious discussion on organ transplantation. However, due to the opposition brought about from the above procedure, with the exception of kidney transplants, the JTS agreed not to perform any organ transplantation. This put Japan in a Dark Age of Organ Transplantation.

3. The General Law for Organ Transplantation

In 1992, the General Law of Organ Transplantation was proposed to The Nation Diet. After spending a long time discussing the matter and making a lot of modifications, the law was concluded on July

16, 1997. This law approved the retrieval of organs, which are to be used for transplantation from brain dead cadavers. The law provides a very strict definition of brain death, diagnostic criteria and diagnostic procedures. The law also covers details of ethical rules, facilities where organs can be retrieved, the equal distribution of organs, and the prohibition of buying or selling organs, etc. Only people who have filled out their last wishes can become organ donors. The doctors who retrieve organs are required to gain an agreement from the donor's relatives as well. It is illegal to retrieve organs from a cadaver without a donor card, even if the relatives has signed an agreement. Under the law, the following organs are legally retrievable: the heart, lung, intestine, pancreas, kidney and cornea. Retrieval of other organs and tissues are prohibited. To comply with the law, within three months, compliance with the rules and guidelines of the organ transplantation to the government offices concerned, a local self-governing body and other relating organisations. A MHW meeting held in July 1997 for the purpose of listing the necessary data that is needed to enforce the law of organ transplantation. The discussions included the role and responsibilities of doctors have in organ transplantation, the manner in which to inform and gain consent of the next of kin, diagnostic criteria of brain death, procedures for record keeping, and the role of the transplantation co-ordinator. These rules and guidelines were set on October 8, 1997, distributed to all organisations involved and have been enforced in The Law for Organ Transplantation since October 16, 1997.

The organ transplantation network was established. The JTS made a system by providing institutes for organ retrieval and transplantation, and actively promoted to popularise the donor card. By having laws in place, more organ transplantation were expected, although in fact, there were few donors. The very first multiple organ transplantation was from a brain dead cadaver on February 25, 1998, a year and a half after the law was implemented. There have been six multiple organ donors since then. The retrieval of hearts, livers, and kidneys transplanted to patients is shown in Table 2. Table 2 shows that there are more transplantation from living donors than from brain dead cadavers. Kidney and liver donors are mostly from living relatives, such as the mother, father, brother or sister. The number of registered donors is very low compared to the number of patients who are willing to take a transplantation.

4. Tissue Graft Issue

There were no discussions about any other tissues or organs under the law stated above. There was an uncomfortable mood among doctors specialising in tissue transplantation, such as heart valve, bone, skin, and so on. Medical groups associated with tissue transplantation, including the JTS, the JOA, and The Japanese Association of Acute Medicine, submitted a petition to the related authorities such as the MHW, members of the Diet, and the Japanese Association of Medical Doctors. As a result, this issue was discussed during a meeting of these specialists. The MHW has added one guideline about tissue transplantation: "The retrieval of tissues from cadavers is legal as long as the act is medically and socially suitable. The retrieval of tissues is only permitted from cardiac dead cadavers. There must also be an agreement, in writing, by the donor or the donor's family." This looks just like Ordinance # 304 in 1954.

In 1996, the JTS formed a working group of professionals composed of medical doctors specialising in the tissue transplantation of heart valves, skin, bone, cornea, etc. for the purpose of tissue transplantation. Prof. Moritoshi Itoman, the director of Kitasato University Hospital Bone Bank (KUBB) participates in this group. The objective of this group is to setup JTS leading guidelines for tissue transplantation and to form the base of a systematic nationwide network. After three years, in 1999, guidelines for tissue transplantation were established.

In December 1999, a specialist meeting for the use of human tissues, as part of the Deliberate Assembly of Health and Welfare Science, was held by MHW. Discussion on framing the guidelines for the governmental point of view was arisen. The participants included the doctors involved in the transplantation of tissue or organs, lawyers, ethicists, corporate members and MHW members.

After several meetings, a report on the guidelines for clinical use of human tissues and organs was completed and submitted to the Prime Minister, because no offical announcement was made on that report, a systematic nation-wide network for tissue transplantation cannot be established. Moreover, there are still members of the Diet opposing to the transplantation. However, based on the JTS guidelines, retrieving tissues and maintaining banks where the tissues are preserved are still being done actively. Better process in providing a safer and better quality tissue and strong support to establish a nation-wide network for these purposes are always be encouraged.

5. Conclusion

This report summarises the current situation of organ and tissue transplantation in Japan. It is good that due to the general law for organ transplantation provided in 1997, organs gained through organ transplantation are equally distributed through the organ transplantation network. Tissue transplantation is mentioned only in the rules for performing organ transplantation and the guidelines from the MHW, and not covered yet in the law. Currently, tissues can be retrieved from brain dead cadavers as well as from cardiac dead cadavers. The number donors will increase. The opportunity to develop tissue transplantation as well as the organ transplantation network in Japan is still explored.

6. Acknowledgement

This paper was presented to APASTB.

7. References

AMAKO, T. (1959). Value and its limit of bone bank, *Gekatiryou* **1**, 324–331.

GOLDHARBER, P. (1961). Osteogenic induction across millipore filters *in vivo, Science* **133**, 2065–2067.

HASHIMOTO, R. (1997). Law No. 104. The general law for organ transplantation, *Government Report* **2181**, 3–5.

ITOMAN, M. *et al.* (1979). Clinical results of frozen allografting and problems of bone bankig, *Seikeigeka* **30**, 905–912.

ITOMAN, M. *et al.* (1988). Banked bone grafting for bone defect repair — clinical evaluation of bone union and graft incorporation, *J. Japan Orthop. Assoc.* **62**, 461–469.

ITOMAN, M. and NAKAMURA, S. (1991). Experimental study on allogeneic bone grafts, *Int. Orthop.* **15**, 1528–1534.

ITOMAN, M. *et al.* (1996). Management of Institutional bone bank, *Teionigaku* **22**, 311–320.

ITOMAN, M. *et al.* (1997). Radiation sterilisation of allografts in Kitasato University Hospital Bone Bank (KUBB). In: *Advance in Tissue Banking*, G.O. Phillips *et al.*, eds., World Science, Singapore, Vol. 1, pp. 323–329.

IWATA, H. *et al.* (1985). Characteristics of AAA bone matrix gelatin and its clinical application, *Seikeigeka* **8**, 33–41.

KITAMURA, S. (1999). *Guidelines for Retrieval, Preservation and Application of Cadaver.* Report from working group 6 (Tissue transplantation) of the Japanese, Society of Transplantation, pp. 1–22.

KNAEPLER, H. *et al.* (1992). Experiments on thermal disinfection and sterilisation of allogeneic bone grafts and the effect of this treatment on biological values, *Chirurg.* **95**, 477–484.

MINISTERY OF HEALTH AND WELFARE (1997). Guidelines for practice of the general law for organ transplantation, *Government Ordinance* No. 1329.

KUROKAWA, T. *et al.* (1999). Revision of "Guidelines for orthopaedic transplantation" and "Frozen bone bank manual", *J. Japan Orthop. Assoc.* **73**, 115–144.

KUROKAWA, T. *et al.* (2000). Femoral head bone bank manual, *J. Japan Orthop. Assoc.* **74**, 135–139.

MASUMI, S. (1981). Retrieval of cadaver bone for banking, *Seikeigeka* **32**, 90–113.

Munting, E. *et al.* (1988). Effects of sterilisation on osteoinduction. Comparison of five methods in demineralized rat bone, *Acta Orthop. Scand.* **59**, 34–38.

NIWA, J. and ITOMAN, M. (1995). Revision of guidelines for orthopaedic transplantation, *J. Japan Orthop. Assoc.* **69**, 71–80.

OGAWA, R. *et al.* (1996). Frozen bone bank manual of the Japanese Orthopaedic Assiciation, *J. Japan Orthop. Assoc.* **70**, 119–128.

OZAWA, T. and ITOMAN, M. (1981). Experimental study on antigenicity of allogeneic bone, *Kitasato Med.* **11**, 482–492.

SIMMONDS, R.J. *et al.* (1992). Transmission of human immunodeficiency virus type-1 from seronegative organ and tissue donor, *N. Engl. J. Med.* **326**, 726–732.

TOMINAGA, T. *et al.* (1999). The influence of sterilisation on the osteoinductive properties of bone in rat bone marrow cell culture, *Kitasato Med.* **29**, 211–219.

UETA, S. (1992). Experimental study on incorporation of various processed allografts and hidroxiapatite, *Kitasato Med.* **22**, 116–123.

URIST, M.R. (1965). Bone formation by autoinduction, *Science* **150**, 893–899.

URIST, M.R. and Hernandez, A. (1974). Excitation transfer in bone. Deleterious effects of cobalt-60 radiation-sterilisation of bank bone, *Arch. Surg.* **110**, 486–493.

WATANABE, T. *et al.* (1998). Experimental study on influence of gamma sterilisation for mechanical properties of the allograft, *Kitasato Med.* 119–126.

WIENTROUB, S. and REDDI, A.H. (1988). Influence of irradiation on the osteoinductive potential of demineralized bone matrix, *Calcif. Tissue Int.* **42**, 255–260.

YAMAMOTO, M. and SUGIOKA, Y. (1991). Guidelines for orthopaedic transplantation, *J. Japan Orthop. Assoc.* **65**, 107–117.

YAMAZAKI, M. and ITOMAN, M. (1992). Effects of preservation prosedures on the biomechanical and osteoinductive properties of rabbit bone, *Kitasato Med.* **22**, 478–484.

YANO, H. *et al.* (1993). Quantitative analysis of revascularisation and new bone formation in experimental bone grafts, *Int. Orthop.* **17**, 109–112.

Advances in Tissue Banking Vol. 6

4 TISSUE DONATION IN A REFERRED HOSPITAL AND IN AN "OUTSIDE HOSPITAL": *THE PAMPLONA MODEL*

ENRIQUE MARAVÍ-POMA, AGUSTINA MARTÍN MONTERO DUE, ENRIQUE MARAVÍ-AZNAR,[1] CARMEN ARENAZ MONTERO DUE, FÁTIMA RODRÍGUEZ FERRO,[2] AMELIA MARAVÍ-AZNAR DUE,[3] RAFAEL TEIJEIRA,[4] MARIAN ALTARRIBA,[2] JOSEFINA RIPOLL and THE VIRGEN DEL CAMINO HOSPITAL TISSUE TRANSPLANT WORKING GROUP

Transplant Co-ordination and DUE-Nurses, [1]MIR of Otorhinolaryngology, [2]MIR of Ophtalmology, of the Virgen del Camino Hospital of the Navarra Health Service — Osasunbidea, [3]DUE-ICU Nurses, San Miguel Clinic, [4]Pathology Service of the Navarra Legal Medicine Institute of the Government of Navarra, Pamplona, Navarra, Spain

Portions of this article were presented at the *9th International Conference of European Association of Tissue Bank* (EATB) and *6th Congress of the Spanish Tissue Bank Association* (STBA) in La Coruña, Spain.

1. Introduction

Human tissue is a valid alternative as a substitute for physiological functions in a human being. One example is the cornea transplant first performed on human beings at the beginning of the last century (Zirm, 1905) by Edward Zirm of the Prague School when the possibility of transplanting organs was still very remote. Since then tissue transplants have become more acceptable and they are now universally accepted as the cornerstone of transplant programmes as far as the number of implants is concerned.

In the last few decades obtaining tissue has been subject to organ transplant co-ordination programmes. These programmes benefited from the development and modernisation of Intensive Care Neurosurgical Unit where locating and obtaining organ donors is a daily mandatory task. From this perspective, hospitals without this characteristic are limited to locating organ donors in their co-ordination efforts. One exception is Catalonia where the Eye Bank of the Barraquer Institute, in Barcelona, promoted donation and performed enucleations since the mid-1950s (Editorial, 2001).

The growing waiting lists of tissue recipients has forced tissue and organ tasks to be distributed (for example in Navarra, Spain). In addition, when the province is small, it is necessary to distribute the medical-assistance tasks amongst the hospital centres, so that each one is a reference for specific tasks. The Virgen del Camino Hospital (VCH) was thus designated as the reference centre for the co-ordination of tissue transplants as a result of a consensus covering the major hospital transplant programmes of the Navarra Health Service — Osasunbidea (SNS-O or NHS-O).

Lastly, the absence of organ donors in this hospital and a mortality rate that prevents the generation of tissue (Maraví-Poma *et al.*, 2000) forced us to develop the search for cadaver donors in outside hospital areas.

The objective is to describe and explain the co-ordination and donation of tissue from cadavers in outside hospital areas.

2. Material and Methods

2.1. Description of the Transplant Co-ordination Programme at the Virgen del Camino Hospital

Pamplona is the capital of the "Foral (having local privileges) Community of Navarra", integrated by a legal pact of 1841 and the "Foral" Enhancement in 1982 with a total area of 10 241 km^2, a population of 523 563 inhabitants and a population growth of 2.10% from 1986–1991 (Comunidad Foral de Navarra, 2001). The so-named "Pamplona Model" is located at the Virgen del Camino Hospital which is a reference for tissue transplants. It was integrated into the Section of Donations and Transplants at the Health Department of the Autonomous Government of Navarra which in turn had been integrated into the Spanish Transplant Organisation (STO or ONT) (Matesanz and Miranda, 1996). The hospital and its transplant programme are based on the Spanish law that regulates tissue transplants (Law 30/1979, Royal Decree 411/1996). The "Pamplona Model" is based on the development of a general protocol supported by the historical background of the transplant programme of the Autonomous Community of Navarra which has been accepted by all the members of the medical establishment who participate in the model.

2.2. Outside of Hospital Tissue Transplant Co-ordination Team at the Virgen del Camino Hospital

Created in 1996, it is composed of a doctor from the Intensive Medicine Unit and two hospital nurses. These members of the medical establishment have successfully taken the National Transplant Co-ordinators' Course organised by the ONT or STO.

2.3. Development of specific protocols

Study and adaptation of use criteria for the detection of potential tissue donors (cornea, bone, skin, osicular tissue, and cardiac valves); infection prevention (bacteriological control and serology); examination of the cadaver as a potential and actual donor; extraction; inter-hospital, extra-hospital and extracommunity transport; storage and transplants.

2.4. Protocol and consensus agreements with the legal authorities

Consensus agreements protocols with judges and forensic pathologists from the different legal courts of Pamplona and the rest of Navarra were developed on the basis of prior agreements.

2.5. Agreements with organisations, institutions, and health centres in Navarra

By means of consensus, audits, and regular meetings with the following organisations: SOS-Navarra, funeral directors, aid and assistance institutions (DYA, Red Cross and medical ambulances), emergency services, intensive care units and nursing management from NHS-O hospitals and private hospitals (University of Navarra Clinic, San Miguel Clinic), meeting minutes and agreements are recorded that are submitted to the provincial co-ordinator for approval and dissemination.

2.6. Specific Teams

The Pamplona Model resulted in a series of teams on permanent duty for each of the links in the co-ordination and logistics of tissue transplant:

- Serological, biochemical and diagnosis team: discussion and analysis of the use criteria formally published by official associations (ONT, 1991). It was designated following an agreement and consensus with the Biochemistry and Microbiological Service Director at the Virgen del Camino Hospital.
- Cornea extraction team.

- Bone extraction team.
- Teams for the diagnosis of natural death due to cardiac arrest, legal and cerebral death.
- Legal assistance team: Motorways Assistants (MA).
- Tissue transport team. rapid transport agency.
- Tissue transplant team: cornea, bone, auditive.

The following are also shown: co-ordination with teams from outside the hospital; transport; ONT or STO; tissue bank, funeral directors, rapid transport agency. Protocols for working with members of the health profession, civilians and relatives from Navarra and outside of Navarra, and, lastly, the economic compensation protocol.

3. Results

In June 1990, transplant co-ordination was created at the Virgen del Camino Hospital, which was directed by an intensive medicine doctor working on a part time basis with the intensive care unit. Its tasks and compensation conformed to the prevailing legal regulations (Royal Decree 426/1980) which were subsequently ratified (Navarra Law Decree, 1994).

In 1996, the hospital was designated as a referred centre for the co-ordination of tissue transplants in Navarra. Shortly thereafter (on December 18, 1996) and in response to a request from the transplant co-ordinator addressed to hospital management, the hospital transplant team was completed with two nurses specialised in transplant co-ordination to support and assist the co-ordinator as overtime work on an "ad-honoren" basis. Management appointed these co-operators in response to the co-ordinator's proposal and as a result of a consensus and authorisation from medical and nursing management, subject to revision every three years. These co-operators were finally incorporated into the programme in January 1997. In this manner, the Virgen del Camino hospital and transplant programme is on alert 24 hours a day including weekdays, holidays and weekends. The entire hospital and outside hospital network and Pathology Service of the Navarra Legal

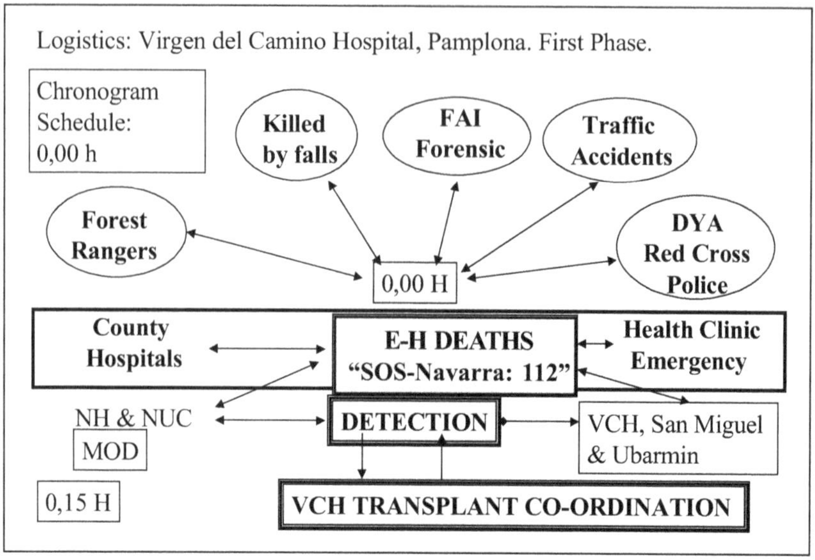

Fig. 1. Logistics: *First phase*: Alerts from SOS-Navarra, Judges and Forensic Pathologists. VCH: Virgen del Camino Hospital; FAI: Forensic Anatomical Institute; DYA: Detente y Ayuda; MOD: Multi-organic Donation; NH: Navarra Hospital; NUC: Navarra University Clinic.

Medicine Institute (NLMI) is covered, where all the tissue donor cadavers in Navarra or the NHS (Fig. 1) involved in legal proceedings are collected.

The transplant co-ordinator's compensation is described in Royal Decree 426/1980. As a result of an agreement amongst general, medical, nursing and hospital personnel management the nurses' compensation is established by means of a medical document. This compensation is covered by the transplant programme revenue from the Government of Navarra as shown in the section corresponding to economic compensation (Table 1).

The three members cooperate in the following manner: The co-ordinator is responsible for the entire tissue transplant programme in Navarra. The two nurses replace the co-ordinator during certain days of the month and his justified absence in accordance with a consensus protocol.

Table 1. Incentive distribution protocol for the Navarra health service — Osasunbidea — Tissue transplant programme. Effective in the year 2001 in the Virgen del Camino Hospital.

Potential tissue donors[1]		Actual tissue donors: corneas			Multi-organ donors: MOD	
		In the VCH	In the FAI	In the HN		
For each donor: 96.16 Euros (16.000 Ptas.)		For each cornea: 147.25 Euros (24.500 Ptas.)			For each MOD: 1190 Euros (198.000 Ptas.)	
Detection	20%	25%[4]	NA	NA	40%	Doctors: 60% DUE-AE: 40%
Extraction and aid[2]	30%	NA	NA	NA	NA	
Serology	NA	10%	10%	NA	5%	
Extraction team	NA	45%	70%	70%	NA	
Co-ordination[3]	50%	20%	20%	30%	20%	
Neurologist					12%	
EEG team					23%	EEG doctors: 55% DUE-EEG: 45%

[1]In this case, every donor is considered a potential one so that the work of all the members of the medical profession involved in the search and detection of potential donors is compensated.
[2]It is defined as assistance in blood sample extraction and assistance provided to the specialist. In the VCH this percentage shall be provided to the person assisting at the time, normally the autopsy room assistant.
[3]Compensation to the VCH transplant programme: This refers to the amounts that do not correspond to anyone but the co-ordinator himself/herself, Due's assigned to the programme, or when one of the members of the medical profession waives his/her rights to the agreed compensation. On the basis of the Hospital Co-ordinator's report these amounts shall be earmarked by the Hospital Managing Director in the Transplant programme activities and for the payment of the two Due assistants in the transplant programme.
[4]In the VCH: A). When the detecting Unit is the ICU, Emergency or hospital patient rooms, the amount (%) corresponding to the DETECTOR shall be distributed 50% between the detecting doctor and the corresponding supervisor (who shall distribute it amongst the non-medical personnel in his/her area). B). When the morgue is involved, the amount is earmarked for the autopsy assistant — hospital porter. C). When the surgical area is involved, the economic assistance is earmarked for the corresponding supervisor who shall distribute it amongst the participating personnel (DUE, assistant nurses, hospital porters, etc.).
VCH: Virgen del Camino Hospital; FAI: Forensic Anatomical Institute; HN: Hospital of Navarra; DUE-AE: University-Graduated Nurses-Assistant Nurses; NA: Not applicable.

4. The Management and Financing of the Transplant Programme at the Virgen del Camino Hospital

The management and financing of the transplant programme organisation devoted expressly to tissue detection, donation, extraction and transplant is based on the fundamental concepts of new institutional economy, property rights, transaction costs and ideological theory (Burgos, Martín and Alonso 1995).

4.1. Consensus protocols for "outside hospital detection" of potential donors

4.1.1. Cadaver detection

A potential tissue donor is defined as any cadaver between 1 and 60 years old that complies with general donor criteria and specific criteria for each tissue. These donors can be found in any hospital, either with cerebral death or cardiac arrest, and in outside hospital areas where these deaths result exclusively from cardiac arrest (mostly traffic accidents).

The protocols worthy of mention due to their special characteristics and originality are described as follows:

Revision and consensus with the affected parties of the prevailing overall and specific protocols and their compliance with the new Tissue Transplant Royal Decree enacted on March 23, 1996, that especially addresses the following:

- The ***Serology Team*** of the VCH transplant programme complies with the new abovementioned law, under the management of the person in charge of the Serology Section of the Clinical and Bacteriological Analysis Service. All the serology is performed on "recipients" and "donors" both alive and as VCH and NHS-O cadavers. The serological process is composed of the following (in addition to the blood group and RH):
  - (a) Emergency Serology: Human Immunodeficiency Virus (HIV) and hepatitis B virus (HBV)
  - (b) Deferred Emergency Serology:

— Cornea donors only: Hepatitis C virus (HCV) and cytomegalovurs (CMV)
— Cornea and bone tissue donors only: HIV-PCR; HCV; CMV; beta-hormone gonadothropic (B-HCG); *Treponema pallidum* (VDRL) and *Brucella.*
— The screening of potential cases of chronic brucellosis is performed in Navarra because this disease was endemic years ago (Díaz, Maraví-Poma and Rivero, 1978; Maraví-Poma *et al.*, 1982; Díaz and Maraví-Poma, 1982)

(c) Multi-organ donor serology (MDS): In addition to determining the blood group and RH factor and analysing hepatic, renal, metabolical and electrolytical functions, the above mentioned tests are performed on an emergency and deferred basis.

The remaining NHS-O hospitals only perform infection serological screening on "*living tissue donors*" for the recovery of spongy bone tissue (femur heads) from orthopaedic surgery for the local hospital bank. It is always "deferred" and it is regularly sent with the serum to the VCH laboratory. The following hospitals are served by the VCH:

— García Orcoyen Hospital in Estella: HIV-PCR; HBV; HCV; CMV and β-HCG.
— Ubarmin Clinic: HIV-PCR and CMV.
— Hospital of Navarra: Only HIV-PCR.

Serotec: Since the beginning of the programme, all the donor serum is collected and stored in a refrigerator chest at the VCH Clinical Analysis Service.

- ***The VCH Cornea Extraction Team***: The service director is responsible for the Eye Bank and a member of the medical staff is responsible for quality, waiting lists, *Keratoplastias* and the search for recipients. Ophthalmological residents (MIR) and VCH surgical specialisation (otorhinolaryngology, maxilofacial surgery, plastic surgery) also participate in this medical team. They act in accordance with an extraction protocol designed by the corresponding specialist.

- ***The VCH Keratoplastia Team***: The Service Director is responsible for cornea transplants and a staff member is responsible for waiting lists and the search for recipients.
- ***The Bone Tissue Programme***:

  — "Spongy" bone tissue extraction teams in each of the NHS-O hospitals. The Traumatology Service Director is responsible for the VCH Bone Tissue Programme.

  — Extraction Team and Regional Bone Bank of the University of Navarra Clinic. Composed of specialists from the Orthopaedics Department, which is not associated with the NHS-O. It operates by means of a special arrangement with our hospital. Person in charge: the service director or delegate.

- ***University Clinic Team***: The person responsible is the transplant co-ordinator. This team acts as an *auxiliary cornea extraction team* and it is authorised to perform *Keratoplastias* and to participate in the bone tissue programme.
- ***San Miguel Clinic Team***: The person responsible for medical management acts as the transplant co-ordinator. The team acts as an auxiliary cornea extraction team and it is authorised to perform *Keratoplastias*.
- ***The SOS-Navarra "112" Team***: It is the co-ordinating centre for civil protection and emergencies in Navarra. It is the organisation where all the calls for help and the detection of "cadaver donors" in all of Navarra are centralised. The person responsible is the director.
- **The *Navarran Legal Medicine Institute (NLMI)***: Its intervention is crucial with cadaver donors involved in legal proceedings, i.e. when deaths have been traumatic and a judge must intervene. Location: NLMI Pathology Service and its members, all of whom can be located by mobile phone. They belong to different legal courts in Navarra. The *Navarran Legal Medicine Institute* Protocol, resulting from a consensus amongst hospitals, and provincial co-ordination, forensic pathologists and SOS-Navarra, is as follows: *Tissue Donor Definition*: All deaths due to accidents for which the doctor or first-aid giver does no detect any heart beat or breathing, thus certifying death. These deaths normally occur in

outside hospital areas and almost all of them are involved in legal proceedings where the judge and forensic pathologist intervene.

4.1.2. NLMI pathology service action protocol

A. Alert from the SOS-Navarra to forensic pathologists:

1. SOS-Navarra, when it is informed of the existence of a cadaver, collects the following data whenever possible: name of the deceased, closest relative or friend, address (street, city and province, country) and the telephone; approximate age and gender; probable cause of the accident or death; existence or non-existence of open fractures in the extremities; and existence of deep eye wounds.
2. SOS-Navarra, with this minimum data and in the case of cadavers involved in legal proceedings, always calls the forensic pathologist (who represents the judge) first and issues *the order to close both eyelids.*
3. Once the judge and forensic pathologist have been informed, SOS-Navarra calls the pager of the *Virgen del Camino Hospital Transplant Co-ordinator,* unless the forensic pathologist has instructed him to the contrary.

B. Summarised selection criteria for tissue donors in outside hospital areas:

1. Age: Between 1 and 60 years
2. Absence of known infectious diseases
3. Absence of open fractures in extremities with cardiac massage
4. No cranial-encephalic traumatism affecting the eyeballs
5. No legal instructions to the contrary
6. The legal autopsy does not prohibit the extraction of bone tissue, unless the intention exists to extract "iliac blades" (agreement amongst the University of Navarra Clinic Bone Extraction Team, orthopaedic surgeons, forensic pathologists and transplant co-ordination).

C. Cadaver inspection protocol on behalf of the VCH Transplant Programme Co-ordinator, with or without the assistance of a forensic pathologist: once the forensic pathologist has been informed, the *Autopsy Assistant (AA)* opens the Forensic Anatomical Institute and facilitates cadaver inspection activities, as well as the collection of blood samples:

  1. Face: Description of open, bleeding wounds, haematomas and their location.
  2. Skull: Description of bruised, incised and bleeding wounds and their location.
  3. Extremities: Description of bruised, incised and bleeding wounds and their location. Especially if fractures exist and whether they are open.
  4. Description of potential unusual findings

4.2. Detection of "potential" cadavers

Alerts from SOS Navarra, Virgen del Camino Hospital, Hospital of Navarra, NLMI Pathology Service (Forensic Anatomic Institute — FAI), Ubarmin Clinic, San Miguel Clinic, informing of the existence of a deceased person in the outside hospital area or in another hospital not associated with the rest (Fig. 1). Once the message has been received, the co-ordinator acts according to the following protocol:

- Call to SOS Navarra expressing an interest in the cadaver.
- Call to the Hospital of Navarra Funeral Director to request that the eyelids be closed with adhesive tape.
- Telephone request to the forensic pathologist on duty for permission to extract tissue, who in turns informs the judge on duty.
- In the case of "spontaneous" family donations, implement the cadaver donation protocol and inform the judge on duty accordingly in case there were any legal objections.
- Cadaver examination. Detect potential fractures and examine the eyeballs to determine whether it is a bone tissue or cornea donor.

- Extract blood and send it to the VCH Emergency Laboratory.
- When the cadaver is between 1 and 60 years old and the corneas are in good condition, advise the VCH Extraction Team. When bone tissue can also be extracted, advise the UNC in order to move the cadaver there for performing an extraction in the orthopaedic surgical room.
- When the cadaver is between 60 and 70 years old, the UNC is advised to examine the corneas, extract complete eyeballs and the bone tissue.
- Once the cadaver extractions have been performed, it is moved to the FAI or the funeral director selected by the family, always through SOS-Navarra. In the case of tissue donors, Navarra legislation authorises the cadaver to be moved by ambulance under strict hygienic conditions (Navarra Law Decree 87/1987), as well as in the rest of Spain (Judicial Authority Council, 1997). This agreement is a result of a consensus amongst Provincial Transplant Co-ordination, the NLMI and SOS-Navarra. The order to move the cadaver is issued through a call from the co-ordinator or forensic pathologist to SOS-Navarra "112".

5. Donations: Forms and Models

Attachment I: Family permission; Attachment II: Request to the magistrate judge on duty; Attachment III: Tissue extraction document.

When the potential donor becomes an actual one, Attachments I, II, and III, signed by the person responsible for the extraction team, are sent to the judge on duty so that this fact is recorded in writing.

5.1. Family permission

This is requested by the co-ordinator directly to the relative or by telephone. If it is done by telephone, permission is requested to the immediate relative and to another relative as witness. This relative is subsequently called to the VCH Intensive Care

Unit or NLMI, for the signature of consent and submission of Attachment I.

5.2. Obtaining of blood samples

5.2.1. Required material

File with transplant forms — VCH. Gloves, gauze, 20 cc syringes with IM needles; long needles for puncturing subclavia veins, the internal yugular or cardiac vein; sample tubes: two with heparine and two without heparine.

Blood samples are then sent to the VCH Emergency Laboratory with a transplant programme letterhead. A tube without heparine showing the name of the donor is left in the FAI for the forensic pathologist.

6. Protocol for "Distribution of Economic Compensation from the Government of Navarra"

The Health Department, by means of a Navarra law decree, earmarks and yearly updates amounts intended to provide incentives for tissue and organ detection, conservation and donation. The compensation shall be paid to hospitals participating in the programme, in response to the proposal of the Section of Associations and Organ Transplant Co-ordination. The subsequent distribution to participating members of the medical establishment in the Donation and Transplant Programme shall be performed by hospital management in proportion to their level of participation in this Programme and following a report from the hospital transplant co-ordinator. The amounts as of January 2001 (Navarra Law Decree 115/1996 and 74/1999) are as follows:

- For a "potential" tissue donor: 96.16 Euros
- For a cornea extraction: 147.25 Euros
- For a "spongy" bone extraction: 96.16 Euros
- For a "cortical" bone extraction: 589 Euros
- For the detection of a Multiorgan donor: 1190 Euros

The compensation distribution (Table 1) at the Virgen del Camino Hospital has been designed exclusively for providing incentives and economic compensation for all "additional" activities derived from tissue and organ detection, serology, donation and extraction, excluding anything related to the "transplant" itself. To this effect a protocol for incentive distribution has been designed, with *a note* that each member of the medical profession pertaining to the different teams and the nurses assigned to the transplant programme receive this compensation by means of a special medical document and on the basis of their assistance to the transplant co-ordinator, in a percentage that is described in Table 1. A consensus agreement that is signed and ratified by the managing director, nursing director, and medical director, following a report by the hospital Virgen del Camino and Navarra provincial transplant co-ordinator.

This compensation may be paid in one of the following ways:

- Compensation to Government personnel: Always through the payroll in the case of Government of Navarra employees.
- Compensation to personnel not employed by the Government of Navarra: Special co-operation of doctors, nurses or auxiliary personnel not covered by this compensation distribution, shall be evaluated and determined by management following a report by the co-ordinator.

7. Logistics: Action Phases

7.1. *First phase*: Alerts from SOS-Navarra, judges and forensic pathologists (Fig. 1)

The police are responsible for alerting the judge on duty and SOS-Navarra in accordance with regulations. The judge in turn informs the forensic pathologist on duty and subsequently the transplant co-ordinator of the death. The first phase begins when SOS-Navarra informs the VCH Co-ordination Team of the existence of a cadaver in accordance with the designed protocol. The "alert" is given and the chronogram begins to run from 00:00 hours.

The alerts come from different places or hospitals where mortal accidents occurred, as reflected in Fig. 1 and in the section on Cadaver Detection. Legal permission is immediately requested from the instructing judge through the forensic pathologist on duty, who authorises the investigation to begin and the cadaver becomes a "potential" tissue donor.

7.2. *Second phase*: Criteria for selection, search and contacts with family members (Fig. 2)

Once the alert is issued, the different protocols on cadaver transport, evaluation and selection and the forensic action are applied. The family is then contacted to obtain their permission, if possible in writing or by telephone, but always in the presence of witnesses, and an interview is arranged for subsequent signatures. Once permission is obtained, blood samples are taken for serological and analytical study and data is collected for the clinical history. Venous access for blood extraction is through the subclavia or

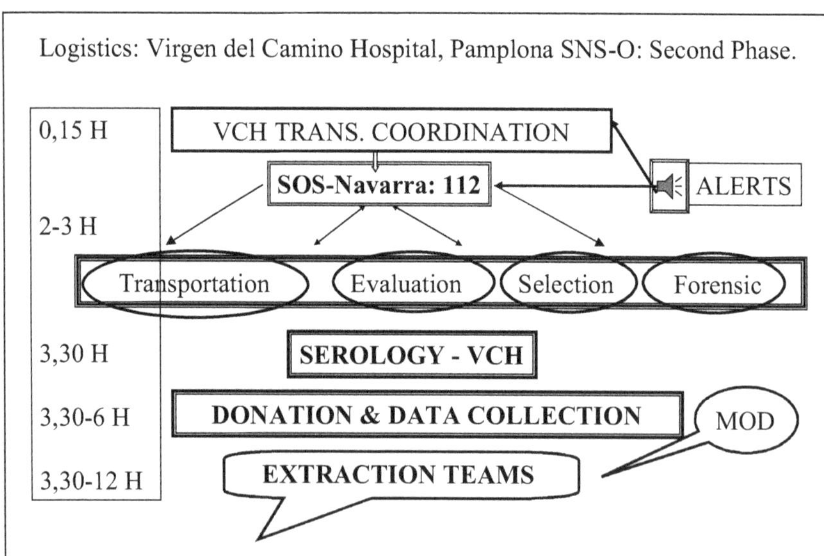

Fig. 2. Logistics: *Second phase*: Criteria for selection, search and contacts with family members.

internal jugular vein, or during the legal autopsy, through the thoracic opening. This phase is completed in 3 1/2 to 12 hours. When tissue from a multi-organ donor (MOD) following a cerebral death is extracted, these two phases are not required.

7.3. *Third phase*: Serology, extraction and transport teams are alerted (Figs. 2 and 3)

When the studies and analyses described above are negative and the cadaver complies with the criteria, it becomes an "actual donor". The different extraction teams are then alerted. The cornea extraction team is the first to act and then the bone and other extraction teams. Following local decontamination, and in accordance with the approved protocols, the tissue is prepared and stored in local or central banks, depending on whether it is cornea or bone tissue. The doctor responsible for each speciality is assigned the task of preparing and locating the "recipient" and scheduling the transplant. When the referred hospital does not have a waiting list, it is passed on to the VCH Transplant Co-ordinator, who is responsible for distributing, preparing and transporting it to the place where the tissue is required (Navarra Province, nearby provinces or the Cornea Bank of the Spanish Transplant Organisation Bank in Madrid–ONT or STO).

7.4. *Fourth phase*: Preservation, storage, distribution and transplant of tissue (Figs. 3 and 4)

Once the "recipient" has been selected, the transplant is performed. The corresponding service is responsible for follow-up, monitoring new developments and updating the waiting list and it reports to the co-ordinator.

7.5. *Fifth phase*: Distribution and agreed incentives

Economical, administrative and relevant regulations. The VCH transplant co-ordinating team informs the judge of all the events,

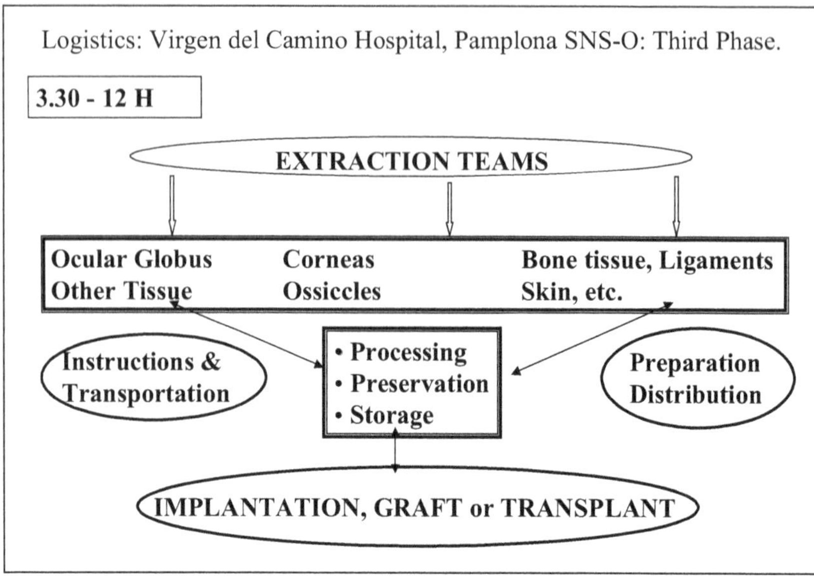

Fig. 3. Logistics: *Third phase*: Serology, extraction and transport teams are alerted.

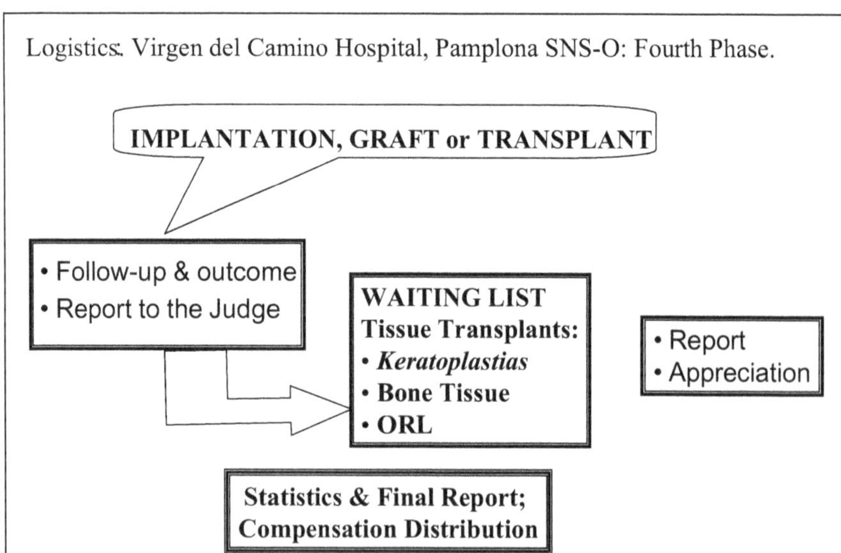

Fig. 4. Logistics: *Fourth phase*: Tissue preservation, storage, distribution and transplant.

writes the report for the provincial co-ordinator, issues the acknowledgements from the Health Councillorship on behalf of the Health Councillor, as well as thanking all the participating teams. The data is stored for statistical purposes and a complete report is written every 6 months. The "incentives distribution" (according to Table 1) for members of the detection, selection, serology and transport teams, excluding the actual transplant, is drawn up.

The entire logistics process of an "outside hospital" co-ordination lasts between 3½ and 12 hours in total when the tissue donor is an "actual" one. When a "potential" donor is involved, the elapsed time is between 30 min and two hours, according to the personal, social and logistics circumstances of the cadaver. The co-ordination team is available 24 hours a day, 365 days a year.

8. Virgen del Camino Hospital Transplant Programme Activity

With 1999 data we describe the donating and transplant activity of a referred hospital for outside hospital tissue. The data corresponds to a referred hospital (VCH), to the Forensic Anatomical Institute, Navarra Health Service-Osasunbidea (NHS-O) and the Spanish Transplant Organisation (ONT or STO).

Since the Pamplona Model began (Fig. 5), it has been observed that the forensic pathologist activity during these years has resulted in 49.5% of potential tissue donors being detected out of 168 cadavers.

The activity of a referred hospital for outside hospital tissue is summarised in Table 2 with 1999 data. The VCH co-ordinating team received 57 alerts from SOS-Navarra with 49 cadavers (86%) becoming "potential tissue donors" from the Virgen del Camino Hospital, Forensic Anatomical Institute (FAI) and the Hospital of Navarra (HN).

In regard to the donation of muscular-skeleton tissue, in 1999, Navarra obtained 380.76 living donors/pmp and cadaver donation was 44.23 donors/pmp (Fig. 6). The total number of parts per million inhabitants was 686.53 pmp.

Table 2. Virgen del Camino Hospital transplant programme activity in 1999.

Nº SOS-NA alerts Potential TD (P-TD): **(86%)**	**57–100%** **49 donors**	Corneas received VCH extraction team:	**0** **39 corneas**
• Family refusal • Medical counterindication • Actual donors	**7–14%** **21–43%** **21–43%**	• In the VCH • In the FAI • In the HN* Waiting list 1999	— **29** **10** **16**
Use of actual TD's:	**21 donors**	Time period:	**57 donors**
Cortical bone extraction: UNC (calote extraction) Cornea extraction: UNC Cornea extraction by the VCH	**17–81%** **1–4.7%** **21–100**	A. 8–15 hours B. 15–20 hours C. 20–24 hours D. 0–8 hours	**31%** **23%** **28%** **18%**
Use of the corneas	**39–100%**	Hours worked:	**Average:**
• *Keratoplastias* in VCH • Sent to the UNC • ONT • For experimental surgery	**16–41%** **9–23%** **7–18%** **7–18%**	Potential-TD Actual-TD	**0.5–2 h** **3.5–12 h**
Serology: Cadaverous and living donors	**179 donors**	**Serology**: Urgene **Serology**: Deferred	**— In cadaverous** **— In living donors**
Living D.: Ubarmin Clinic Living D.: Estella Hospital	**60–33.5%** **30–16.7%**	**Serology**: Complete with cadaverous, and Living D: Hospital of Navarra	**49–27.4%** **40–22.4%**

*Corneas extracted in the NH Surgical Room following organ extraction; TD: Tissue donor; P: Potential; UNC: University of Navarra Clinic; VCH: Virgen del Camino Hospital; FAI: Forensic Anatomical Institute; HN: Hospital of Navarra; ONT: Spanish Transplant Organisation; Tx: Transplants.

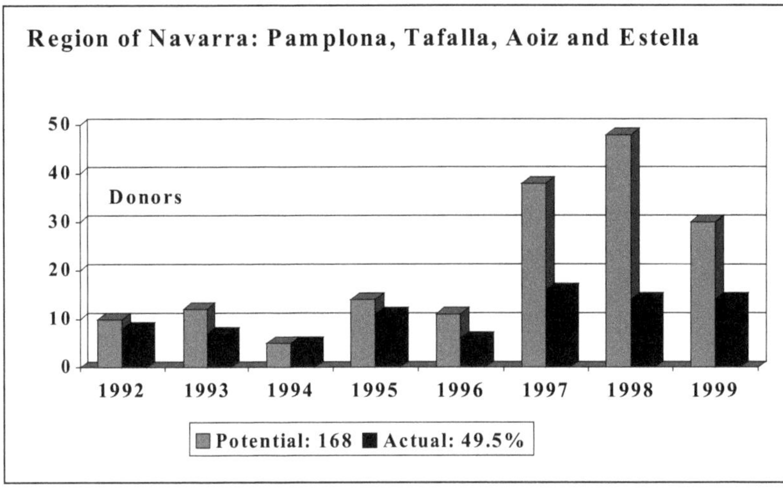

Fig. 5. Forensic medicine activity.

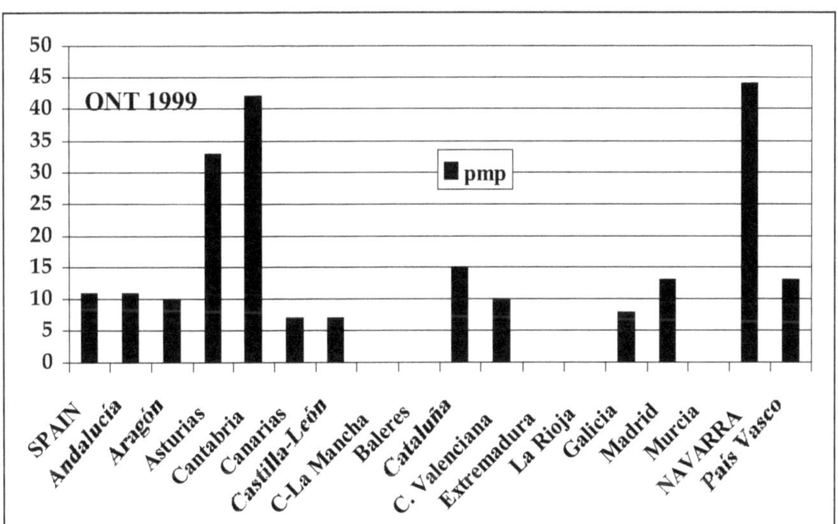

Fig. 6. Cadaveric donation activity. Skeleton-muscular tissue (1999 STO/ONT data).

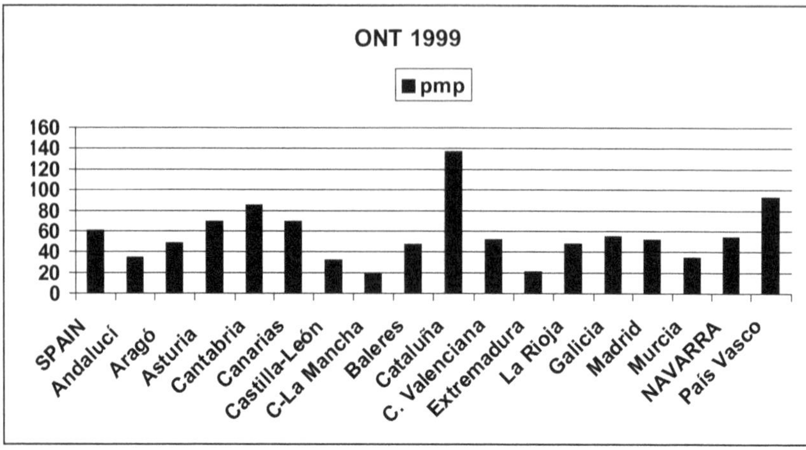

Fig. 7. Transplant activity using cadavers. Corneal trasplants by regions in Spain (1999 STO/ONT data).

4018 corneas were extracted in 1999 for transplant purposes in Spain. However, 39.2% of tissue obtained was ultimately rejected for Keratoplastia, double the rejection figure at the VCH that was 18% (Table 1). The cornea transplant rate was 47 per million inhabitants, slightly below the average Spanish rate of 61.6/pmp (Fig. 7). The waiting list rate for cornea transplants in Navarra was 2/pmp compared to the Spanish average of 52.45.

9. Discussion

Transplant activity in Navarra began in 1969 with the first renal transplant. Multi-organ transplants were first performed in 1984. In 1990 the post of tissue transplant co-ordinator was created in Navarra that has always been held by an intensive medicine specialist. Since 1992, when the VCH was designated a referred centre for cornea transplants, there have been no cerebral deaths donors and almost no cadaver donors. Consequently, transplant co-ordination implemented cadaver donor detection in outside hospital areas in accordance with the corresponding recommendations.

(Aranzabal *et al.*, 1996). In 1996 the VCH is defined as a referred centre for the tissue programme in Navarra.

The Tissue Transplant Outside Hospital Co-ordination Model, the so-named *Pamplona Model* rests on three pillars: 1) The referred hospital (VCH) *Co-ordination*, 2) *SOS-Navarra* "112" and 3) Forensic Anatomical Institute. This method does not have any known precedents, because all the hospitals detect cadavers almost exclusively in their own hospital (Escudero *et al.*, 1997). In this respect, the little experience we have (Maraví-Poma *et al.*, 2000) consists of extracting corneas in the house of the deceased (Alonso and Burgos, 1996). Consequently, the *Pamplona Model* is based on the "centralisation" of "outside hospital" cadaver detection in a public service: *SOS-Navarra*, that covers all of Navarra and is accessible and profitable.

As in the rest of Spain (Matesanz and Miranda, 1996; Alonso-Pulpón, 2000), the success of the tissue transplant programme rests on three basic pillars:

1) Adequate organisation which is an exact reflection of and is based on the ONT or STO.
2) Incentives which are partly based on salaries and exclusively covers detection, donation and extraction activities. It is also true that a health system like that of Navarra, as in the rest of Spain (Alonso-Pulpón, 2000), which in our opinion is the best public health system in the world (ONT or STO), offers few incentives to the members of the medical profession. Transplants are the only activity that offers direct incentives in a regulated form in Spain and Navarra, with advantages and disadvantages in comparison to the rest of the health system.
3) A high traffic and labour accident rate. Most of the cadavers come from traffic accidents for which Navarra has a high rate. More specifically, 2514 accidents occurred in the year 2000, in which 92% were traffic accidents. Seventy-nine of these accidents resulted in 100 sudden death and 15 people subsequently died, where 58% of these 115 people were below 50 years in their perfect health when the accidents occurred. This mortality

figure is the highest in the last seven years in Navara (SOS-Navara, 2000). According to the same source in Navarra, and in proportion to Spanish population, the traffic death rate is 82% higher than the overall average and more than double of the European average. Since 1998, Navarra has a systematic plan for the identification and elimination of "dangerous stretches of road" in Spain motorway network and recently the Government of Navarra and the Councillorship of Health and Public Works has, amongst other priorities, the goal of preventing traffic and labour accidents.

Co-ordination and logistics cannot be scheduled in an urgent manner, because it can "vary in time" throughout the day. It operates with professional and economic incentives. This is another of the characteristics that distinguishes the "*Pamplona Model*". Incentives for donation and distribution, are intended to promote and support donations along its entire logistics chain by providing the stipulated amounts through a medical document considered "special and over-time". It is handled separately from the assistance provided by the members of the health profession and this is the approach of the health authorities and the co-ordinator. This assistance specifically excludes "transplant" action that is clearly medical assistance covered by the Spanish Health System and in this case by the NHS-O (Royal Decree 1996; 1999). The economic quantities are established yearly by the Parliament of Navarra, and they are debated, approved and included in the General Budgets.

The percentage of tissue donors registered at the NHS-O is much higher than the average registered at the INSALUD area in Spain. Family refusals (14%) are relatively frequent, but they are much lower than the average observed in Spain: 22.1% in the year 2001.

10. Conclusion

The *Pamplona Model* with the Virgen del Camino Hospital as a reference for the co-ordination of outside hospital tissue donors, is efficient. Scientific, organisational, and administrative progress and the contributions of different teams have made tissue transplant from outside hospital donors a daily practice in Vergen del Camino Hospital, with a growing number of donors and patients with access to this therapeutic method.

11. Acknowledgements

The VCH Transplant Co-ordination wishes to thank Mr. Robert Kimball, Liberal Arts graduate and member of the sales staff at VAW-INASA for translating and revising the English version of this article.

12. Virgen del Camino Hospital Tissue Transplant Working Group

1. *Hospital Virgen del Camino*. Transplant Co-ordinator: Dr. Enrique Maraví-Poma ICU. DUE Nurses: Dña. Agustina Martín Montero and Carmen Arenaz Erburu.
2. *ICU*. Co-ordinator Dr. Enrique Maraví-Poma, Director Dr. Alfonso Manrique Larralde. DUE Nursing: Mª Jesús Elizondo and ICU team.
3. *Hospital de Navarra*. Transplant Co-ordinator: Dr. José Elizalde.
4. *Ophthalmology*. Dr. Rafael Iturralde, in-charge of the Eye Bank and Dra. Marian Altarriba Freire, in-charge of quality, *Keratoplastia* waiting lists, and the search for recipients.
5. *VCH Residents*. Ophthalmology Resident: Dra. Fátima Rodríguez Ferro; Maxilofacial Resident: Dr. David Pellicer Artigot; Plastic Surgery Resident: Dr. Ignacio Palacios Ortega, Dr. Javier Montón Echeverría, Dr. Pedro Roldan Velez; Otorhinolaryngology Resident: Dr. Enrique Maraví-Aznar.
6. *Serology*. Person-in-charge: Dr. Víctor Martínez de Artola; Resident: Dra. Lucía Herrero Lobo, Dr. Gregorio Aisa Rivera.
7. *Traumatology*. Director: Dr. Manuel Martínez Grande; Person-in-charge: Dr. Javier Heras Elizalde, Dr. José Javier López Blanco; DUE Nursing: María Yerro Cruz.
8. *Emergencies*. Person-in-charge: Dr. Javier Sesma and emergency team.

9. *Haematology*. Director: Dr. Francisco Oyarzabal.
10. *Navarra Legal Medicine Institute. Team of the Government of Navarra*. Forensic Pathology, Director of the NLMI: Dr. Rafael Teijeira; Motorways Assistants (MA): Messrs. Iñaki Fadrique Garde, Javier Cirez Ortega, José Joaquín Garrido Villalvilla.
11. *University of Navarra Clinic (UNC)*. Co-ordinator: Dr. Pedro Errasti; Ophtalmology: Dr. Javier Moreno; Orthopaedics and Bone Bank: Dr. Mikel San Julían.
12. *San Miguel Clinic*. Co-ordinator: Dr. Yolanda Yoldi.
13. *SOS-Navarra*. Person-in-charge: Mr. Francisco Fernández Elizalde.

13. References

ALONSO-PULPÓN, L. (2000). Cardiac transplants in Spain. Organisation and results, *Rev. Esp. Cardiol.* **53**(Suppl 1), 39–52.

ALONSO, M. and BURGOS RODRÍGUEZ, R. (1996). The transplant co-ordinator in obtaining and transplanting tissue, *Rev. Esp. Trasp.* **5**, 153–157.

ARANZABAL, J., ALONSO, M., CABRER, C., FELIPE, C., GETINO, A., MARAVÍ-POMA, E., NAVARRO, A. and OTERO, J. (1996). The transplant co-ordinator and activities related to the use of human tissue, *Rev. Esp. Trasp.* **5**, 150–152.

BURGOS, R., MARTÍN, J. and ALONSO, M. (1995). Management models and organ transplants. In: *Co-ordination and Transplants, the Spanish Model*, R. Matesanz and B. Miranda, eds., Grupo Aula Medica, S.A. Madrid (España), pp. 177–184.

CANBAY, A.E., ELLA, R. and WEBER, F. (2001). Attitudes towards cadaveric organ donation: Comparison of German and Turkish citizens in Germany, *Organs Tissues* **4**, 29–35.

CONSEJO GENERAL DEL PODER JUDICIAL (1997). Agreement 49 of the Permanent Commission: "Take note and acknowledge the agreement of the Intra-territorial Council of the Spanish Health System to the Minister of Health and Consumption which submits recommendations to the corresponding judicial associations for facilitating organ and tissue transport and avoiding autopsy practices under the conditions described in Article 785.8ª f) of the Criminal Law Code". This INSALUD agreement shall be submitted to the Presidents of the High Justice Courts for their knowledge and dissemination to the courts under their jurisdiction for their knowledge only. *Registro General del Poder Judicial,* 25 Junio: Salida Nº 1/14762.

DÍAZ R,, MARAVÍ-POMA, E. and RIVERO, A. (1976). Comparison of counter-immunoelectrophoresis with other serological tests in the diagnosis of human brucellosis. (Summary of Mr. Enrique Maraví-Poma's doctoral thesis for obtaining a PhD.), *Bull World Health Organ* **64**, 417–424.

DÍAZ, R., MARAVÍ-POMA, E., FERNÁNDEZ, J.L., GARCÍA-MERLO, S. and RIVERO-PUENTE, A. (1982). Brucellosis: Study of 222 cases. Part IV: Diagnosis of human Brucellosis, *Rev. Clin. Esp.* **166**: 107–110.

EDITORIAL (2001). Is it time for tissue? *Trasplantament* 17 Junio, 1.

ESCUDERO, MD., OTERO J. and GAGO, E. (1997). Tissue donor. In: *Organ and Tissue Donor: Evaluation and Assistance,* López Navidad *et al.*, eds., Springer Verlag Ibérica. Barcelona, Spain, pp. 508–520.

LAW 30/1979 (1979). Law 30 dated October 27, 1979 on Organ Extraction and Transplant. Refer to the implementation of Royal Decree 1980.

MARAVÍ-POMA, E., MURIE, M., GAMBOA, J., DÍAZ, R. and RIVERO-PUENTE, A. (1982). Brucellosis: Study 222 cases. Part III: Chronic Brucellosis clinic, *Rev. Clin. Esp.* **166**, 101–105.

MARAVÍ-POMA, E., MARTÍN, A., ARENAZ, A. and LIZARRAGA, J. (2000). Hospital mortality and tissue donations in the referred hospital: Interhospital comparative data. The *9th International Conference of the European Association of Tissue Banks* (EATB) and *6th Congress of the Spanish Association of Tissue Banks* (AEBT), November 30–December 2, La Coruña, Spain.

MATESANZ, R. and MIRANDA, B. (1996). *Organ Donation for Transplants. The Spanish Model.* Grupo Aula Medica, S.A., Madrid, España.

NAVARRA LAW DECREE (ORDEN FORAL) 37/1992. Moving a cadaver involved in legal proceedings. Dated January 27. *Official Navarra Bulletin* of February 5.

NAVARRA LAW DECREE (ORDEN FORAL) 115/1996, (1996). Of the Health Councillor, dated May 7, that promotes the yearly updating of economic compensation for organ and tissue detection and extraction. Government of Navarra, Health Department, May 12.

NAVARRA LAW DECREE (ORDEN FORAL) 74/1999 (1999). Of the Health Councillor, dated 10th March, that updated the amounts of economic compensation for organ and tissue detection and extraction. Government of Navarra, Health Department, March 12, Salida n° 2295.

NAVARRA LAW DECREE (ORDEN FORAL) (1994). Navarra Law Decree dated February 21, of the Health Councillor, that regulates technical and administrative co-ordination of the transplant programme for organs, tissue and other anatomical parts in Navarra. *Official Navarra Bulletin.* March 14, **BON n° 31**, 1091.

ONT (2000). *Reports and Consensus Documents.* Spanish Transplant Organisation, Ministry of Health and Consumption ed., Madrid (España), Editorial Complutense, S. A.

ROYAL DECREE (REAL DECRETO) (1980). Royal Decree 426/1980, dated February 22, that implements Law 30/1979, October 27, on organ extraction and transplant. *Official Spanish Bulletin,* March 13, **BOE n° 63**, 41–50.

ROYAL DECREE (REAL DECRETO) (1996). Royal Decree 411/1996, dated March 1, that regulates activities related to the use of human tissue. *Official Spanish Bulletin,* March 23, **BOE n° 4**, 123–134.

ROYAL DECREE (REAL DECRETO) (1999). Royal Decree 2070/1999, dated December 30, that regulates the activities for obtaining and clinically utilising human organs and territorially co-ordinating organ and tissue donation and transplant. *Official Spanish Bulletin*, January 4, **BOE nº 3**, 179–190.

SOS-NAVARRA "112" (2000). Report on emergency health assistance, excluding assistance at home, which is covered by SOS Navarra, by mobilising the ambulance network and transportation to hospitals in the year 2000.

ZIRM (1995). Http://www.cirugiaocularcr.com/indicaciones.htm

SECTION II:

MUSCULOSKELETAL

Advances in Tissue Banking Vol. 6

5 ALLOPLASTY IN CONJUNCTION WITH METALLOOSTEOSYNTHESIS FOR BONE DEFECTS REPLACEMENT IN PAEDIATRICS OF TUMOUR-LIKE PATHOLOGIES OF THE SKELETON

S.P. MIRONOV, A.I. SNETKOV, M.V. LEKISHVILI, A.S. SAMKOV, I.A. KASYMOV, M.G. VASILIEV and S.Y. BATRAKOV

Central Institute of Traumatology and Orthopaedics (CITO), Moscow, Russia

1. Introduction

Transplantation of bioimplants with the aim to replace post-incision osseous defects is an urgent problem not only in trauma and orthopaedics but also in dental, maxillary-facial surgery, otolaryngology, etc. In paediatric orthopaedics dysplastic diseases of skeleton bones demand the most attention. Extensive damages as well as heavy deformations of skeleton long bones require complicated reconstructive interventions directed to the elimination of dysplastic tissue, replacement of the defects by a certain plastic material and fixation of damaged segments (Urist, 1979; Snetkov, 1988; Kasymov, 2000). An alloplasty with grafts of allogenic origin is the most common technique for osseous pathology relief in paediatric orthopaedics. These allografts, being of high plasticity, are not actually limited by size, shape and amount of the material required. Additional trauma, being the case in autotransplantation (Kumta *et al.*, 2000), is avoided using these grafts. Those are the reasons for this plastic technique to be in advance in paediatrics of skeleton dysplasia (Volkov and Biser, 1969; Volkov and Berezhnoi, 1983; Berezhnoi *et al.*, 1997; Volkov, 1985). A tendency of more extensive surgical intervention, conducting of radical and organo-protecting oncology incisions, a desire to improve the results of medical treatment inevitably lead to a search for more efficient plastic material for osseous defects restoration. Demineralised grafts prepared by means of various techniques are the materials of such kind. In recent years, these bioimplants have been rather well developed at theoretic, experimental and clinical levels (Urist, 1979; Saveliev, 1996; Reddi, 1998; Solheim, 1998; Russel and Block, 1999). The material has shown to be excellent in various fields of plastic surgery due to its high osteoconductive and osteoinductive properties (Harakas, 1984; Gendler, 1986; Bolander and Balian, 1990).

2. Results and Discussion

Partially demineralised allografts with various extent of demineralisation, prepared by special technology, have been also used to activate reparative processes in the field of plastics in CITO. A bulk bone tissue sampled from cadavers of 19–45 years old was used. Over this age donor group a percentage of mineral phase in bulk (compact) bone is equal to 62.10 ± 0.61% total weight (Matveichuk, 1998). The decrease of the sample weight under demineralisation (up to 5% of the initial) corresponded to low extent of demineralisation. The grafts of moderate and high extent of matrix demineralisation showed respectively up to 35–40% and 70–75% weight loss.

The technology of demineralised graft preparation developed in the Tissue Bank of CITO includes a number of stages (Lekishvili and Kasymov, 1999). The demineralisation was carried out by 0.6–1.8 N hydrochloric acid at 18–20°C. Lyophilisation was conducted using LZ 9.2 installation (Chech Rep.) over 48–72 hours with subsequent sterilisation by high-velocity electron beam of 20–25 kGy dosage using LUE-8-5M and U003 MV emitting technological equipment (Russia) of 8–9 MeV radiative energy and 1.0 kGy/s dose power. This sterilising technique has been demonstrated to be very efficient. At the same time, osteoinductive properties of biomaterials were rather well preserved. These observations were also in agreement with other studies (Dziedzic-Goclawska *et al.*, 1991; Munting *et al.*, 1998).

In a decade from 1990 till 2000 there have been 110 children with dysplastic diseases with extensive damages and complex deformations of skeleton bones treated in the Clinic of Paediatric Osseous Pathology and Orthopaedics at CITO. Among them 74 suffered with polyostotic form of fibrous dysplasia, 28 dyschondroplasia and 8 osseous cysts. The patient's age ranged from 8 to 16 years old.

Alloplasty combined with metalloosteosynthesis of various kinds was used for surgical relief. The necessity to apply the alloplasty combined with metalloosteosynthesis was related to following reasons: Fixation of bone crushed fragments occurred after correcting osteotomies, correction of deformation and shortening of an extremity bone under application of metallic holder as a long-term endoprosthesis.

All of the variants of the combination of alloplasty with metalloosteosynthesis were separated into three general groups depending on type of the metallic construction used:

1. On-bone metalloosteosynthesis.
2. Intramedullar metalloosteosynthesis.
3. Compressive distraction metalloosteosynthesis.

The on-bone metalloosteosynthesis combined with alloplasty has been applied to 72 patients with polyostotic form of fibrous dysplasia associated with extensive damage of long bones of lower extremities. Angular holders (ASIF, Germany) with the assigned neck-diaphysis angle of 130° and 200–320 mm cover plate were used.

3. Operational Technique

The thigh bone was disclosed by a longitudinal entrance on lateral surface of the femur over a distance from large trochanter to diastole metaphysis. The resection of cortical layer was performed over the anterior surface of the bone allowing fixation of the holder along its lateral surface. After the damaged tissue was removed and the bone partitions were treated by rotary cutters from the inside, the correcting osteotomies at the upper deformation of the segment were carried out (Figs. 1 and 2).

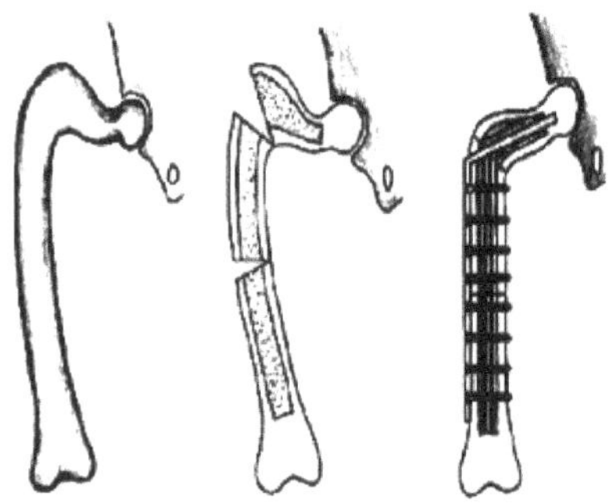

Fig. 1. Scheme of the correction and substitution of a femur defect with allografts in combination with on-bone metalloosteosynthesis.

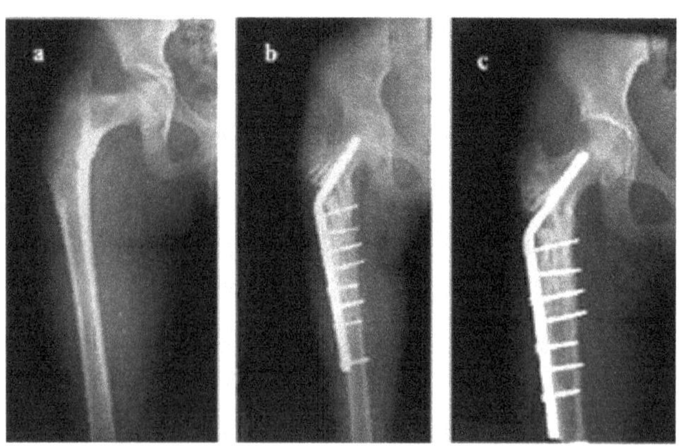

Fig. 2. Alloplastics combined with on-bone metalloosteosynthesis: (a) before operation; (b) after the operation; (c) 6 months after the operation.

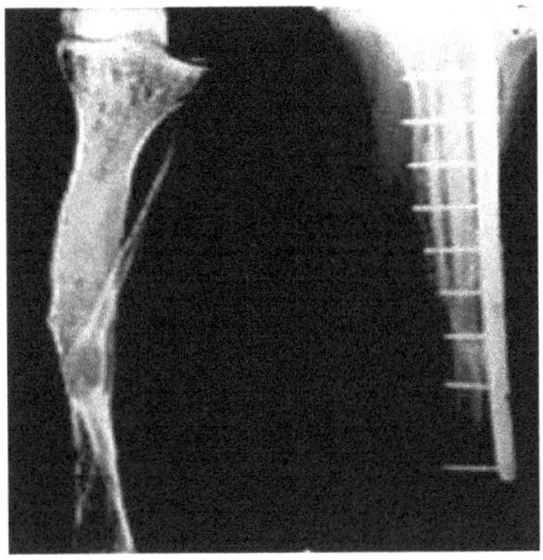

Fig. 3. On-bone metalloosteosynthesis combined with alloplasty by demineralised allografts.

Intraosseous alloplasty with bulky cortical implants having surface layer of demineralisation was the second stage. As the alloplasty was complete, the blade of the slab was intruded to cervix of the femur. The cervix defect was filled with cortical implanting material of finer composition with high to moderate extent of demineralisation. The cover plate was fixed by screws, which were drawn through the allografts (Fig. 3).

As a significant tapering of the cortical layer over the inner bone surface was the case, the on-bone stacking of the allograft (matrix) over the surface using screw fixation was obligatory for stability of the graft.

Similar operational technique was also applied to treat extensive damages of tibia. Strait metallic constructions were used in these cases.

Intramedullar metalloosteosynthesis with bone alloplasty with partially demineralised implants has been applied to treat localised form of dysplasia and tibia deformation for restricted number of patients.

The resection of dysplastic nidi did not lead to any considerable decrease of the bone strength, whereas its fixation by metallic pin along with allografts of high or moderate extent of demineralisation provided sufficiently stable osteosynthesis.

Compressive distraction osteosynthesis combined with alloplasty was applied to 28 patients suffered with dyschondroplasia, 2 fibrous dysplasia and 8 osseous cysts.

Following criteria for the application of the external fixing apparatus have been used:

- above 6 years old age of the patient
- more than 5 cm shortening of damaged segment
- more than 10° angular deformation of the segment
- status of contiguous articulation.

Under all the above-indicated nosological forms, the reasons to apply this surgical technique were as follows: presence of pathological nidus, shortening and deformation of extremity segment.

The following treatment techniques were used:

- method of single-action alloplasty and elongation of the segment by compressive distraction osteosynthesis
- method of double-stage action (peripheral resection of the bone, alloplasty of the defect, elongation of the segment by means of compressive distraction osteosynthesis at the background of organotypical reconstructing of the bone in the plastics region).

The method of single-action alloplasty and elongation of the segment was of most frequent use (29 patients). In this technique, cortical allografts of low extent of demineralisation (superficially demineralised were used). These grafts provide additional supporting points for the pins of external fixation apparatus.

3.1. Single-action alloplasty with compressive distraction osteosynthesis

Soft tissues in a projection of pathological nidus were resected. After a pathological tissue was removed an osteotomia at the top of the deformation of the bone was carried out. The alloplasty with split cortical implants of low demineralisation extent was performed to substitute the osseous defect. Therewith, a single bulky allograft of the length 3–4 cm exceeding the expected elongation value was introduced intramedullary. Osteosynthesis using a compressive-distraction apparatus was followed by the bone plastics. Axial deformation of a segment below 20° was corrected by the single-action, larger deformation — in course of correction of the extremity shortening. The distraction in the apparatus was maintained to 1 mm per day. In course of the distraction the status of the segment axis and the regenerate was monitored by X-rays. Stabilisation in the apparatus was conducted during 1.5–2 month up to an ageing of the regenerate followed by total compensation of the extremity shortening (Figs. 4 and 5).

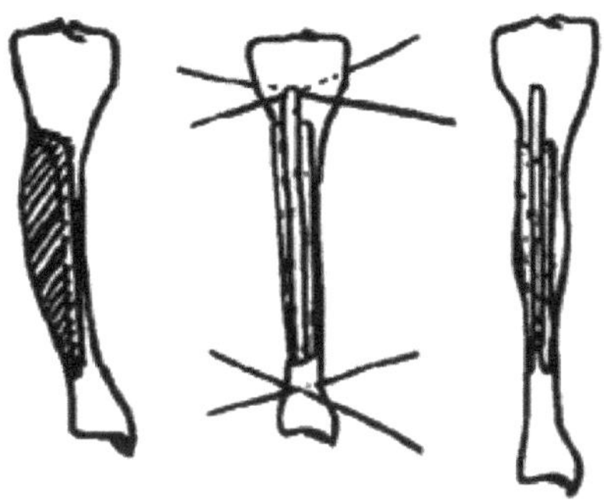

Fig. 4. Scheme of the plastics with simultaneously conducted lengthening of the segment.

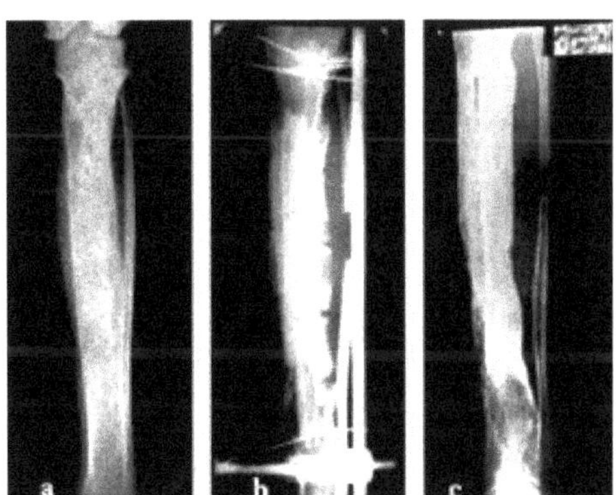

Fig. 5. Two-stage treatment of skeleton fibrous dysplasia: (a) before the operation; (b) 3 months after the operation; (c) 1 year after the operation.

3.2. Double-stage treatment of tumour-like diseases

This technique was applied to 9 patients with pronounced deformations of long bones and total damage of the segments by dysplasia process. In the first stage the pathologic tissue was removed, the bone was corrected and the defect was substituted by cortical split implants. A combined material composed by allografts of low and high demineralisation was used. Dense low-demineralised implants served as intermedullar holders providing preservation of appropriate axis of the segment. The second stage was the elongation of the extremity under conditions of organotypical reconstructing of the implants (during 1.5–2 years) using general technique.

Subsequent complications after bone-plastic operations combined with metalloosteosynthesis were observed for 17 patients, being 15% of the total. Suppuration in the incision region was observed for 4 patients (3 after bone plastics with on-bone metalloosteosynthesis at fibrous dysplasia and 1 with Olie's disease subjected to bone plastics combined with compressive distraction osteosynthesis). In 8 cases (6 patients with fibrous dysplasia and 1 with Olie's disease) relapses were indicated. That required further repeated surgical incisions. Fractures of the on-bone holders were observed for 5 patients with fibrous dysplasia. The holders were replaced.

In the other observations carried out over a long-term monitoring (up to 10 years), there were no complications. Total anatomic and functional reparation of the operated extremities has been indicated within usual schedule (Fig. 6).

Therefore, plunging metalloosteosynthesis with alloplasty using cortical allografts of different extent of demineralisation provides a specific set of conditions for radical elimination of pathological tissue and substitution of extensive osseous defects at diffusive bone damage by fibrous dysplasia. After resorption of the graft is complete, the holder takes a part of loading upon itself acting as a long-term endoprosthesis providing prophylactics against deformation and protecting the bone under variable loading.

Compressive-distraction osteosynthesis in combination with alloplasty amplifies the latitude of revision orthopaedics considerably

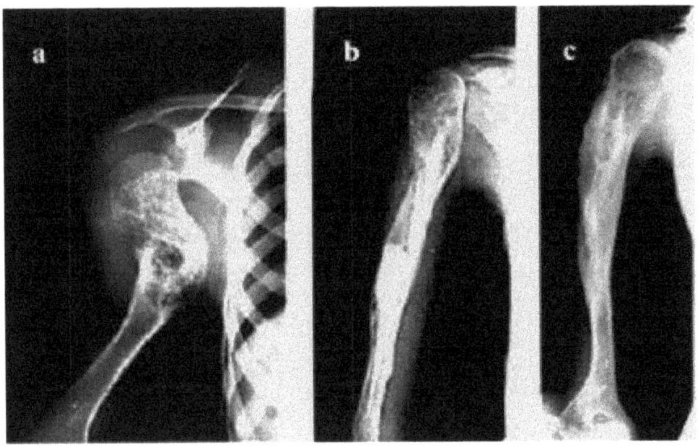

Fig. 6. Two-stage treatment of skeleton dysplasia. Olie's disease: (a) before the operation; (b) 10 months after the operation; (c) 2 years after the operation.

and ensures not only positive "oncological" result but also restoration of shape and length of damaged extremities.

4. References

BEREZHNOI, A.P., SNETKOV, A.I. and KASYMOV, J.A. (1997). Alloplasty techniques in paediatric bone pathology. In: *Proc. VI Conference of Orthopaedic-Traumatologists of Russia,* pp. 638–640.

BOLANDER, M.E. and BALIAN, G. (1990). Use of demineralized bone matrix in the repair of segmental defects. *US Patent* 4 902 296 from 20.02.1990.

DZIEDZIC-GOCLAWSKA, A., OSTROWSKI, K., STACHOWICZ, W. *et al.* (1991). Effect of radiation sterilization on the osteoinductive properties and the rate of remodelling of bone implants preserved by lyophilization and deep-freezing, *Clin. Orthop.* **272,** 30–37.

GENDLER, E. (1986). Perforated demineralized bone matrix: A new form of osteoinductive bio-material, *J. Biomed. Mat. Res.* **20,** 687–697.

HARAKAS, N.K. (1984). Demineralized bone-matrix-induced osteogenesis, *Clin. Orthop.* ***188***, 239–251.

KASYMOV, J.A. (2000). Alloplasty for operational treatment of bone pathology in paediatrics, Ph.D. Thesis.

KUMTA, S.M., LEUNG, P.C., GRIFFITH, J.F. *et al.* (2000). Vascularised bone grafting for fibrous dysplasia of the upper limb, *J. Bone Joint Surg. (Br).* **82**, 409–412.

LEKISHVILI, M.V. and KASYMOV, J.A. (1999). A technique for preparation of bone allografts. *Patent of Russian Federation* No. 2147800 from 17.02.1999.

MATVEICHUK, I.V. (1998). Structural-functional adaptation of bone tissue as composite material regarding species, age and functional peculiarities, Ph.D. Thesis.

MUNTING, E., WILMART, J.F., WIJNE, A. *et al.* (1998). Effect of sterilization on osteoinduction: Comparison of five methods in demineralized rat bone, *Acta Orthop. Scand.* **59**, 34–38.

REDDI, A.H. (1998). Initiation of fracture repair by bone morphogenetic proteins, *Clin. Orthop.* **355S**, S66–S72.

RUSSEL, J.L. and BLOCK, J.E. (1999). Clinical utility of demineralized bone matrix for osseous defects, arthodesis, and reconstruction: impact of processing techniques and study methodology, *Orthopedics* **22**, 524–531.

SAVELIEV, V.I. (1996). Preparation and storage of demineralized osseous tissue for clinical applications. *Transactions: Demineralised Osseous Allografts and Its Application in Reconstructive Surgery.* St. Peterburg, pp. 3–12.

SNETKOV, A.I. (1988). Operational treatment of polyosseous form of fibrous dysplasia in paediatrics, *I.I. Grecov Vestnik Khirurgii.* **6**, 85–89.

SOLHEIM, E. (1998). Osteoinduction by demineralised bone, *Int. Orthopaed.* **22**, 335–342.

URIST, M.R. (1979). A bone morphogenetic system in of bone matrix with a mixed aqueous and non-aqueous solvent, *Proc. Soc. Exp. Biol. Med.* **162**, 48–53.

VOLKOV, M.V. and BISER, V.A. (1969). *Homotransplantation of Osseous Tissue in Pediatrics.* Meditzina, Moscow, p. 216.

VOLKOV, M.V. and BEREZHNOI, A.P. (1983). Principe aspects of dystrophic osseous cyst treatment in paediatrics, *Vestnik Khirurgii* **131**, 68–73.

VOLKOV, M.V. (1985). *Osseous Diseases in Pediatrics.* Meditzina, Moscow, pp. 88–100.

Advances in Tissue Banking Vol. 6

6 CLINICAL STRATEGY FOR APPLICATION OF DEEP FROZEN-RADIATION STERILISED BONE ALLOGRAFTS

WOJCIECH MARCZYǸSKIJANUSZ KOMENDER

Institute of Traumatology, Orthopaedics and Neurosurgery of Central Clinical Hospital Military School of Medicine in Warsaw, Poland

JANUSZ KOMENDER

Bank of Human Tissues Medical Academy in Warsaw, Poland

1. Introduction

Orthopaedic diseases of osseous tissue and its post-traumatic damage usually require surgical treatment which often consists of reconstruction. In many cases the quantity of osseous tissue of the place of intervention is insufficient for adequate reconstruction. In these patients autogenic or allogenic bone grafts should be used (Dziedzic-Goclwska, 1977; Goldberg and Stevenson, 1987; Komender, 1977; Marczynski, 1991, Mazess, 1988; Ostrowski, 2000; Wlodarski, 1991). Obtaining autografts requires an additional surgery and it is often impossible to gain a sufficient quantity of tissue (Nusbickel *et al.*, 1989). For these reasons, our attention has shifted towards frozen allografts which we have successfully used for many years.

The use of allogenic biostatic frozen and radiation sterilised bone grafts has become commonplace in orthopaedics and traumatology. Bone grafts stabilise bone structure in places where healing is disturbed or osteogenesis decreased. Grafts consist of either spongy or compact bone and have different forms. Bone allografts can be placed in many different ways. In some cases they are inserted intramuscularly and connected to the bone only at their ends. As a result an osseous bridge is formed (Alho *et al.*, 1989; Anderson, 1989; Malinin *et al.*, 1985; Marczynski and Tylman, 1991; Mazurkiewucz and Trelinska, 1985; Tylman, 1986).

2. Methods and Results

Allogenic biostatic frozen and radiation sterilised bone grafts were transplanted to 1376 patients in the Institute of Traumatic Surgery, Orthopaedics and Neurosurgery in Warsaw between 1981 and 1998. 731 patients (53%) were female, 645 (47%) were male. There were 827 adults (60%) and 549 children (40%). The age of patients was 2 to 70 with a mean of 36 years (Fig. 1).

Most grafts were used in patients aged 11 to 19 (426, 31%) and patients in their 20 (298, 21%). Bone transplantation was performed in patients with up to 32 different diagnoses. The most frequent indications were scoliosis (206 cases, 15%), hip joint hypoplasia

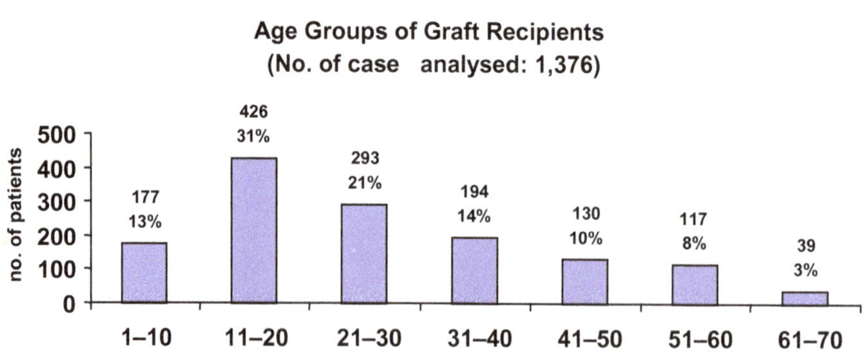

Fig. 1. Age groups of graft recipients.

(177 cases, 13%), and solitary cysts (179 cases, 13%). All diagnoses were divided into six groups: bone union failure (489 cases, 35%), congenital anomalies (374, 27%), benign neoplasms (173, 12%), prosthesis reimplantation (43, 3%), arthrosis (80, 7%) and miscellaneous cases (91, 7%) (Fig. 2).

Various types of grafts were used, depending on indications. Spongy bone grafts where used in 1085 (79%) and compact in 295 (21%) patients. Slivers were used in 839 cases (61%), bars in

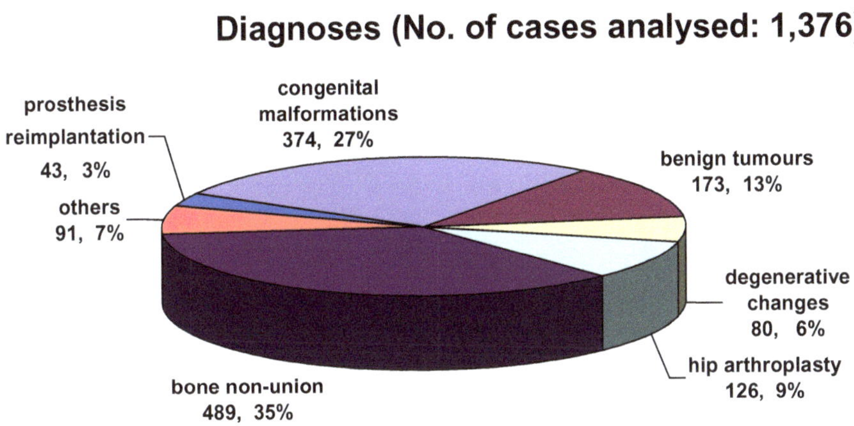

Fig. 2. The indications for the bone transplantation.

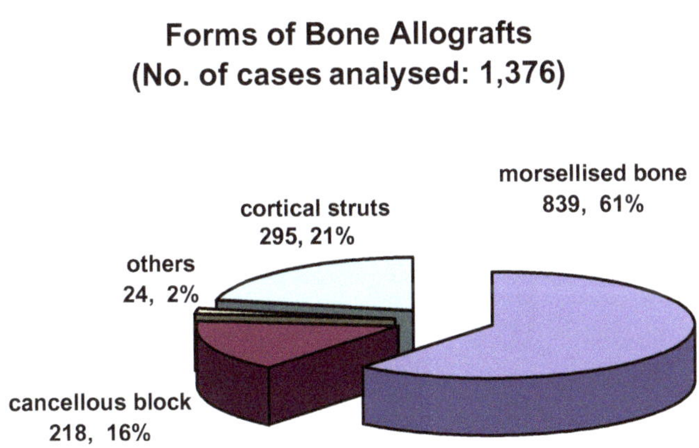

Fig. 3. Forms of bone allografts.

295 cases (21%), plasters in 218 cases (16%) and solid grafts in 24 cases (2%) (Fig. 3).

The way in which the graft was implanted depended on the diagnosis and the location of the affected osseous tissue. The grafts were placed periosteally in 1206 cases (88%) and subperiosteally in 170 cases (12%). Adhesion apposition was used in 661 cases (48%), incomplete intraosseus implantation in 335 cases (24%), intraosseus implantation in 316 cases (23%) and intramuscular implantation in 64 cases (5%). The grafts were implanted in the spine in 394 cases (28%), in the femur in 284 cases (21%), in the hip in 271 cases (20%), in the tibia in 229 cases (17%), in the arm in 126 cases (9%), in the forearm in 42 cases (3%), in the foot in 18 cases (1%) and in the hand in 12 cases (1%) (Fig. 4).

In the spine slivers were used in 217 cases (55%) and bars in 177 (45%). In the femur slivers were used in 224 cases (79%), bars in 31 cases (11%) and plasters in 29 cases (10%). In the hip surgery plasters were used in 144 cases (53%), slivers in 123 cases (46%) and bars in 4 cases (1%). In crural bones slivers were used in 114 cases (50%), plasters in 98 cases (43%) and bars in 17 cases (7%). In the arm slivers were used in 83 cases (66%), plasters in 26 cases (21%) and bars in 17 cases (13%). In the forearm slivers were used in 37 cases (88%) and plasters in 5 cases (12%). The result of the treatment was analysed 2 to 15 years after the surgery.

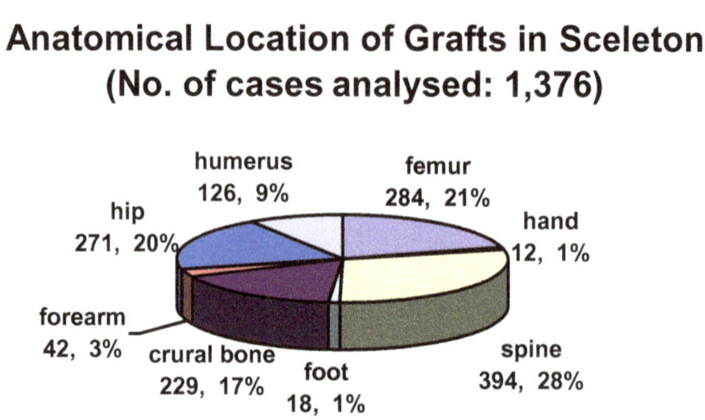

Fig. 4. Anatomical location of grafts in the skeleton.

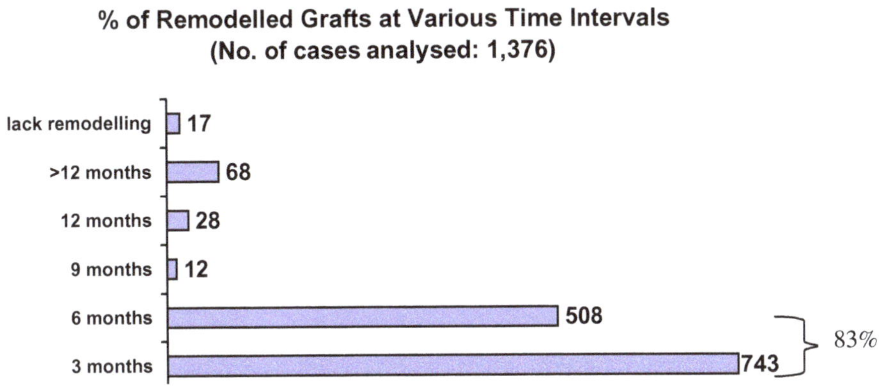

Fig. 5. Percentage of remodelled grafts at various time intervals.

The progress of the bone graft remodelling was evaluated on the basis of physical and X-ray examinations. The examinations were carried out at regular intervals. Three months after the surgery, 745 grafts (46%) were incorporated, 509 (37%) six months after the surgery, 123 (9%) nine months after the surgery and 28 (2%) twelve months after the surgery. After this period, 68 more grafts (5%) were incorporated and 17 grafts (1%) became sequesters (Fig. 5).

Among our patients, there were six cases of infection and one case of congenital pseudoarthrosis. All 316 intraosseus grafts (100%) were incorporated. 661 adhesive insertion grafts were incorporated (90%). The grafts which were incompletely embedded in the bone were incorprated in 335 cases (86%). All grafts which had been placed subperiosteally were incorporated where 1085 of the grafts apposed outside periosteum (90%) were incorporated. All grafts placed in muscles were partially or completely resorbed. The total number of grafts which had either completely incorporated or were in the process of incorporation amounted to 1238 (90%).

The time of graft incorporation depended on many factors such as diagnosis, site of implantation, patient's age, the type and size of the graft as well as the method of its implantation. Spongy bone grafts were incorporated after 3 to 6 months, compact bone grafts

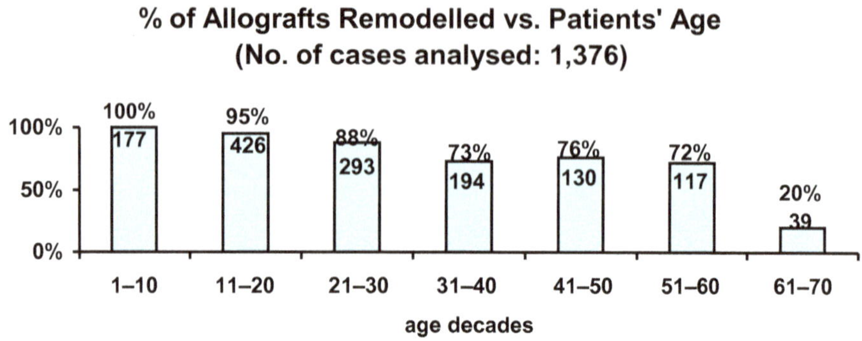

Fig. 6. Percentage of allografts remodelled versus patient's age.

in the form of bone bars were incorporated after 6 to 24 months. By correlating patients' age with the efficacy of the procedure, we have shown that the treatment was highly effective in patients aged 2 to 10 (100%). In patients aged 11 to 19 the efficacy of the treatment was 95% and in patients in their 20s it was 88%. In older age groups the efficacy of bone graft implantation decreased (Fig. 6).

High or satisfactory efficacy of the treatment was achieved in 1249 cases (90%). In 126 cases (9%), the procedure was not effective or it was not possible to assess the graft. We have achieved high overall efficacy of graft implantation in all six groups created on the basis of the diagnosis.

3. Examples of Graft Use in Specific Clinical Situations

Prosthesis reimplantation of the hip joint is performed on a large scale all over the world. Aseptic loosening of the hip joint prosthesis often occurs many years after implantation. The process of loosening is caused by surgical errors as well as mechanical and biological factors. Free particles of polyethylene, metal and acrylic cement initiate the process. Loosening is often accompanied by bone destruction. When reconstruction of the damaged bone is attempted ways of reconstructing acetabulum and femur and achieving functional stability of the hip joint prosthesis must be determined. Surgical technique used during hip arthroplasty

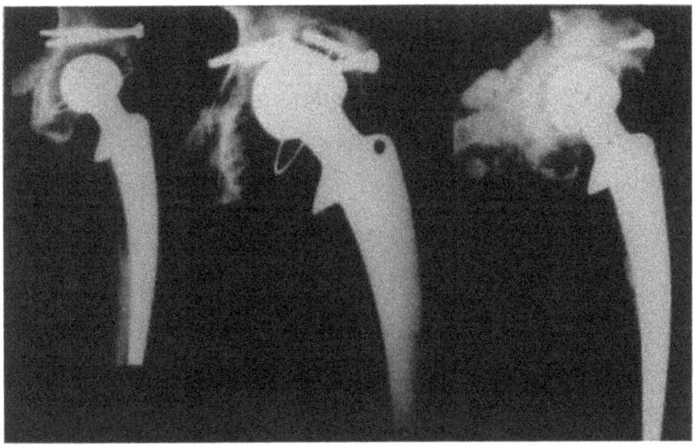

Fig. 7. X-ray of the hip after total arthroplasty. Radiograph showing picture after surgery, after loosening prothesis, and after revision with used bone allografts.

enables precise reconstruction of bone stock both around the acetabulum and femur metaphysis. For this purpose, we used morselised, frozen allogenic spongy bone grafts and 43 patients (3%) underwent hip arthroplasty. Radiograms were taken immediately after the surgery and 3, 6 and 12 month later. The grafts were incorporated 3 to 6 months after the surgery. The patients were allowed to exert full body-weight on the joint 6 month after the surgery (Fig. 7).

Post-traumatic spinal fractures require surgical treatment in 15–20% of patients. Post-traumatic instability and spinal canal stenosis with or without neurological symptoms are indications for surgical treatment. Surgical methods include decompression of spinal canal structures, internal stabilisation along with immobilisation of the fewest possible number of segments, which enabled us to bring patients early into the erect position without the necessity of external immobilisation. For the surgery, we used frozen allografts of spongy morselised or compact bones in the form of bars. After the bone grafts had been incorporated the mechanical stabilisation was replaced by the biological stabilisation. We treated 68 patients (5% of all patients) with thoracic and lumbar spine fractures. Primary neurological complications of various degrees were diagnosed in 56 patients (82% of patients with spinal fractures). We treated 8 patients with fractures in the thoracic section Th_1 to Th_{10} (11%). The fractures occurred most commonly in the transitional section Th_{11} to L_2 (50 patients, 74% of patients with spinal fractures). 10 fractures occurred in the lumbar section L_3 to L_5 (15% of patients with spinal fractures). The bone grafts were incorporated 3 to 6 months after surgery (Fig. 8).

Congenital and developmental scolioses were treated in 102 children between 1990 and 1997. Clinically and radiologically confirmed progression of the scoliosis was an indication for surgical treatment. Two surgical methods were used: a one-stage correction

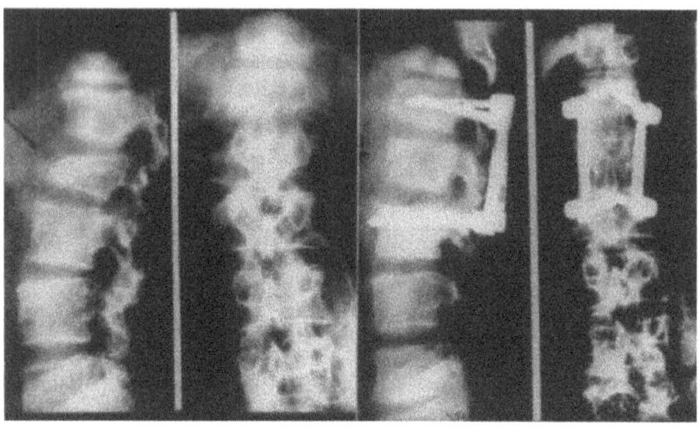

Fig. 8. Radiograph showing instability fracture L_1 after injury and after transpedicular stabilisation, with used frozen allografts.

Fig. 9. Telescope rod with dystraction instrument.

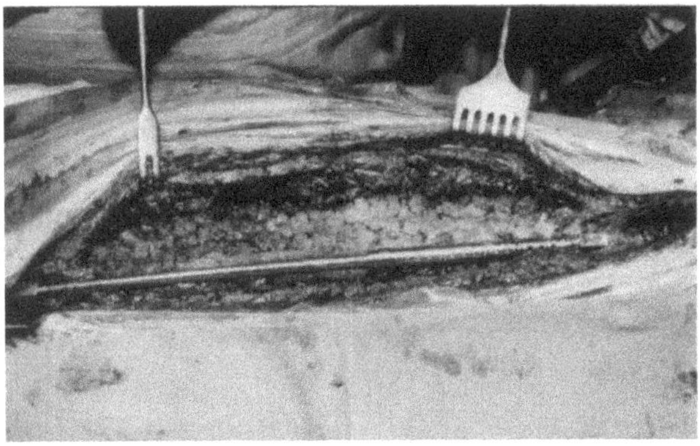

Fig. 10. Spondylodesis with use morsellised, frozen bone grafts during last dystraction of the rod.

and a multiple-stage correction. One-stage correction was achieved with the aid of one or two distraction rods. In scolioses which increased with the child's age, multiple-stage treatment was carried out with the use of a telescopic rod type RRC-2 (Fig. 9). Fifty three children, whose bone development had not been completed according the Risser test, were treated this way. Apart from mechanical correction and stabilisation, allogenic biostatic radiation sterilised frozen bone grafts were implanted in all operated patients to ensure additional biological stabilisation (Fig. 10).

The choice of a particular type of graft depended on the clinical situation. If the graft was to be placed in the lamina and the spinosus process of vertebra, morsellised bone grafts were used. We recently used morsellised bone grafts which were mixed with crushed garamycin sponge before implantation.

Among the multiple-stage treated scolioses the angle before the surgery was 40 to 100 degrees, average 70 degrees, and after the surgery 13 to 40 degrees, average 27 degrees. The percentage of post-surgical correction was 56 to 74%. In each child we treated the loss of the correction due to fast growth. On average we performed 5 correction surgeries on one child. For these surgeries, we used frozen bone grafts which were applied in two stages. During the implantation of the telescope rod the grafts were placed in the area

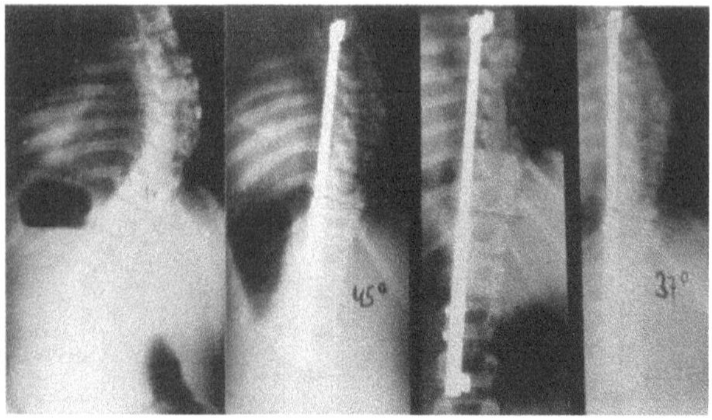

Fig. 11. Pre-operative and post-operative radiographs of the scoliosis. Example of multiple-stage treatment using a telescope rod in the scolioses increasing with the growth of the child.

where the hooks had been implanted. In children undergoing the final correction, morselised bone was implanted on the decorticated surface of the lamina and the spinoous process of the veretebra in the space between the telescope rod and the spine. The bone grafts were incorporated 3 to 6 months after the surgery (Fig. 11).

Bone cysts cause pain and fractures but they have no specific symptoms. They are usually identified by a casual X-ray of a bone or because of a pathological fracture. Frozen and radiation-sterilised bone allografts were implanted in 179 patients (13%) with cysts resulting from benign tumors. Most of them were children (137 cases, 76% of all cyst cases). Adults constituted 24% (42 cases). The most common location of bone cysts was the humerus (53 patients, 30%), the femur (47 patients, 26%), the tibia and the fibula (47 cases, 26%). Other bone cysts were found in the hip (10 cases, 6%), hands and feet (9 cases, 5%) and forearm (4 cases, 2%). The most common type was a solitary bone cyst (127 cases, 71%).

Other types of cysts were cartilaginous tumors (12 cases, 7%), fibrocellular tumors (9 cases, 5%), osteoid osteomas (6 cases, 4%) and aneurysmal bone cysts (9 cases, 5%).

During surgical treatment the bone cysts were excised and examined, its sites electrocoagulated and mechanically and chemically cleaned. Then they were filled with morselised spongy or

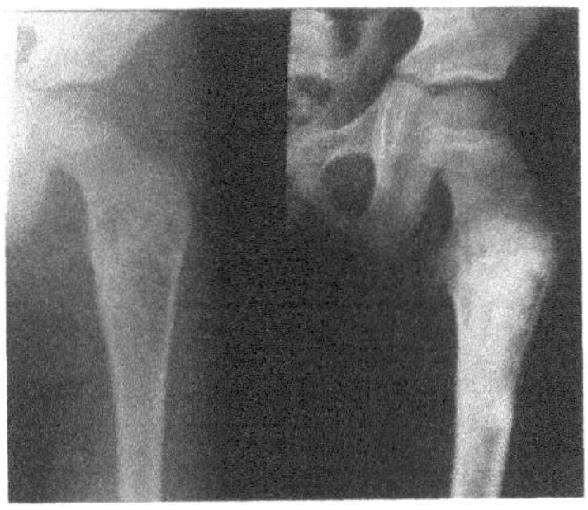

Fig. 12. Radiograph showing multiloculare cyst of the femur pre-operative and 3 months post-surgery with use frozen bone grafts.

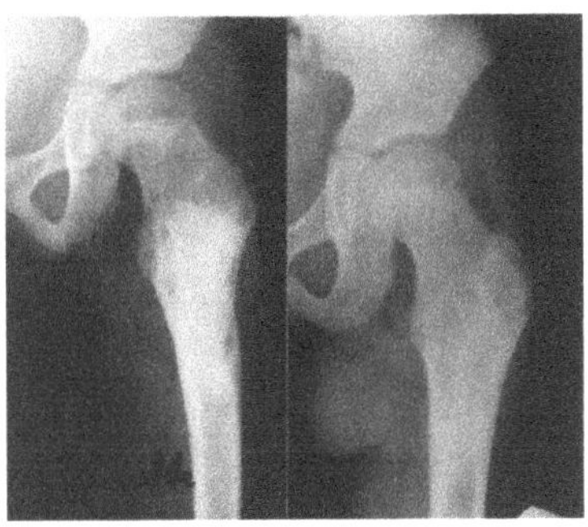

Fig. 13. Radiograph showing rebuilding bone allografts after 6 and 12 months post-surgery.

cortical bone grafts. The cyst tissue underwent histopathological examination. Physical examination and X-rays were evaluated every 6–8 weeks. Allograft incorporation was observed in 82 patients (46%) within three months and in 66 more patients (37%) six months after the surgery (Figs. 12 and 13).

Frozen bone grafts were used in 489 patients (35% of all patients) with a post-traumatic non-union bone. The age of the patients was 12 to 69 with a mean of 34 years. Frozen bone grafts were used in 306 patients (62.5% of miscellaneous group patients) with a delayed-union bone, in 34 patients (5%) with primary post-traumatic bone defects, in 117 patients (23.9%) with pseudo-arthrosis and in 42 patients (8.6%) with habitual dislocation of the shoulder joint. The specific diagnosis and the indications for the surgery determined the method of graft implantation which was intraosseus, incompletely intraosseus, adhesive or intramuscular. In 459 cases (94%) spongy bone grafts were used. In 30 cases (4%) cortical bone grafts were used, in 288 cases (58.9%) morselised bone grafts, in 171 cases (35.1%) cancellous blocks and in 30 cases (6%) cortical struts. The process of the bone graft remodelling was evaluated on the basis of physical and X-ray examinations, which were carried out every 3 months after the surgery. In 225 patients (46%) the grafts were incorporated 3 months after the surgery, in 181 cases (37%) after six months, in 44 cases (9%) after nine months, in 10 cases (2%) after twelve months. In 24 cases (5%) the remodelling process lasted over a year. In 5 cases (1%) the grafts became sequesters due to infection. In 42 cases (8.6%) partial resorption of the grafts was observed. Seventeen patients required a second bone grafting (Fig. 14).

Surgical filling of defects after intra-articular fracture of proximal tibia was performed in 15 patients between 1995 and 1998, In which 70% of fractures involved the lateral tibia condyle, 10% medial condyle and 20% both condyles. All patient were diagnosed with an X-ray and CT in order to establish the location and extent of the bone defect. For the tibia reconstruction we used morsellised bone grafts and screw or plate stabilisation. Clinical assessment of functions of the operated knee joints was carried out according to

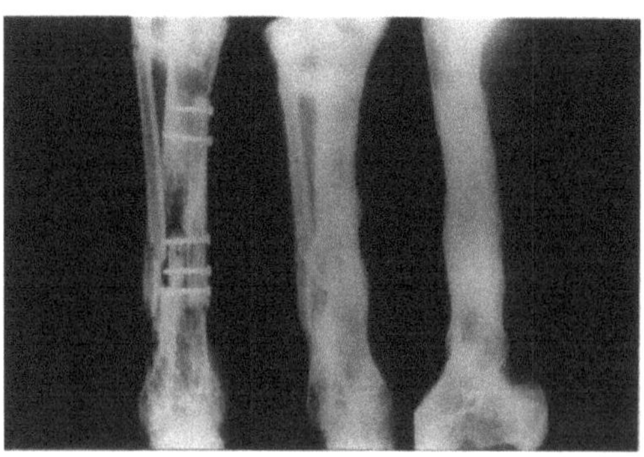

Fig. 14. Bone defect of the tibia post inflammatory process. Osteosynthesis by bone bar and screw. Defect was fulfilled by frozen morsellised bone grafts.

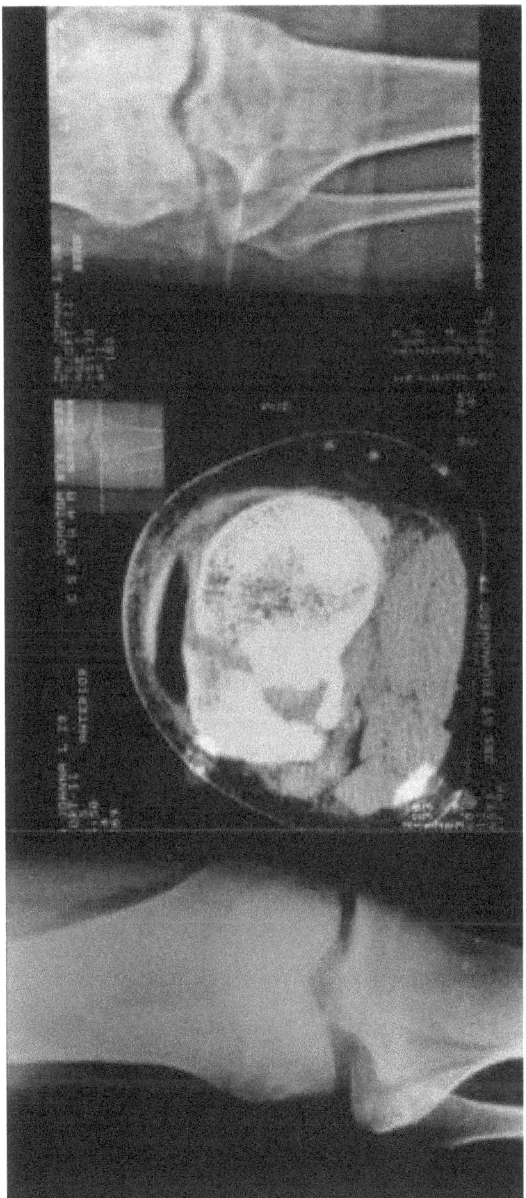

Fig. 15. Anero-posterior X-ray and CT of fracture condylus lateralis of the tibia.

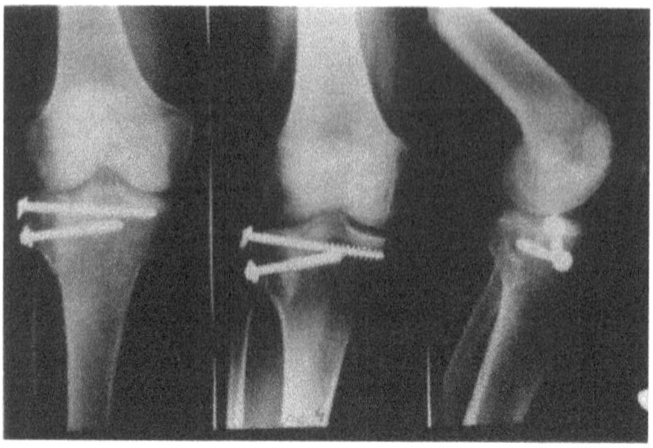

Fig. 16. Operational filling of defects after intra-articular fracture of proximal tibia with stabilisation by screws; immediately after surgery and 3 months late.

the Lysholm-Gillquist score and IKDS score. Frozen bone grafts were incorporated 3 to 6 months after the surgery (Figs. 15 and 16).

4. Conclusions

1. The application of allogenic, biostatic frozen grafts makes it possible to reduce the extent and duration of operations.
2. The allogenic, biostatic frozen and radiation sterilised bone grafts undergo the "creeping substitution" (incorporation) in 3 to 6 months.
3. The allogenic grafts in various forms (granulate, bone wreckage, plasters, solid grafts and bone bars) suit different surgical requirements.
4. The progress of the graft substitution (incorporation) depends on its structure and the site of implantation. It can be assessed by an X-ray examination.
5. Morselised frozen allografts of spongy bone enable the reconstruction of osseous tissue damaged in the process of hip joint prosthesis loosening.
6. Frozen bone grafts are advantageous osteogenic material. Bone graft transplantation enables spine stabilisation after the removal of instruments.

7. Application of frozen bone grafts to produce a biological spondylodesis in the surgical treatment of scolioses shortens the operation time, limits the blood loss and the risk of complications. Moreover, it does not cause any additional untoward effects and scars.

5. References

ALHO, A., KARAHARJU, E.O., KORKALA, O., LAASONEN, E.M., HOLMSTROM, T. and MULLER, C. (1989). Allogenic grafts for bone tumor. 21 cases of osteoarticular and segmental grafts, *Acta Orthop. Scand.* **60**, 143–153.

ANDERSON, W.J. (1989). Allograft bone for arthrodesis and repair of skeletal hand problems, *J. Hand Surg.* **14**, 332–335.

DZIEDZIC-GOCLAWSKA, A. (1977). Radiation sterilisation of the tissue. In: *Biostatic Grafts. cz.I*, J. Komender, ed., PZWL, Warszawa, p. 50.

GOLDBERG, V.M. and STEVENSON, S. (1987). Natural history of autografts and allografts, *Clin. Orthop.* **225**, 7–16.

KOMENDER, J. (1977). *Biostatic Allografts.* Publ., Co., PZWL, Warsaw, pp. 24–46.

MALININ, T.I., MARTINEZ, O.V. and BROWN, M.D. (1985). Banking of massive osteoarticular and intercalary bone allografts-12 years experience, *Clin. Orthop.* **197**, 44–57.

MARCZYÑSKI, W. and TYLMAN, D. (1991). Operating treatment of idiopatic necrosis of the head of the femur with use allogenic bioststatic frozen bone grafts, *Military Health Service Sci. J.* **11**(12), 692–696.

MARCZYÑSKI, W, (1995). *Treatment of Non-union and Defects of Bone.* Publ. Co., Bellona, Warsaw, pp. 10–28.

MAZURKIEWICZ, H. and TRELINSKA, A. (1985). Role of bone allografts in reconstruction defects of acetabulum and femur in replacement of the hip, *Chir. Org. Motus Orthop. Polonica* **50**, 403–408.

MAZESS, R.M. (1988). Bone densitometry (letter), *Am. J. Roentgenol.* **150**, 207–208.

NUSBICKEL, F.R., DELL, P.C., MCANDREW, M.P. and MOORE, M.M. (1989). Vascularized autografts for reconstruction of skeletal defects following lower extremity trauma. A review, *Clin. Orthop.* **243**, 65–70.

OSTROWSKI, K. (2000). Infections transferred by transpants. *Section of Medical Scientes Polish Academy of Scientes*, Publ. Co., Warsaw, pp. 5–15.

TYLMAN, D. (1986). Bone union of fracture of bones — Biological aspect, and influence physical agent. *Chir. Org. Motus Orthop. Polonica* **51**, 433–446.

LODARSKI, K.H. (1991). Bone histogenesis mediated by nonosteogenic cells, *Clin. Orthop.* **272**, 8–15.

Advances in Tissue Banking Vol. 6

7 FROM BONE ALLOGRAFTS TO SYNTHETIC BONE GRAFTS: BONE MORPHOGENETIC PROTEINS AND OSTEOINDUCTION

A.H. REDDI

Center for Tissue Regeneration and Repair
Department of Orthopedic Surgery
University of California Davis, School of Medicine
Sacramento, CA 95817

1. Introduction

Transplantation of bone to unite recalcitrant non-union of fractures is well known in orthopaedic surgery practice. The use of fresh autogenous bone grafts while desirable, results in donor site morbidity and associated complications. In view of this, allografts prepared by tissue banks and bone banks is gaining widespread acceptance. Furthermore, in massive bone defects the use of adjunct fortification with bone morphogenetic proteins (BMPs) from bone may improve the potency of a bone grafts. BMPs are endogenous morphogens in bone morphgenesis and development (Reddi, 1998). Morphogenesis is the multistep sequential cascade of pattern formation, organogenesis and bilateral mirror-image symmetry of musculoskeletal structures including bone. Proteins initiating bone morphogenesis are called BMPs and were initially identified, purified, isolated, from demineralised bone matrix cloned and expressed in a recombinant form.

The aim of this article is to convey the current excitement and recent research progress in BMPs and how it might aid tissue banking, with the ultimate goal of help design and optimised allografts. In addition, improvements in allografts will lead to design of a totally synthetic bone graft using principles of tissue banking and tissue engineering. Tissue engineering is the science of fabrication of new tissues for functional restoration of lost parts due to cancer, diseases, and trauma. The three key ingredients are bone inductive BMPs, responding mesenchymal stem cells and an extracellular bone matrix scaffolding. The confluence of the fields of tissue banking and tissue engineering ensures an optimal graft for the patient.

2. Bone Morphogenesis is a Cascade

Over a century ago decalcified bone was used in treatment of osteomyelitis implying a role for bone matrix (Senn, 1889). Lacroix (1945) in Belgium hypothesised the role of an osteogenic inducer, osteogenin, in bone formation. A key discovery was made by Urist (1965) that demineralised, lyophilised segments of rabbit bone induced new bone in intramuscular sites. Endochondral bone morphogenesis was induced by subcutaneous implantation of particles of demineralised bone matrix and the multiple steps in bone morphogenesis were dissected (Reddi and Huggins, 1972). Bone morphogenesis is a sequential cascade with three key phases: chemotaxis, mitosis and differentiation of mesenchymal cells initially into cartilage and replacement by bone (Reddi, 1981; Reddi and Anderson, 1976). The sequential cascade begins with the binding of plasma fibronectin, a cell adhesive protein to implanted demineralised matrix, permitting mesenchymal cell attachment (Weiss and Reddi, 1980) and maximal proliferation as monitored by (^{3}H) thymidine incorporation on day 3. Chondroblast differentiation is evident on day 5. Maximal chondrogenesis was observed on days 7–8. On day 9 hypertrophy of cartilage was observed with concominant mineralisation of cartilage matrix. Angiogenesis and vascular invasion is a prerequisite for osteoblast differentiation and is maximal on days 10–11. The newly formed endochondral bone is remodelled and is the site of haematopoiesis (Reddi, 1994; 1998). The sequential bone morphogenesis in response to demineralised bone matrix mimics early stages of skeletal morphogenesis in limb bud in embryos and fracture healing in adults recapitulating embryonic bone development and morphogenesis. Thus, it is possible to isolate the key signals for bone morphogenesis from demineralised bone matrix. We next describe the purification, cloning and expression of BMPs.

3. Purification, Cloning and Expression of Recombinant BMPs

In general a requirement for purification of an osteoinductive BMP is a bonafide bioassay. The demineralised bone matrix is insoluble and is in the solid state. Dissociative extractants such as 4M guanidine hydrochloride or 8M Urea or 1% sodium dodecyl sulfate at pH 7-4 were used extract 2% to 3% of the proteins. The solubilised proteins in the extract and the insoluble collagenous residue were not osteoinductive in an ectopic site in a 28-day old rat. However, combining the extract with the collagenous matrix residue rendered the recombined matrix osteoinductive. Thus, an operational reconstitution of the solubilised proteins in the extract with the insoluble collagenous matrix was achieved (Sampath and Reddi, 1981; 1983). Further, this critical experiment allows one to enunciate the concept of a collaboration between a soluble signal in the extract and the insoluble extracellular matrix substratum to initiate new bone formation. This experimental finding in our laboratory was a quantum advance in the isolation and purification of bone morphogenetic proteins. This bioassay led to the molecular cloning and recombinant expression of BMPs by biotechnology firms.

In order to scale up the purification procedures a switch was made from rat diaphyseal bone to bovine bone. However, demineralised bovine, human, monkey and porcine bone matrices were not osteoinductive in rats and variable response seen in immunodeficient nude rats and mice. The underlying biochemistry was unraveled by preparing extracts of demineralised bone from xenogeneic species such as bovine, human, monkey and porcine and partially purified by molecular seize chromotography. Molecular fractions of less then 50 kDa were consistently osteoinductive when assayed with allogeneic collagenous matrix in rats. Thus, the soluble osteoinductive proteins are homologous in bovine, human, monkey, porcine and rat and can elicit bone formation in rats and are not species-specific (Sampath and Reddi, 1983). The yield of protein was about 1 μg from 1 kg of bovine bone requiring well over a ton of bovine bone to obtain enough purified protein to determine the amino acid sequence (Luyten *et al.*, 1989). The amino acid sequence data permitted design of cognate oligonucleotide probes permitting molecular cloning of the BMPs (Reddi, 1998; Wozney *et al.*, 1988). At last count there are over 15 BMPs. This list does not include BMP-1 as it is an enzyme, procollagen c-proteinase involved in proteolytic processing of mature collagen.

Bone morphogenetic proteins are dimeric molecules with two polypeptide chains held together by a single disulfide bond. Each monomer is biosynthesised as a polypeptide chain of over 400 amino acids which includes the precursor including the pro-region and the mature BMP. The crystal structure of BMP 7 was determined by X-ray crystallography and highlights the characteristic cysteine knot of members of the BMP family (Griffith *et al.*, 1996).

In the bone morphogenesis cascade cartilage differentiation, hypertrophy and cell death is followed by bone formation by the typical endochondral bone formation cascade. Therefore, all BMPs are cartilage morphogenetic proteins since cartilage is the first albeit transient phenotypic transformation state as in epiphyseal growth plate. It is noteworthy that extracts of bovine articular cartilage contain distinct cartilage-derived morphogenetic proteins (CDMPs) that are also known as growth/differentiation factor, (GDFs) 5 through 7. CDMPs/GDFs are members of the BMP family (Chang *et al.*, 1994; Storm *et al.*, 1994).

4. BMPs and Chondrogenesis, and Cartilage Maintenance

Recombinant human BMP 4 (BMP-2B) and purified BMP3 stimulate chondrogenesis in limb bud mesodermal cells (Chen *et al.*, 1991). Thus, BMPs might play a role in early chondrogenesis. In addition BMPs have a profound role in maintenance of articular cartilage phenotype. It is common knowledge that in monolayer cultures chondrocytes progressively undergo dedifferentiation and lose their cartilage-specific type II collagen matrix and is replaced by type I collagen. Unlike monolayers of chondrocytes, in explant cultures of articular cartilage as the extracellular matrix (ECM) encases the chiondrocytes the dedifferentiation is prevented and/or delayed. In serum-free chemically defined medium recombinant BMP4 and BMP7 stimulate proteoglycan synthesis in bovine and porcine articular cartilage explants (Lietman *et al.*, 1997; Luyten *et al.*, 1992).

5. BMPs: Pleiotropy and Thresholds

Pleiotropy is the property of a single gene or gene product such as protein to have a multiplicity of different biological actions. Since, BMPs govern the three key steps in bone induction cascade such as chemotaxis (migration of cells), mitosis (multiplication of cells by division) and differentiation initially of cartilage and replacement by bone, BMPs are true pleiotropic morphogens. In addition, BMPs regulate haematopoiesis, stimulate extracellular matrix synthesis, influence cell survival maintenance and cell death (apoptosis). Recombinant BMP4 stimulates chemotaxis of human peripheral blood monocytes at femtomolar concentrations (Cunningham *et al.*, 1992). The mitogenic action of BMP 4 on mesenchymal cells is at picomolar concentration. At a slightly higher concentration BMP4 initiates *in vitro* chondrogenesis. It is important to emphasise that these concentrations are in solution *in vitro* and *in vivo* BMPs are bound to extracellular matrix components such as collagens I and IV, heparan sulphate, heparin and the bone mineral hydroxyapatite. Thus, when, BMPs are bound to extracellular matrix, the local concentration is difficult to determine and may offer an optimal molecular conformation for biological actions. This aspect of the molecular cell biology of morphogens has considerable physiological significance for local contact-mediated short range actions in bone formation by osteoblasts and remodelling by osteoclasts. BMPs are pleiotropic morphogens that act in concentration dependent thresholds which are critical in local cellular environments and are dependent on context and microenvironment.

6. BMP Receptors and Smads

The biological actions of BMPs are transduced via specific BMP receptors (Heldin *et al.*, 1997). BMP receptors are of two types, I and II, and are serine/threonine protein kinases. These kinases are enzymes that phosphorylate substrate Smads and activate them. The Smads are cytoplasmic substrates for cell membrane-associated BMP type I receptor. BMP type II receptor phosphorylates type I receptor (Reddi, 1998). There are eight different Smads. Smads 1, 5 and 8 are substrates for BMP receptors. Phosphorylation of either Smad 1, 5 or 8 activates them to interact with common partner Smad 4 and this hetereomeric complex enters the nucleus to activate and turn on BMP-responsive genes. There are two inhibitory Smads 6 and 7 which are normally resident in nucleus which act as a relay to inhibit and turn off BMP receptor I mediated phosphorylation of Smad 1, 5 and 8. Thus, there is an intricate homeostatic control between BMP receptor activated turning on of genes and their turning off by Smads 6 and 7 by inhibiting type I BMP receptor kinase. BMPs regulate cell cycle progression, mitosis and differentiation of progenitor stem cells in macrow into bone.

The availability of BMPs to interact with BMP receptors and activation of BMP-response genes is determined by the bioavailability of BMPs. Recent work has identified BMP-specific antagonists such as noggin (Zimmerman *et al.*, 1996), chordin, and members of DAN family such as gremlin (Hsu *et al.*, 1998). These antagonists have the same affinity to BMPs and the BMP receptors. Thus, these antagonists could be used therapeutically in pathological to inhibit conditions excessive bone formation. It is noteworthy that these antagonists may play a role in fibrous nonunions of fractures. In addition, the concentration of BMP binding proteins in bone graft may be prognostic utility in bone banking.

7. BMPs: Clinical Applications

As BMPs were purified, cloned and expressed based on a functional bone formation assay it is certainly possible to use recombinant human BMPs in orthopaedic surgery. Recombinant BMP2 and BMP7 (also by the two biotechnology companies known as osteogenic protein-1) are the two BMPs that have been studied extensively due to the obvious financial and clinical implications. BMPs have been evaluated in preclinical models and in the clinic (Bostrom *et al.*, 1996; Boyne *et al.*, 1997; Cochran *et al.*, 1999; Cook *et al.*, 1994a; 1994b; Hollinger *et al.*, 1996; Yamaguchi *et al.*, 1991). Until recently autogenous bone grafts were considered as "gold standard" for bone grafts. However, the harvesting of bone from iliac bone results in donor site morbidity. The imminent completion of clinical trials with recombinant BMP2 and BMP7 in tibial non-unions is awaited with much excitement by patients and biotechnology industry. Progress in this area is dependent on strides in carriers for delivery of recombinant BMPs. Collagen appears to be an optimal carrier. Additional challenges include synthesis of novel carriers (Ripamonti, 1992; 1996) with defined BMP-release kinetics optimised for different clinical segments for example diabetes and patients with advanced osteoporosis. Despite these advances the challenge of scaffolds for massive

Table 1. Steps in evolution of synthetic bone grafts based on bone morphogenetic proteins

1. Allografts from Bone Bank
2. Allografts fortified with BMPs
3. Synthetic scaffolding of collagen and BMPs
4. Mineralised collagen matrix and BMPs
5. Mineralised gollagen with BMPs and stem cells

segmental defects of the sort faced by orthopaedic oncologists and trauma surgeons requires novel approaches and research. Tissue engineering is the science of manufacture of skeletal spare parts for human skeleton based on morphogens such as BMPs, responding stem cells and biomimetic biomaterial scaffolds to mimic extracellular matrix components (Reddi, 1998) and the proof of concept for tissue engineering using BMPs and molds with vascularised muscle flops was presented (Khouri *et al.*, 1991). In this regard tissue banking scientists and surgeons can be satisfied that tissue banking practice is paving the path for an accelerated progress towards the goal of a synthetic bone graft fortified with recombinant BMPs based on principles of tissue banking.

8. Acknowledgements

I wish to thank Rita Rowlands for truly outstanding help with bibliography and preparation of the manuscript. This work is supported by the Lawrence J. Ellison Chair in Musculoskeletal Molecular Biology.

9. References

BOSTROM, M., LANE, J.M., TOMIN, E., BROWNE, M., BERBERIAN, W., TUREK, T., SMITH, J., WOZNEY, J. and SCHILDHANER, T. (1996). Use of bone morphogenetic protein-2 in the rabbit ulnar nonunion model, *Clin. Orthop. Rel. Res.* 272–282.

BOYNE, P.J., MARX, R.E., NEVINS, M. TRIPLETT, G., LAZARO, E., LILLY, L.C., ALDER, M. and NUMMIKOSKI, P. (1997). A feasibility study evaluating rhBMP-2/absorbable collagen sponge for maxillary sinus floor augmentation, *Int. J. Periodon. Restorative Dent.* **17**, 11–25.

CHANG, S.C., HOANG, B., THOMAS, J.T., VUKICEVIC, S., LUYTEN, F.P., RYBA, N.J., KOZAK, C.A., REDDI, A.H. and MOOS, M. Jr. (1994). Cartilage-derived morphogenetic proteins.

New members of the transforming growth factor-beta superfamily predominantly expressed in long bones during human embryonic development, *J. Biol. Chem.* **269**, 28227–28234.

CHEN, P., CARRINGTON, J.L., HAMMONDS, R.G. and REDDI, A.H. (1991). Stimulation of chondrogenesis in limb bud mesoderm cells by recombinant human bone morphogenetic protein 2B (BMP-2B) and modulation by transforming growth factor beta 1 and beta 2, *Exp. Cell Res.* **195**, 509–515.

COCHRAN, D.L., SCHENK, R., BUSER, D., WOZNEY, J.M. and JONES, A.A. (1999). Recombinant human bone morphogenetic protein-2 stimulation of bone formation around endosseous dental implants, *J. Periodontol.* **70**, 139–150.

COOK, S.D., BAFFES, G.C., WOLFE, M.W., SAMPATH, T.K., RUEGER, D.C. and WHITECLOUD, T.S. 3rd. (1994a). The effect of recombinant human osteogenic protein-1 on healing of large segmental bone defects, *J. Bone Joint Surg. Am.* **76**, 827–838.

COOK, S.D., DALTON, J.E., TAN, E.H., WHITECLOUD, T.S. 3rd and RUEGER, D.C. (1994b). *In vivo* evaluation of recombinant human osteogenic protein (rhOP-1) implants as a bone graft substitute for spinal fusions, *Spine* **19**, 1655–1663.

CUNNINGHAM, N.S., PARALKAR, V. and REDDI, A.H. (1992). Osteogenin and recombinant bone morphogenetic protein 2B are chemotactic for human monocytes and stimulate transforming growth factor beta 1 mRNA expression, *Proc. Natl. Acad. Sci. USA* **89**, 11740–11744.

GRIFFITH, D.L., KECK, P.C., SAMPATH, T.K., RUEGER, D.C. and CARLSON, W.D. (1996). Three-dimensional structure of recombinant human osteogenic protein 1: Structural paradigm for the transforming growth factor beta superfamily, *Proc. Natl. Acad. Sci. USA* **93**, 878–883.

HELDIN, C.H., MIYAZONO, K. and TEN DIJKE, P. (1997). TGF-beta signalling from cell membrane to nucleus through SMAD proteins, *Nature* **390**, 465–471.

HOLLINGER, J., MAYER, M., BUCK, D., ZEGZULA, H., RON, E., SMITH, J., JIN, L. and WOZNEY, J. (1996). Poly(alpha-hydroxy acid) carrier for delivering recombinant human bone morphogenetic protein-2 for bone regeneration, *J. Controlled Rel.* **39**, 287–304.

HSU, D.R., ECONOMIDES, A.N., WANG, K., EIMON, P.M. and HARLAND, R.M. (1998). The Xenopus dorsalizing factor Gremlin identifies a novel family of secreted proteins that antagonize BMP activities, *Mol. Cell.* **1**, 673–683.

KHOURI, R.K., KOUDSI, B. and REDDI, H. (1991). Tissue transformation into bone *in vivo* — A potential practical application, *JAMA-J. Am. Med. Assoc.* **266**, 1953–1955.

LACROIX, P. (1945). Recent investigations on the growth of bone, *Nature* **156**, 576.

LIETMAN, S.A., YANAGISHITA, M., SAMPATH, T.K. and REDDI, A.H. (1997). Stimulation of proteoglycan synthesis in explants of porcine articular cartilage by recombinant osteogenic protein-1 (bone morphogenetic protein-7), *J. Bone Joint Surg. Am.* **79**, 1132–1137.

LUYTEN, F.P., CUNNINGHAM, N.S., MA, S., MUTHUKUMARAN, N., HAMMONDS, R.G., NEVINS, W.B., WOODS, W.I. and REDDI, A.H. (1989). Purification and partial amino acid sequence of osteogenin, a protein initiating bone differentiation, *J. Biol. Chem.* **264**, 13377–13380.

LUYTEN, F.P., YU, Y.M., YANAGISHITA, M., VUKICEVIC, S., HAMMONDS, R.G. and REDDI, A.H. (1992). Natural bovine osteogenin and recombinant human bone morphogenetic protein-2B are equipotent in the maintenance of proteoglycans in bovine articular cartilage explant cultures, *J. Biol. Chem.* **267**, 3691–3695.

REDDI, A.H. (1981). Cell biology and biochemistry of endochondral bone development, *Coll Relat. Res.* **1**, 209–226.

REDDI, A.H. (1994). Bone and cartilage differentiation, *Curr. Opin. Genet. Dev.* **4**, 737–744.

REDDI, A.H. (1998). Role of morphogenetic proteins in skeletal tissue engineering and regeneration, *Nature Biotech.* **16**, 247–252.

REDDI, A.H. and ANDERSON, W.A. (1976). Collagenous bone matrix-induced endochondral ossification hemopoiesis, *J. Cell Biol.* **69**, 557–572.

REDDI, A.H. and HUGGINS, C. (1972). Biochemical sequences in the transformation of normal fibroblasts in adolescent rats, *Proc. Natl. Acad. Sci. USA* **69**, 1601–1605.

RIPAMONTI, U. (1996). Osteoinduction in porous hydroxyapatite implanted in heterotopic sites of different animal models, *Biomaterials* **17**, 31–35.

RIPAMONTI, U., MA, S. and REDDI, A.H. (1992). The critical role of geometry of porous hydroxyapatite delivery system in induction of bone by osteogenin, a bone morphogenetic protein, *Matrix* **12**, 202–212.

RIPAMONTI, U., VAN DEN HEEVER, B., SAMPATH, T.K., TUCKER, M.M., RUEGER, D.C. and REDDI, A.H. (1996). Complete regeneration of bone in the baboon by recombinant human osteogenic protein-1 (hOP-1, bone morphogenetic protein-7), *Growth Factors* **13**, 273–289.

SAMPATH, T.K. and REDDI, A.H. (1981). Dissociative extraction and reconstitution of extracellular matrix components involved in local bone differentiation, *Proc. Natl. Acad. Sci. USA* **78**, 7599–7603.

SAMPATH, T.K. and REDDI, A.H. (1983). Homology of bone-inductive proteins from human, monkey, bovine, and rat extracellular matrix, *Proc. Natl. Acad. Sci. USA* **80**, 6591–6595.

SENN, N. (1889). On the healing of aseptic bone cavities by implantation of antiseptic decalcified bone, *Am. J. Med. Sci.* **98**, 219–240.

STORM, E.E., HUYNH, T.V., COPELAND, N.G., JENKINS, N.A., KINGSLEY, D.M. and LEE, S.J. (1994). Limb alterations in brachypodism mice due to mutations in a new member of the Tgf-Beta-superfamily, *Nature* **368**, 639–643.

URIST, M.R. (1965). Bone: Formation by autoinduction, *Science* **150**, 893–899.

WEISS, R.E. and REDDI, A.H. 1980. Synthesis and localization of fibronectin during collagenous matrix-mesenchymal cell interaction and differentiation of cartilage and bone *in vivo*, *Proc. Natl. Acad. Sci. USA* **77**, 2074–2078.

WOZNEY, J.M., ROSEN, V., CELESTE, A.V., MITSOCK, L.M., WHITTERS, M.J., KRIZ, R.W., HEWICK, R.M. and WANG, E.A. (1988). Novel regulators of bone formation: Molecular clones and activities, *Science* **242**, 1528–1534.

YAMAGUCHI, A., KATAGIRI, T., IKEDA, T., WOZNEY, J.M., ROSEN, V., WANG, E.A., KAHN, A.J., SUDA, T. and YOSHIKI, S. (1991). Recombinant human bone morphogenetic protein-2 stimulates osteoblastic maturation and inhibits myogenic differentiation *in vitro*, *J. Cell Biol.* **113**, 681–687.

ZIMMERMAN, L.B., DE JESúS-ESCOBAR, J.M. and HARLAND, R.M. (1996). The Spemann organizer signal noggin binds and inactivates bone morphogenetic protein 4, *Cell* **86**, 599–606.

Advances in Tissue Banking Vol. 6

8 THE INFLUENCE OF STERILISATION ON THE OSTEOINDUCTIVE PROPERTIES OF BONE IN RAT BONE MARROW CELL CULTURE

TOSHIYUKI TOMINAGA, TOSHIHIRO IZUMI and MORITOSHI ITOMAN

Department of Orthopaedic Surgery
Kitasato University School of Medicine
1-15-1 Kitasato, Sagamihara
Kanagawa 228-8555, Japan

HITOSHI ITO

Takasaki Radiation Chemistry Research Establishment
1233 Watanuki, Takasaki, Gunma, Japan

1. Introduction

In patients with bone tumours or open comminuted fractures, some substitutes are required to supplement the bone and reconstruct the bone defects. Although fresh autogenous bone is an ideal graft material, availability is limited and the patient must accept the inconvenience of an additional surgical incision, increased post-operative immobility and the weakness of the donor site (Younger and Chapman, 1989). Therefore, bone allografts or biomaterials are often utilised. Although allografts and biomaterials have both advantages and disadvantages relative one to another, one of the advantages of allografted bones over biomaterials is that grafted bones have osteoinductive activity (Redondo *et al.*, 1997; Thielemann *et al.*, 1982). Those activities are explained by the action of growth (Hauschka *et al.*, 1986; Urist *et al.*, 1979; 1983). On the other hand, disease transmission or infection is an important issue in bone allografting. Although the most common methods for preserving bone for grafting are freezing and freeze-drying, there is growing concern about the sterility and safety of bone allografts since the report of a case of HIV transmission following bone grafting (Simonds *et al.*, 1992).

Several methods are currently used for the preservation and sterilisation of bones. There include ethylene oxide sterilisation, heat treatment of frozen bones and irradiation of freeze-dried bones (Aspenberg *et al.*, 1990; Knaepler *et al.*, 1992; Komender *et al.*, 1991; Zasacki, 1991). It is reported that some bacteria or viruses can be inactivated by heat treatment at 80°C for 10 min, or by irradiation (Garrel and Knaepler, 1993; Spire *et al.*, 1985; Turner *et al.*, 1956).

Biological and biomechanical studies on preserved and sterilised bones showed that the processing procedure resulted in a significant loss of osteoinductive properties and strength (Pelker *et al.*, 1983; Voggenreiter *et al.*, 1994). There are studies on the influence of these processes (freezing, freeze-dying, irradiation and heat treatment) on the biological activity of the bone matrix (Buring and Urist, 1967; Goclawska and Stachowicz, 1997; Goldberg and Stevenson, 1987; Turner *et al.*, 1956; Urist *et al*, 1967; Yang *et al.*, 1994), and irradiation is reported to decrease the osteoinductive *activity in vivo* as well as *in vitro* (Urist and Hernandez, 1974).

To investigate the effect of several bone sterilisation processes, we employed irradiation or heat-treatment for sterilisation, and examined the osteoinductive activity of processed bones by assessing cell proliferation, alkaline phospatase (ALP) activity and ALP staining in a rat bone marrow cell culture system.

2. Materials and Methods

2.1. Preparation of irradiated bones

Under anaesthesia, tibiae, femora and humeri were harvested from 12-week-old Wistar rats. The bones were freed of adhering soft tissue, the epiphyses were discarded and the marrow was removed by washing with phosphate-buffered saline (PBS). Bones were extracted in a chloroform/ethanol (1:1) mixture for 4 hours and died in air overnight. They were then lyophilised, cut into uniform 70 mg (dry weight) fragments, and irradiated at either 10, 20, 25, 30, 40, or 50 kGy by γ-irradiation at the Japan Atomic Energy Research Institute. The irradiated bones were stored at −20°C until use in the following experiments.

2.2. Preparation of heat-treated bones

The bones were harvested and the marrow was removed as described above. They were heated in PBS using a water bath at 60°C for 10 hours or 80°C for 10 min. After heat treatment, the bones were cut into uniform 100 mg (wet weight) fragments because dry weight of 70 mg bone weighed 100 mg when wet. They were stored at −80°C until use. Frozen bones were stored at −80°C without heat treatment.

2.3. Isolation and culture of bone marrow cells

Bone marrow cells were isolated from 6-week-old Wistar rats. Tibiae and femora were excised aseptically. After the epiphyses were removed, the marrow was flushed out with a stream of culture medium, i.e. α-minimum essential medium (α-MEM; GIBCO, Rockville, USA) supplemented with 10% foetal bovine serum (GIBCO), 50 mg/ml ascorbic acid (WAKO, Osaka, Japan), and antiobiotics — streptomycin-penicillin-fungizone (GIBCO), through a 20-gauge needle attached to a syringe. Cells were released by disrupting the marrow by repeated pipetting, and plated in a 25-cm^2 tissue culture flask (Falcon, Becton Dickinson, New Jersey, USA)

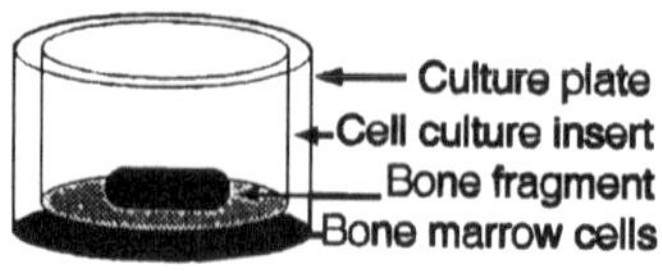

Fig. 1. A cell culture well with a culture insert. A bone fragment can be seen in the insert.

containing 10 ml of medium. After about 14 days when primary cultures reached confluence, the cells were detached with 1 mg/ml trypsin (WAKO) and 0.2 mg/ml EDTA (Sigma, St. Louis, USA) in PBS, and seeded at 4×10^4 cells/35 mm well (Falcon tissue culture plate) for subculture. Cultures were maintained in a humidified atmosphere cosisting of 95% air and 5% CO_2 at 37°C. Two ml of the culture medium was added and changed twice a week.

At subconfluence, a cell culture insert (8.0 μm pore size: Falcon 3090) was set in the well, and a bone fragment was placed in it (Fig. 1). At this point, the medium was supplemented with 10 mM β-glycerophosphate (Sigma) and 1×10^{-8} M dexamethasone (Sigma). The day when the bone fragment was placed in culture was designated day 0. The cells were cultured for 14 days, and the culture medium was changed twice a week. Cell culture inserts without bones were employed as controls.

2.4. Analysis of cell proliferation

Following the culture with bone fragments, the medium was discarded, culture wells were rinsed twice with PBS and cells were dispersed with trypsin/EDTA. Cell numbers were counted in duplicate using a haemocytometer (metalised counting chamber: Becton Dickinson) after stained with 0.4% trypan blue (Sigma). All of the assays were done in triplicate.

2.5. ALP histochemistry

The histochemical staining for ALP was performed by a modification of the Pease method (Drury and Wallington, 1967).

Culture wells were examined by ALP staining on days 0, 7 and 14. Cells were rinsed twice with PBS, fixed for 15 min in 10% formaldehyde neutral buffer solution, rinsed with distilled water, and left in distilled water for 15 min. Substrate 0.1 mg/ml naphtol AS MX-PO_4 (Sigma) in 4 µl/ml N, N-dimethyl formamide (WAKO), tris-HCl (0.2 M, pH 8.3) and 0.6 mg/ml red violet LB salt (Sigma). Add this into the well and incubate it for 45 min at room temperature. The wells were then washed with distilled water and the stained preparation was photographed.

2.6. Measurement of ALP activity

ALP activity was determined by a modification of the Lowry method (Rosen *et al.*, 1988). Briefly, cells were cultured in 6-well plates as described above. Cell monolayers were rinsed twice with PBS, and scraped off in 1.5 ml tris-HCl buffer (pH 7.2) containing 0.1% triton X-100 (WAKO) on days 0, 7 and 14. ALP was released from the cells by 3 cycles of freezing and thawing. The cells were centrifuged and the supernatant was kept frozen at −80°C until use. ALP activity was measured using p-nitrophenyl phosphate as the substrate in 0.1 M aminomethyl propanol containing 2 mM $MgCl_2$, pH 10.5 and 2 mM p-nitrophenol was used as a standard. The content of cellular protein was measured using a BioRad DC protein assay kit (Bio-Rad laboratories, California, USA) and bovine serum albumin served as a standard. ALP activities were determined by absorption at 420 nm and expressed as nanomoles/min/mg protein. All assays were done in triplicate with bone marrow cells of the same origin. Several experiments with bone marrow cells of independent origin gave similar results.

2.7. Statistical analysis

All assays were performed in 3 wells and expressed as means ± standard deviations (SD). Statistical significance was determined by analysis of variance (ANOVA) and subsequently by Fisher's protected least significant difference (PSLD) multiple comparison test.

3. Result

3.1. The effects of different doses of irradiation on bone marrow cells

The addition of 0–50 kGy-irradiated bones to bone marrow cell cultures led to an increase in cell numbers with a maximal effect of unirradiated bone on day 7 (179% compared with control). Irradiation at 50 kGy resulted in numbers that reached 136% of the control on day 14 (Fig. 2). The number of cells obtained when using 10–30 kGy-irradiated bones showed a significant increase compared with controls, but the increases with 40–50 kGy-irradiated bones failed to reach significance compared with controls on day 7. Although there were no significant differences in cell numbers among the unirradiated and irradiated bone groups, those groups did show significant increases over controls on day 14.

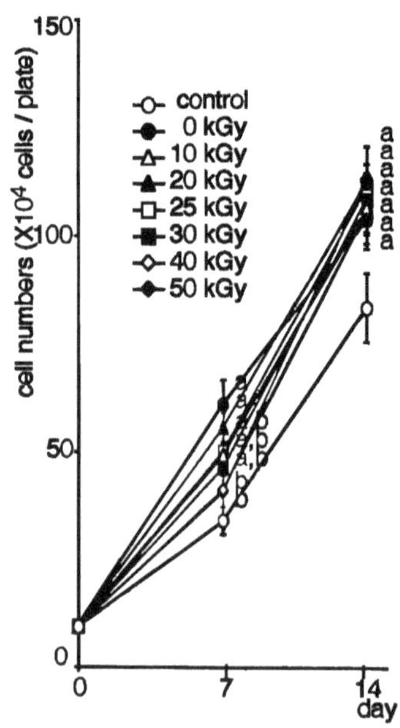

Fig. 2. Cell numbers in bone marrow cultures with or without irradiated bone fragments. Data are presented as means ± SEM for $n = 3$ cultures. "a" denotes a significant difference from controls at $p < 0.05$. "b" denotes a significant difference from 0 kGy at $p < 0.05$.

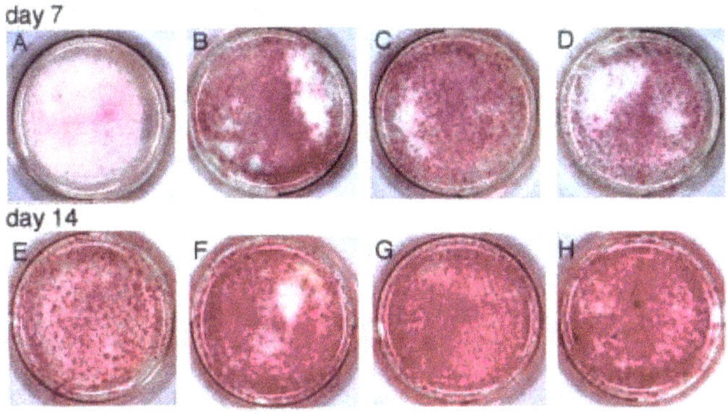

Fig. 3. Photograph of the culture dishes after staining for ALP. **A, E**: Cells without bone fragments on days 7 and 14. **B, F**: Cells with unirradiated bone fragments on days 7 and 14. **C, G**: Cells with irradiated bone fragments (25 kGy) on days 7 and 14. **D, H**: Cells with irradiated bone fragments (50 kGy) on days 7 and 14.

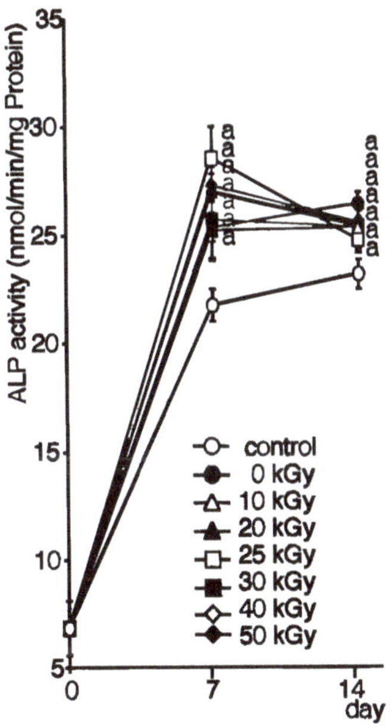

Fig. 4. ALP activity in bone marrow cells with or without irradiated bone fragments. Data are presented as means ± SEM for $n = 3$. "a" denotes a significant difference from controls at $p < 0.05$.

When the cell culture wells were stained for ALP, the staining appeared different in the presence or absence of bone. ALP staining of the cultures with bone (unirradiated or irradiated up to 50 kGy) was denser than controls both on days 7 and 14 (Fig. 3).

When ALP activities were measured, the addition of irradiated bones to bone marrow cells appeared to result in an increase in ALP activity. Although there was no difference among the bone groups, there was a maximum increase of 32% at 25 kGy on day 7, and 10% at 50 kGy on day 14 compared to controls (Fig. 4).

3.2. Effects of the different bone sterilisation processes on bone marrow cell growth

Sterilisation of bones by heating at 60°C for 10 hours, heating at 80°C for 10 min, or by irradiation at 25 kGy were used to assess the effect of bone treatment on the proliferation and differentiation of bone marrow cells.

The addition of bone to bone marrow cells led to increases in the cell numbers in all of the bone groups compared with controls both on days 7 and 14. However, the use of bone fragments sterilised by any of the 3 different methods showed suppression in cell numbers by 18% compared with the effects of frozen bone on day 7. The numbers of bone marrow cells obtained with the 3 different bone sterilisation methods showed no significant differences between bone groups on day 7, but cell numbers in the 80°C group showed an increase compared with other sterilisation methods on day 14 (Fig. 5).

The differences in ALP staining of the culture dishes were apparent between the control and the bone groups, and ALP stains using any of the sterilised bones were denser than controls (Fig. 6).

The increase in ALP activity reached 40% above the control in sterilised bone groups on day 7, and there was no difference among the groups. On day 14, the increase in ALP activity was 13% compared with controls in 60°C group (Fig. 7). ALP activities in 80°C and 25 kGy groups suppressed by 10% compared with frozen on day 14.

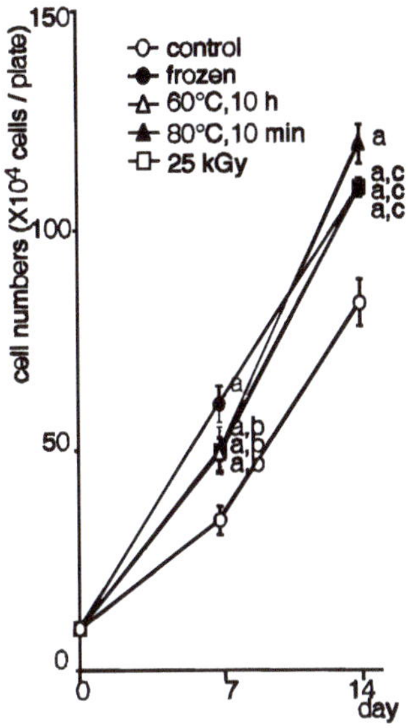

Fig. 5. Cell numbers in bone marrow cultures with or without processed bone fragments. "frozen" indicates the unprocessed frozen bone. Data are presented as means ± SEM of triplicates. "a" denotes a significant difference from controls at $p < 0.05$. "b" denotes a significant difference from frozen bone at $p < 0.05$. "c" denotes a significant difference from 80°C at $p < 0.05$.

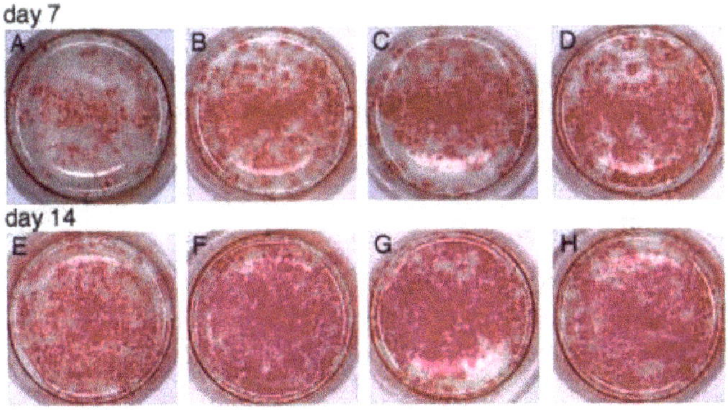

Fig. 6. Photograph of the culture dishes after staining for ALP. **A, E**: Cells without bone fragments on days 7 and 14. **B, F**: Cells with the heat-treated bone fragments on days 7 and 14. **C, G**: Cells with the heat-treated bone fragments (60°C for 10 hours) on days 7 and 14. **D, H**: Cells with irradiated bone fragments (25 kGy) on days 7 and 14.

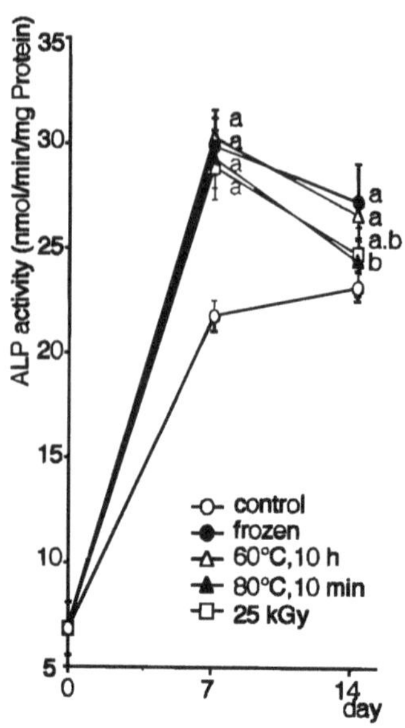

Fig. 7. ALP activity in bone marrow cells with or without processed bone fragments. Data are presented as means ± SEM of triplicates. "a" denotes a significant difference from controls at $p < 0.05$. "b" denotes a significant difference from frozen bone at $p < 0.05$.

4. Discussion

We have investigated the effect of sterilisation-processed bone on cell proliferation and differentiation in cultures of Wistar rat's bone marrow cells. Although some differences between sterilisation methods were observed, the addition of processed bone to the cultures resulted in increases in cell numbers and ALP activity, independent of the processing methods.

Our bone marrow cell culture system includes osteogenic cells that form the mineralised tissue, in which osteocalcin is expressed in the presence of β-glycerophosphate and dexamethasone (Maniatopoulos *et al.*, 1988; Milen *et al.*, 1998; Peter *et al.*, 1998). In this study, the bone fragments placed in the cell culture insert enhanced ALP activity as well as accelerating cell proliferation. Since the culture system we used separated the bone marrow cells from the added bone fragment by a culture insert with sieve, it is hypotisised that humoral factors such as growth factors are involved in the action of processed bones. In previous *in vitro* and *in vivo* studies on the demineralised bone, proteins in the bone matrix are reported to promote cell proliferation and differentiation (Becerra *et al.*, 1996; Huang *et al.*, 1988; Reddi and Huggins, 1972; Urist, 1965). It is know that there are many groeth factors in bone matrix. Among them are transforming growth factor-β (TGF-β) and bone morphogenetic protein (BMP), which play important roles in osteogenesis. Furthermore, bone is the richest source of TGF-β, containing more than 200 μg/kg of wet weight (Finkelman *et al.*, 1991; Mohan and Baylink, 1991; Seyedin *et al.*, 1986). BMP-2 promotes cell proliferation and differentiation in rat bone marrow cell culture (Puleo, 1997; Rickard *et al.*, 1994), and basic fibroblast growth factor has some effects similar to BMP-2 (Nahada *et al.*, 1997; Locklin *et al.*, 1995; Pitaru *et al.*, 1993). BMP-2 and TGF-β also promote cell proliferation and/or differentiation in human bone marrow cell culture (Fromigue *et al.*, 1998; Lecanda *et al.*, 1997). These growth factors in bone might be involved in the effects of bone fragments in our culture system.

Irradiation is used for sterilisation of graft bones. There is some controversy concerning the osteoinductive capacity of bone after irradiation. Some investigators report that irradiation decreases the osteoinductive capacity of bone (Munting *et al.*, 1988; Urist and Hernandez, 1974; Zhang *et al.*, 1997), whereas others reported a lack of influence on the osteoinductive capacity after 22 kGy (Hallfeldt *et al.*, 1995; Katz *et al.*, 1990; Wientroub and Reddi, 1988). Wientroub and Reddi (1988) suggested that the reported decreased osteoinductive capacity of irradiated bone could be explained by the denaturation of protein by the heat generated by irradiation. We employed long-duration, low-rate bone irradiation in order to prevent heat generation. Our result showed that bone irradiated at 25 kGy retained the original osteoinductive effect on bone marrow cells.

In addition to irradiation, several other methods are used for the serilisation of graft bones in clinical practice. These include the use of ethylene oxide gas, autoclaving and heat treatment.

We examined two methods of heat treatment, which included treatments at 60°C for 10 hours (Shikata *et al.*, 1978) or at 80°C for 10 min (Knaepler *et al.*, 1992), compared with irradiation sterilisation. Urist *et al.* (1967) reported that the osteoinductive activity was preserved by heat treatment at 50–70°C, while it was decreased at temperature above 80°C and was completely lost at 100°C. Ito *et al.* (1995) reported that the activity was preserved by heat treatment at 60°C for 10 hours or at 70°C for 1 hour, while it was decreased at 70°C or 80°C for 10 hours. In the present study, very little deterioration could be found after sterilisation under the conditions we used (60°C for 10 hours, 80°C for 10 min, irradiation at 25 kGy), regarding either cell proliferation or ALP activity in our culture system. Considering that hepatitis B virus can be inactivated by heat treatment at 80°C for 10 min (Mauler *et al.*, 1987) or by irradiation more than 20 kGy (Goclawska *et al.*, 1991), sterilisation of allograft these methods could be useful in retaining osteoinductive properties.

6. Acknowledgement

The authors wish to thank Dr. Y. Takagaki for useful comments and NAI (Tokyo, Japan) which provides English editing services for manuscript preparation.

7. References

ASPENBERG, P. *et al.* (1990). Dose-dependent reduction of bone inductive properties by ethylene oxide, *J. Bone Joint Surg.* **72**(B), 1030–1037.

BECERRA, J. *et al.* (1996). Demineralized bone matrix mediates differentiation of bone marrow stromal cells *in vitro*: Effect of age of cell donor, *J. Bone Miner. Res.* **11**, 1703–1714.

BURING, K. and URIST, M.R. (1967). Effects of ionizing radiation on the bone induction principle in the matrix of bone implants, *Clin. Orthop.* **55**, 225–234.

DRURY, R.A.B. and WALLINGTON, E.A. (1967). *Carleton's Histological Technique, Fourth*, Oxford University Press, Toronto, pp. 244–245.

FINKELMAN, R.D. *et al.* (1991). Vitamin D deficiency causes a selective reduction in deposition of transforming growth factor in rat bone: Possible mechanism for impaired osteoinduction, *Proc. Natl. Acad. Sci. USA* **88**, 3657–3660.

FROMIGU, O. *et al.* (1998). Bone morphogenetic protein-2 and transforming growth factor 2 interact to modulate human bone marrow stromal cell proliferation and differentiation, *J. Cell. Biochem.* **68**, 411–426.

GARREL, T.V. and KNAEPLER, H. (1993). Disinfection of allogenic bone grafts with low heat, *Transfusion* **33**(7), 615.

GOCLAWSKA, A.D. *et al.* (1991). Effect of radiation sterilization on the osteoinductive properties and the rate of remodeling of bone implants preserved by lyphization and deep-freezing, *Clin. Orthop.* **272**, 30–37.

GOLDBERG, V.M. and STEVENSON, S. (1987). Natural history of autografts and allografts, *Clin. Orthop.* **225**, 7–16.

HAUSCHKA, P.V. *et al.* (1986). Growth factors in bone matrix, *J. Biol. Chem.* **261**(27), 12665–12674.

HALLFELDT, K.K.J. *et al.* (1995). Sterilization of partially demineralized bone matrix. The effects of different sterilization techniques on osteogenetic properties, *J. Surg. Res.* **59**, 614–620.

HANADA, K. *et al.* (1997). Stimulatory effects of basic fibroblast growth factor and bone morphogenetic protein-2 on osteogenic differentiation of rat bone marrow-derived mesenchymal stem cells, *J. Bone Miner. Res.* **12**, 1606–1614.

HUANG, D.S.T. *et al.* (1988). Mitogenic response of cells in culture to demineralized bone matrix, *J. Oral Maxillofac. Surg.* **46**, 460–463.

ITO, T. *et al.* (1995). Sensitivity of osteoinductive activity of demineralized and defatted rat femur to temperature and duration of heating, *Clin. Orthop.* **316**, 267–275.

KATZ, R.W. *et al.* (1990). Radiation-sterilized insoluble collagenous bone matrix is a functional carrier of osteogenin for bone induction, *Calcif. Tissue Int.* **47**, 183–185.

KNAEPLER, H. *et al.* (1992). Experiments on thermal disinfection and sterilization of allogenic bone grafts and the effect of this treatment on biological value, *Unfallchirung* **95**, 477–484.

KOMENDER, J. *et al.* (1991). Therapeutic effects of lyophilized and radiation-sterilized, allogeneic bone, *Clin. Orthop.* **272**, 38–49.

LECANDA, F. *et al.* (1997). Regulation of bone matrix protein expression and induction of differentiation of human osteoblasts and human bone marrow stromal cells by bone morphogenetic protein-2, *J. Cell. Biochem.* **67**, 386–398.

LOCKLIN, R.M. *et al.* (1995). *In vitro* effects of growth factors and dexamethasone on rat bone marrow stromal cells, *Clin. Orthop.* **313**, 27–35.

MANIATOPOULOS, C. *et al.* (1988). Bone formation *in vitro* by stromal cells obtained from bone marrow of young rats, *Cell. Tissue Res.* **254**, 317–378.

MAULER, R. *et al.* (1987). Inactivation of HLTV-III/LAV, hepatitis B and non-A/non-B viruses by pasteurisation in human plasma preparations, *Dev. Biol. Stardard* **67**, 337–351.

MILEN, M. *et al.* (1998). Dexamethasone stimulates osteogenic differentiation in vertebral and femoral bone marrow cell cultures: Comparison of IGF-1 gene expression, *J. Cell Biochem.* **71**, 382–391.

MOHAN, S. and BAYLINK, D.J. (1991). Bone growth factors, *Clin. Orthop.* **263**, 30–48.

MUNTING, E. *et al.* (1988). Effect of sterilization on osteoinduction. Comparison of five methods in demineralized rat bone, *Acta Orthop. Scand.* **59**(1), 34–38.

PELKER, R.R. *et al.* (1983). Biomechanical properties of bone allografts, *Clin. Orthop.* **174**, 54–57.

PETER, S. J. *et al.* (1998). Osteoblastic phenotype of rat bone marrow stromal cells cultured in the presence of dexamethasone, glycerophosphate, and L-ascorbic acid, *J. Cell Biochem.* **71**, 55–62.

PITARU, S. *et al.* (1993). Effect of basic fibroblast growth factor on the growth and differentiation of adult stromal bone marrow cells: Enhanced development of mineralized bone-like tissue in culture, *J. Bone Miner. Res.* **8**, 919–929.

PULEO, D. A. (1997). Dependence of mesenchymal cell responses on duration of exposure to bone morphogenetic protein-2 *in vitro*, *J. Cell Physiol.* **173**, 93–101.

REDDI, A. H. and HUGGINS, C. (1972). Biochemical sequences in the transformation of normal fibroblasts in adolescent rats, *Proc. Natl. Acad. Sci. USA* **69**, 1601–1605.

REDONDO, L. M. *et al.* (1997). Repair of experimental mandibular defects in rats with autogenous, demineralized, frozen and fresh bone, *J. Oral Maxillofac. Surg.* **35**, 166–169.

RICKARD, D. J. *et al.* (1994). Induction of rapid osteoblast differentiation in rat bone marrow stromal cell cultures by dexamethasone and BMP-2, *Dev. Biol.* **161**, 218–228.

ROSEN, D. M. *et al.* (1988). Transforming growth factor-beta modulates the expression of osteoblast and chondroblast phenotypes *in vitro*, *J. Cell. Physiol.* **134**, 337–346.

SEYEDIN, S. M. *et al.* (1986). Cartilage-inducing factor-A. Apparent identity to transforming growth factor-beta, *J. Biol. Chem.* **261**(13), 5693–5695.

SHIKATA, T. *et al.* (1978). Incomplete inactivation of hepatitis B virus after heat treatment at 60°C for 10 hours, *J. Infect. Dis.* **138**(2), 242–244.

SIMONDS, R. J. *et al.* (1992). Transmission of human immunodeficiency virus type 1 from a seronegative organ and tissue donor, *N. Engl. J. Med.* **326**, 726–732.

SPIRE, B. *et al.* (1985). Inactivation of lymphadenopathy-associated virus by heat, gamma rays, and ultraviolet light, *Lancet* **II**, 188–189.

THIELEMANN, F. W. *et al.* (1982). Osteoinduction. Part I: Test model and comparative long term observation of allogenic and xenogenic matrix implants, *Arch. Orthop. Traumat. Surg.* **99**, 217–222.

TURNER, T. C. *et al.* (1956). Sterilization of preserved bone grafts by high voltage cathode irradiation, *J. Bone Joint Surg. (Am)* **38**(4), 862–884.

URIST, M. R. (1965). Bone: formation by autoinduction, *Science* **150**, 893–899.

URIST, M. R. and HERNANDEZ, A. (1974). Excitation effects of cobalt 60 radiation-sterilization of bank bone, *Arch. Surg.* **109**, 486–493.

URIST, M. R. *et al.* (1979). Solubilized and insolubilized bone morphogenetic protein, *Proc. Natl. Acad. Sci.* **76**(4), 1828–1832.

URIST, M. R. *et al.* (1986). Human bone morphogenetic protein, *Proc. Soc. Exp. Biol. Med.* **173**, 194–199.

URIST, M. R. *et al.* (1967). The bone induction principle, *Clin. Orthop.* **53**, 243–283.

VOGGENREITER, G. *et al.* (1994). Effects of preservation and sterilization on cortical bone grafts. A scanning electron microscopic study, *Arch. Orthop. Trauma Surg.* **113**, 294–296.

WIENTROUB, S. and REDDI, A. H. (1988). Influence of irradiation on the osteoinductive potential of demineralized bone matrix, *Calcif. Tissue Int.* **42**, 255–260.

YANG, C. Y. *et al.* (1994). The healing grafts combining freeze-dried and demineralized allogeneic bone in rabbits, *Clin. Orthop.* **298**, 286–295.

YOUNGER, E. M. and CHAPMAN, M. W. (1989). Morbidity at bone graft donor sites, *J. Orthop. Trauma* **3**(3), 192–195.

ZASACKI, W. (1991). The efficiency of application of lyophilized, radiation-sterilized bone graft in orthopedic surgery, *Clin. Orthop.* **272**, 82–87.

ZHANG, Q. *et al.* (1997). Ethylene oxide does not extinguish the osteoinductive capacity of demineralized bone, *Acta Orthop. Scand.* **68**(2), 104–108.

Advances in Tissue Banking Vol. 6

9 BONE GRAFTING FOR ACETABULAR DEFICIENCIES IN TOTAL HIP ARTHROPLASTY

TIIT HAVIKO

Clinic of Traumatology and Orthopaedics
Tartu University Clinics, Tartu, Estonia

PEETER LAASIK

Elva Hospital, Elva, Estonia

1. Introduction

Reconstruction of acetabular deficiencies may present difficult problems in patients with segmental and cavitary acetabular bone loss, but especially in the case of combined defects and pelvic discontinuity. Impaction bone grafting with morselised bone tissue and cementation of the acetabular cup, as described by Slooff *et al.* (1984), is useful for primary hip replacement in acetabular protrusion. This procedure showed good early results (Slooff *et al.*, 1996) as well as good long-term outcome with a survival rate of 90% at 12 years (Schreurs *et al.*, 1998). Compressed allograft chips for acetabular reconstruction also yielded promising results in acetabular cavitary cysts and in revision hip arthroplasty (Azuma *et al.*, 1994). However, in serious central cavitary deficiencies the above mentioned technique, appears to be insufficient for obtaining good long-term results. A stress analysis of acetabular reconstruction in protrusio acetabuli revealed that reduction in stress on the medial wall of the acetabulum happened when a reinforcement ring — with cementation of a polyethylene cup into it — was used (Crowninshield *et al.*, 1983). In the case of peripheral segmental acetabular deficiencies with a considerable loss of the bone stock, conventional techniques with plain cementation have resulted in poor long-term outcome. The bone graft, usually taken from the removed femoral head, is necessary, especially in osteoarthrosis secondary to congenital dysplasia. The use of the femoral head auto- or allografts restores the bone stock, which allows implantation of the acetabular component (Harris and Gothers, 1976; Becker *et al.*, 1999).

We performed a clinical and radiological evaluation of patients with bone grafting of the acetabulum both in primary and revision hip arthroplasty. We also studied incorporation of bone grafts and their consolidation with the mature bone of the acetabulum, as well as radiolucencies around the acetabular cup and cup failure.

2. Patients and Methods

2.1. Pre-operative data

Between January 1993 and December 1998, we performed 121 operations of the bone grafting of the acetabular component in total hip arthroplasties in 111 patients (86 women and 35 men). The mean age of the patients was 62.4 (32–83) years. The initial diagnoses were 85 coxarthrosis, 15 rheumatoid arthritis, 6 posttraumatic conditions and 5 other secondary osteoarthrosis. Primary total hip arthroplasty was performed in 89 patients. In 22 patients, revision surgery was made for the aseptic loosening of the acetabular component with a major deficiency of the acetabular bone stock 2 to 6 years after primary surgery. The mean age of the patients at revision operations was 65.4 (47–83) years. The hips had been revised one to three times. The acetabular deficiencies were classified according to the American Academy of Orthopaedic Surgeons Committee on the hip (D'Antonio *et al.*, 1989) as follows: 59 hips were classified as type I segmental deficiencies, 49 hips as type II cavitary deficiencies (protrusio acetabuli in 15 hips), 11 hips as type III combined deficiencies and 2 hips as type IV pelvic discontinuities. The estimation of the acetabular defects was based on the pre-operative radiographs on the AP and lateral views, and on the intraoperative findings described in operative reports.

2.2. Operative technique

The operations were performed by four orthopaedic surgeons, while posterolateral approach was mostly used. Trochanteric osteotomy was not performed even in revision cases. In the case of primary total hip arthroplasty (89 patients), four types of total hip prostheses were used: Lubinus SP II and Lubinus IP, as well as Müller straight and Spotorno (one case). The acetabulum was prepared in the usual way employing reamers of adequate size. Subchondral bone was preserved partially. In the reconstruction of peripheral segmental deficiency (41 patients), a structural massive auto- or allograft of the femoral head was used in the postero-superolateral aspect of the acetabulum. In the case of allograft reconstruction, frozen femoral heads were thawed in a warm physiological solution for about 30 min and prepared for bone grafting according to the reguired size/shape. Microbiological samples were taken from the allograft. Fixation of bone grafts was done with 2–4 3.2 mm cortical screws. The screws were placed in the posterior-superior or in the posterior-inferior quadrant of the acetabulum. Further, the acetabulum was once again reamed until the test cup was sufficiently covered by the bone stock. A cemented cup of 44 to 52 mm in diameter was inserted, and Palacos bone cement was used.

On the femoral component, a cemented stem with the 28 mm or 32 mm head was implanted. Reconstruction of central segmental acetabular defects (16 hips) was performed by filling the defect with an impaction of morselised bone autografts. For routine implantation of the socket, we used a cementing technique with anchoring of the socket to the posterosuperior part of the acetabulum. In 2 patients with central segmental acetabular deficiencies with total absence of the medial wall, a massive allograft socket was fixed to the bone defect, followed by implantation of the acetabular component. Reconstruction on peripheral cavitary defects was mostly performed by impaction morselised autogenous bone grafting. In case cavitary cysts were present in the acetabulum, all soft tissue was removed and bone grafts were prepared from the resected femoral head and neck. For central cavitary defects (protrusio acetabuli), impacted morselised bone allografts (9 hips) or autografts (6 hips) with a chip size of 4-6 mm were used. The bottom of the acetabulum was filled with bone chips and was then thightly impacted with bone packers. To anchor the acetabular component, 3–5 holes with a diameter of 5 mm were drilled in the peripheral part of the acetabulum nonimpacted with bone chips. The standard Lubinus or Müller acetabular component was inserted.

Morselised bone grafts in association with an acetabular support cup (10 hips) were used in severe cases of central cavitary deficiency, mostly in revision operations (7 hips). After the impaction of bone chips the Kerboul (Sulzer) or the Ewerwahn (Link) acetabular support cup was fixed to the acetabulum, followed by the implantation of the cemented acetabular component. In 2 cases of pelvic discontinuity, a massive bone allograft was used in combination with an acetabular support cup. Prophylactic antibiotics and low molecular heparin were given routinely.

2.3. Post-operative management

The patients were kept in bed for 1–2 days, after which they started to walk, using a walking frame or crutches, with partial weight bearing until 3 months after an operation for peripheral cavitary deficiencies. In the case of segmental deficiencies, central cavitary deficiencies and pelvic discontinuity, partial weight bearing lasted for 6 months and full weight bearing was adopted thereafter.

2.4. Clinical and radiographic assessments

Clinical and roentgenological evaluation of the patients was carried out regularly at 3, 6, 12, 18, 24 months, and at 3, 5 and 6 years after the operation. Radiographically, the position of the cup was defined normal if the socket abduction angle in the AP radiograph was measured at 30°–50° and anteversion in the lateral radiograph was measured at 10°–30°.

Radiolucencies, migration, tilting of the cup, and breakage of the screws were taken notice of. The width of the radiolucent zone at the bone-cement interface in each of the acetabular zones of DeLee and Charnley (1976) was measured. Migration of the cup or its tilting was determined with the X-ray template, on comparison of the postoperative and the follow-up AP radiographs (Zehntner and Ganz, 1994). Radiographic failure of the acetabulum was defined if radiolucent lines, exceeding 2 mm in all three zones, or migration of the cup by more than 5 mm, were observed (van der Linde and Tonino, 2001).

3. Results

We did not observe any early complications related to the technical aspects of surgery. There was one dislocation of the hip and one infection among 111 operated patients. All patients were followed up early, 3 months after surgery, and regularly up to 6 years after the operation.

3.1. Segmental deficiency

In the group of segmental acetabular deficiency, 54 patients (55 hips) were available for follow-up in the period of one to six years after reconstruction operation.

Massive bone grafting with screw fixation was analysed in 41 patients (42 hips), the mean age of patients was 55.7 (32–75) years (Table 1). Poor operation results were recorded in six cases. In five cases of them, resorption of auto- or allograft was seen at 1 to 2.5 years after surgery, which was followed by a migration of

Table 1. Characteristics of 42 hips treated with massive bone grafting of peripheral segmental acetabular deficiencies.

A	B	C	D	E	F	G	H	I	J	K
1	F	39	OA	Lu	Lu	Au	2	Y	N	N
2	F	40	OA	Lu	Lu	Au	5	Y	N	N
3*	F	47	OA	Lu	Lu	Al	5	Y	N	N
4	F	34	CL	Lu	Lu	Au	6	Y	N	N
5	M	64	OA	Lu	Lu	Au	5	Y	N	N
6	M	40	CL	Lu	Lu	Au	5	Y	N	N
7	F	49	CL	Lu	Lu	Au	6	Y	N	N
8	F	49	CL	Lu	Lu	Au	6	Y	N	N
9	F	44	OA	Lu	Lu	Au	4	Y	N	N
10	F	58	OA	Lu	Lu	Al	4	Y	N	N
11	F	64	OA	Lu	Lu	Au	2	Y	N	N
12	F	83	OA	Lu	Lu	Al	3	Y	N	N
13*	F	54	CL	Lu	Lu	Al	5	Y	N	N
14*	F	56	CL	Lu	Lu	Al	3	Y	N	N
15	F	42	CL	Lu	Lu	Au	2	Y	N	N
16	F	70	OA	Lu	Lu	Al	2	N	Y	Y
17	F	60	OA	Lu	Lu	Au	2	Y	N	N
18	M	51	CL	Lu	Lu	Al	2	Y	N	N
19	F	58	OA	Lu	Lu	Au	3	Y	N	N
20	F	51	OA	Lu	Lu	Au	5	Y	N	N
21	F	66	OA	Lu	Lu	Au	2	Y	Y	Y
22	F	32	CL	Zi	Sp	Au	3	Y	N	N
23	F	42	OA	Lu	Lu	Au	3	Y	N	N
24	F	67	OA	Lu	Lu	Au	1.5	Y	N	N
25	M	73	OA	Lu	Lu	Au	1	Y	N	N
26	F	36	OA	Lu	Lu	Au	1	Y	N	N
27	F	64	OA	Lu	Lu	Au	4	Y	N	N
28	F	56	OA	Lu	Lu	Au	2.5	N	Y	Y
29	F	57	OA	Lu	Lu	Au	2.5	Y	N	N
30*	F	66	OA	Lu	Lu	Al	2	N	Y	Y
31	F	40	OA	Lu	Lu	Au	2	Y	N	N
32	F	41	OA	Lu	Lu	Au	1.5	Y	N	N
33*	M	66	OA	Lu	Lu	Al	1.5	N	N	Y
34*	F	68	OA	Lu	Lu	Al	3	Y	N	N

Table 1. (*Continued*)

A	B	C	D	E	F	G	H	I	J	K
35	F	50	CL	Lu	Lu	Au	2	Y	N	N
36	F	75	OA	Lu	Lu	Au	1	N	Y	Y
37*	F	75	OA	Lu	Lu	Al	1	Y	N	N
38	F	53	OA	Lu	Lu	Au	2	Y	N	N
39	F	56	CL	Lu	Lu	Au	2	Y	N	N
40*	M	71	OA	Lu	Lu	Al	2	Y	N	N
41*	M	59	RA	Lu	Lu	Al	1.5	Y	N	N
42	F	75	OA	Lu	Lu	Au	1	Y	N	N

A: Case number; B: Sex; C: Age at operation; D: OA = Osteoarthrosis, RA = Rheumatoid arthritis, CL = Congenital luxation; E: Type of cup, Lu = Lubinus, Zi = Zimmer; F: Type of stem, Lu = Lubinus, Sp = Spotorno; G: Type of graft, Au = Autograft, Al = Allograft; H: Follow-up in years; I: Consolidation, Y = Yes, N = No; J: Resorption, Y = Yes, N = No; K: Failure, Y = Yes, N = No; *Revision operation

the acetabular cup. In one case (a 66-year-old man) a migration of the cup was seen despite the consolidation of the bone graft and the absence its resorption. In two cases migration of the acetabular cup was stopped at 2 and 2.5 years and no revision operation was needed. In other four cases, a revision operation was done in the following period. In other 36 cases of massive bone grafting, as well as in 13 cases of central segmental deficiency, the evidence of incorporation of bone grafts after one to six years was confirmed rontgenologically. Regarding the results of massive bone grafting of peripheral segmental deficiencies, positive results were achieved in 86% of cases (Fig. 1).

3.2. Cavitary deficiency

In the group of cavitary deficiencies 43 patients (46 hips), mean age 69.4 (60–83) years, were observed one to six years after operation. In peripheral cavitary defects, bone autografts without a support acetabular cup were used. Roentgenologically, incorporation of

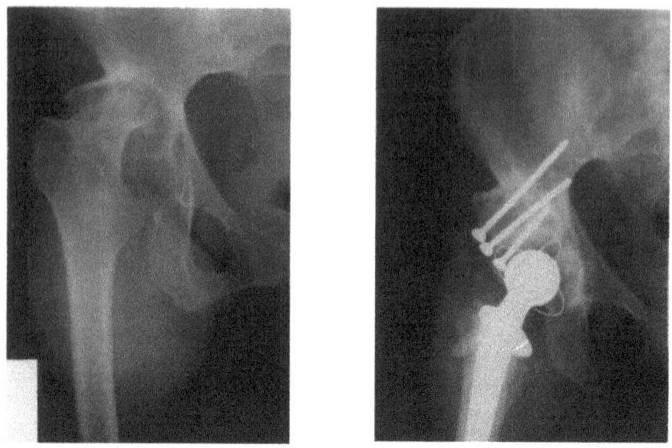

(a) (b)

Fig. 1. (a) Pre-operative radiograph of a 42-year-old female patient with rheumatoid arthritis. (b) Radiograph made 3 years after total hip arthroplasty with restoration of the acetabular bone stock. Consolidation of the allograft with host bone.

bone grafts in all cases occurred after one post-operative year. In 15 cases of central cavitary defects (protrusio acetabuli), migration of the acetabular cup occurred only in one case. This 68-year-old female patient had undergone a bone autografting for protrusio acetabuli followed by cementation of the acetabular component. Normal incorporation of bone autografts took place after two years. Despite that, a moderate migration of the socket with increase in the abduction angle was evident, which was still not an indication for revision surgery. In other cases of central cavitary defects, roentgenologically confirmed incorporation of morselised bone grafts, including ten hips operated with the use of the acetabular support cup, took place after 1 to 6 years (Fig. 2). One patient, a 69-year-old woman, was not included in the analysis due to the too short follow-up period (Table 2).

3.3. Pelvic discontinuity

Two patients (a 76-year-old man and a 60-year-old woman) with pelvic discontinuity following repeated previous operations on the

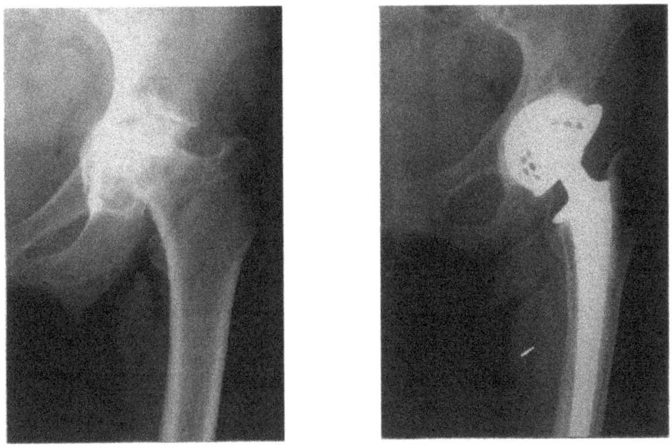

Fig. 2. (a) Pre-operative radiograph of a 63-year-old female patient with rheumatoid arthritis. (b) After reconstruction of the bone stock with morselised allografts, an Ewerwahn acetabular support cup was implanted. Allogenous bone reveals definite radiological incorporation after 4 years.

Table 2. Characteristics of ten hips treated with bone grafting and an acetabular support cup.

A	B	C	D	E	F	G	H	I	J	K
1*	F	76	OA	1	Si	Ke	5	Y	N	N
2*	F	61	OA	2	Ae	Ke	2	Y	N	N
3*	M	83	OA	1	Si	Ke	5	Y	N	N
4*	M	76	OA	3	Mü	Ke	1.5	Y	N	N
5*	F	69	OA	1	Lu	Ke	1	Y	N	N
6	F	69	OA	0	Lu	Ew	0.5	Not analysed		
7	F	70	OA	0	Lu	Ew	1.5	Y	N	N
8*	F	60	OA	1	Lu	Ke	1.5	Y	N	N
9*	F	60	RA	1	Mü	Ke	6	Y	N	N
10	M	59	RA	0	Lu	Ke	1	Y	N	N

A: Case number; B: Sex; C: Age at operation; D: OA = Osteoarthrosis, RA = Rheumatoid arthritis; E: Number of previous operations; F: Primary prostheses, Si = Sivash, Ae = Aesculap, Mü = Müller, Lu = Lubinus; G: Type of cage, Ke = Kerboul, Ew = Ewerwahn; H: Follow-up in years, I: Incorporation, Y = Yes; J: Radiolucencies, N = No; K: Failure, N = No; *Revision operation hip were observed two years after a new revision operation with massive bone allografting and the use of the Kerboul support cup. In both cases the resorption of the pelvis and the support function of the lower limb were achieved after a long period (2 years) of partial weight bearing.

4. Discussion

Total hip replacement usually yields good or excellent outcomes both clinically and roentgenologically. The success of arthroplasty depends upon many factors, one of them being the quality of the acetabular bone stock prior to operation. In the mid-1970s, Harris and Grothers (1976) presented a surgical technique describing autogenous bone grafting with the use of the femoral head for correcting severe acetabular deficiency in total hip arthroplasty.

In our practice, we have used the AAOS classification of acetabular deficiencies, which facilitates the planning of surgery for different failed acetabular components.

Acetabular reconstruction with morselised autogenous bone grafts usually yields good results in the case of peripheral cavitary deficiencies, as was also shown in our investigation. This method is applicable also in central cavitary defects with promising early results (Slooff *et al.*, 1996) and long-term results, with a survival rate of up to 90% at 12 years (Schreurs *et al.*, 1998). In our analysis of 15 cases of protrusio acetabuli, a failure of the cup occurred only in one case after two years. In 10 cases of more severe bone loss, acetabular reconstruction was achieved with bone grafting followed by implantation of the acetabular support cup. No failure of the acetabulum was recorded. The results obtained with this method were promising also at 3 years of follow-up in a series of Dartee *et al.* (1988).

Biological augmentation of the acetabular roof is extremely important in peripheral segmental acetabular deficiencies, especially in revision arthroplasty. Therefore, bone grafting is considered necessary in the aspect of long-term results (Morsches *et al.*, 1989). Massive bone grafting of the acetabular roof in our series was done in 42 cases with screw fixation. Good results were achieved after one to six years in 86% of patients with segmental acetabular deficiency.

The use of allografts is still controversial. We used massive allografts in nine revision arthroplasties. Resorption of the allograft was seen in one case, 2 years after the operation. The failure of the cup was evident in two cases, 1.5–2.5 years after the operation. At the same time, some authors have reported an overall success rate as low as 28% at 2.5 years (Hoikka *et al.*, 1993).

Assessment of radiological results is necessary for detecting early changes in the acetabular component. In the standard case, it should be done earlier than two years after operation. Since radiolucent lines or zones usually correspond to bone resorption, and since increasing radiolucencies correlate with mechanical loosening (Thanner, 1999), we have performed a regular radiographic examination of the patients in the described series six months to six years after the operation.

5. Acnowledgements

This study was supported by the Estonian Science Foundation grant No. 3945.

6. References

AZUMA, T., YASUDA, H., OKAGAKI, K. and SAKAIK, H. (1994). Compressed allograft chips for acetabular reconstruction in revision hip arthroplasty, *J. Bone Joint Surg. (Br)* **76**, 740–744.

BECKER, R., URBACH, D., GRASSHOFF, H. and NEUMANN, H.W. (1999). Structural bone grafting in arthroplasty for congenital hip dysplasia, *Acta Orthop, Scand.* **70**(5), 430–434.

CROWNINSHIELD, R.D., BRAND, R.A. and PEDERSEN, D.R. (1983). A stress analysis of acetabular reconstruction in protrusio acetabuli, *J. Bone Joint Surg. (Am)* **65**, 495–499.

D'ANTONIO, J.A., CAPELLO, W.N., BORDEN, L.S., BARGAR, W.L., BIERBAUM, B.F., BOETTCHER, W.G., STEINBERG, M.E.,

STULBERG, S.D. and WEDGE, J.H. (1989). Classification and management of acetabular abnormalities in total hip arthroplasty, *Clin. Orthop.* **243**, 126–137.

DARTEE, D.A., HUIJ, J. and TONINO, A.J. (1988). Bank bone grafts in revision hip arthroplasty for acetabular protrusion, *Acta Orthop. Scand.* **59**(5), 513-515.

DE LEE, J.G. and CHARNLEY, J. (1976). Radiological demarcation of cemented sockets in total hip replacement, *Clin. Orthop.* **121**, 20–32.

HARRIS, W.H. and GROTHERS, O.D. (1976). Autogenous bone grafting using the femoral head to correct severe acetabular deficiency for total hip replacement. In: *The Hip: Proceedings at the 4th Open Scientific Meeting of the Hip Society,* Mosby, St. Louis, pp. 161–185.

HOIKKA, V., SCHLENZKA, D., WIRTA, J., PAAVILAINEN, T., ESKOLA, A., SANTAVIRTA, S. and LINDHOLM, S. (1993). Failures after revision hip arthroplasties with threaded cups and structural bone allografts. Loosening of 13/18 cases after 1-4 years, *Acta Orthop. Scand.* **64**, 403–407.

MORSCHER, E., DICK, W. and SEELIG, W. (1989). Revisions-Arthroplastik des Hüftgelenkes mit autologer und homologer Spongiosa, *Orthopäde* **18**, 428–437.

SCHREURS, B.W., SLOOFF, T.J., BUMA, P., GARDENIERS, J. and HUISKES, R. (1998). Acetabular reconstruction with impacted morsellised cancellous bone graft and cement, *J. Bone Joint Surg. (Br)* **80**, 391–395.

SLOOFF, T.J., HUISKES, R., VAN HORN, J. and LEMMENS, A.J. (1984). Bone grafting in total hip replacement for acetabular protrusion, *Acta Orthop. Scand.* **55**, 593–596.

SLOOFF, T.J., BUMA, P., SCHREURS, B.W., SCHIMMEL, J.W., HUISKES, R. and GARDENIERS, J. (1996). Acetabular and femoral reconstruction with impacted graft and cement, *Clin. Orthop.* **323**, 108–115.

THANNER, J. (1999). The acetabular component in total hip arthroplasty. Evaluation of different fixation principles, *Acta Orthop. Scand.* **70**(Suppl 286), 41.

VAN DER LINDE, M. and TONINO, A. (2001). Acetabular revision with impacted grafting and a reinforcement ring, *Acta Orthop. Scand.* **72**(3), 221–227.

ZEHNTNER, M.K. and GANZ, R. (1994). Midterm results (5.5–10 years) of acetabular allograft reconstruction with the acetabular reinforcement ring during total hip revision, *J. Arthroplasty* **5**, 469–479.

Advances in Tissue Banking Vol. 6

10 *IN VITRO* ENGINEERED CARTILAGE

V. MIRABET and P. VACA

Tissue Bank, Transfusion Centre of Valencia
Valencia, Spain

J.F. PERTUSA

Department of Animal Biology
University of Valencia, Burjassot, Spain

J.L. LEQUERICA and E. SANZ

Institute of Biomedicine of Valencia
Spanish Council for Scientific Research, Valencia, Spain

1. Introduction

Articular cartilage is a tissue which demonstrates a limited capacity for self-repair after injury. Thus, reconstructive surgery on cartilage defects includes techniques to replace the loss of tissue, either with small plugs from a less weight bearing region of the joint (autograft) or with a full allograft from a tissue bank (Bugbee and Convery, 1999; Fitzpatrick and Morgan, 1998; Friedlaender *et al.*, 1999; Tavakol *et al.*, 1993).

Although the use of autologous tissue is the *gold standard*, it is limited by the small amount of cartilage available in the body for transplantation to other sites. Moreover, the effects of damage to the donor site must be considered.

On the other hand, allografts can be stored by freezing (at −80°C or below) or fresh (between 1 to 10 °C). Fresh allografts only can be stored for a few days, with culture medium, then microbiological culture results will not have developed fully due to shortened incubation times. Nevertheless, chondrocytes from fresh stored cartilage have been shown to retain cell viability over 60% after 60 days of storage, with periodic refresh of culture medium. Frozen cartilage grafts provide more time for screening of the tissue, but freezing reduces significantly cell viability (Csönge *et al.*, 1998).

Since few years ago, transplantation of cell suspension *in vitro* cultured chondrocytes has been used as an alternative to restore lost tissue (Brittberg *et al.*, 1994; Katsube *et al.*, 2000). This technique has been considered to be successful in clinical application in combination with a periosteal flap. However, the surgical manipulation of the chondrocyte suspension is difficult, because the risk that transplanted cells may leak out of the defect with joint motion. It must also be considered that cultured chondrocytes as monolayer tend to assume a fibroblastic dedifferentiated phenotype and synthesising predominantly type I collagen instead of type II. Nevertheless, re-differentiation of these cells cultured *in vitro* as a monolayer has been established using three-dimensional culture systems. In order to stabilise the chondrocyte phenotype during culture, various biological substances have been used as scaffold: collagen, fibrin glue, alginate, etc. (Rudert *et al.*, 1998; Katsube *et al.*, 2000; Wakitani *et al.*, 1998; Sims *et al.*, 1998; Solchaga *et al.*, 1999).

Our work seeks in the first place to study the most appropriate cultivation conditions to expand the cellular population and then to observe to the cellular behaviour within fibrin or alginate as cell carrier.

2. Materials and Methods

2.1. Chondrocyte isolation and monolayer culture

Human articular cartilage specimens were provided from our tissue bank (Tissue Bank of Valencia) from six multiorgan donors ranging in age from 16 to 50 years (mean 33 ± 13). Cartilage biopsies were taken from femoral condyle ($n = 4$) and meniscus ($n = 2$), and then stored (1–10°C) in sterile saline isotonic solution until tissue processing (< 24 hours).

Before mechanical dissection, the cartilage remained 1 hour in an antibiotic solution (50 μg/ml each of tobramycin, vancomycin, trimethropin/sulfamethoxazole and amphotericin B, in Hank's balanced saline solution, HBSS) at room temperature. Then, the biopsies tissues were minced and the fragments (≈1 mm^3) were subjected to extracellular matrix digestion in a solution of collagenase II (2 mg/ml) and hyaluronidase (0.1 mg/ml) in Dulbecco's modified Eagle medium (DMEM). The tissue fragments embedded in this enzymatic solution were incubated at 37°C for 12–16 hours in 50 ml conical centrifuge tubes. The cell suspension was then filtered through a sterile gauze, rinsed with HBSS and centrifuged for 10 min at 580 g. The resulting pellet was resuspended in basal culture medium as specified below and then seeded into 25 cm^2 culture flasks, and incubated at 37°C, 5% CO_2 and 95% humidity. Cell density in the final suspension was determined by a hemacytometer and cell viability was assessed by trypan blue staining. The medium was replaced every 2 to 3 days and the cells grown near to confluence. These primary cells were harvested with a solution of 0.06% trypsin and 0.02% ethylene diamine tetra-acetic acid (EDTA) in HBSS, for 5 min at 37°C. The resulting cell suspension was centrifuged as above, resuspended in fresh medium and replated in 25 cm^2 culture flasks (split 1:3), again grown to confluence and harvested. These cells were used for (i) three-dimensional scaffolds, (ii) cell proliferation assays, or (iii) stored by freezing (1°C/min) in DMEM with 5% dimethylsulphoxide (Me_2SO_3) and 30% foetal bovine serum.

2.2. Culture medium

A basal nutrient solution was used consisting of DMEM with 4 g/l of glucose to which the following were added: L-glutamine (2 mM), insulin (10 μg/ml), transferrin (5.5 μg/ml), sodium selenite (6.7 μg/l), penicillin (100 U/ml), streptomycin (100 mg/ml), amphotericin B (2.5 μg/ml), and ascorbate 2-phosphate (35.5 μg/l).

In addition, this basal medium was supplemented with different types of serum (10%) for the assay of cellular proliferation: human serum (HS), human umbilical cord serum (UCS), foetal bovine serum (FBS) and Ultroser® (GIBCO) (ULT, used at a final concentration of 2%, equivalent to 10% FBS). Sera were heat inactivated by mixing it in a water bath at a thermostatically controlled temperature of 56°C for 30 min.

2.3. Cell proliferation assay

To optimise culture conditions of chondrocytes, four different culture media regarding the serum supplement (HS, UCS, FBS or ULT) were assayed. The experiments were carried out in parallel for the four groups. At least 245 cultures were performed for each group and donor.

The Cell Proliferation Reagent WST-1® (Boehringer Mannheim) was used. It is a colorimetric assay for the quantification of cell proliferation, based on the cleavage of the tetrazolium salt WST-1 by mitochondrial dehydrogenases in viable cells.

Chondrocytes were cultured as monolayer in 96-well-plates at a density of 500 cells/well, incubated at 37°C, 5% CO_2 and 95% humidity. After 5 days in culture, the Cell Proliferation Reagent WST-1® was added (10 μl/well) in a 100 μl/well final volume of culture medium and then, the plates were incubated for 4 hours in culture conditions. Subsequently, plates were shacked thoroughly for 1 min on a shaker at room temperature. A microtiter plate (ELISA) reader quantified the formazan produced. The absorbance was measured to 450 nm of wavelength. Wells without cells were used as blank, with the same volume of culture medium and reagent WST-1, to know non-specific absorbance due to the type of culture medium, incubation time or light exposure. Mitochondrial activity was expressed in units of optical density (OD_{450}). At the same manner the effect of the percentage of serum supplemented to the culture medium was determined, then data of proliferation were obtained at time when culture supplemented with 20% HS reached 100% confluence. Results were analysed according to statistics methods.

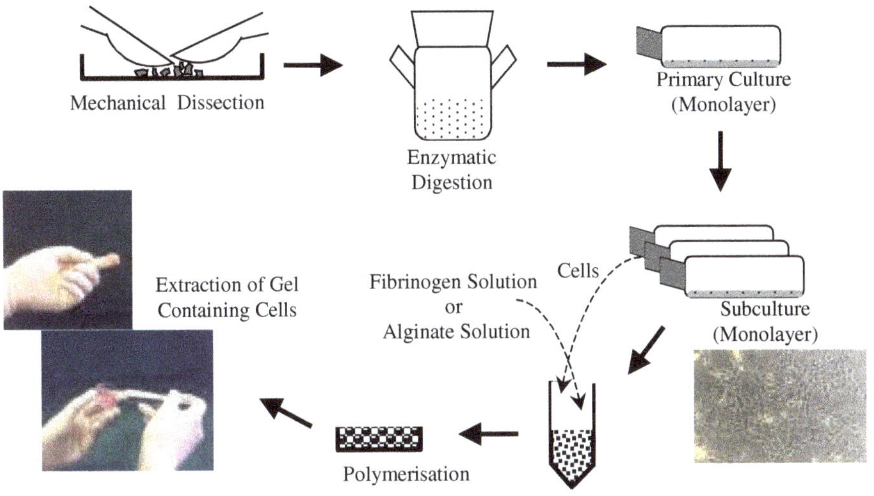

Fig. 1. General scheme of the proceeding to obtain a three-dimensional gel containing chondrocytes from a human cartilage biopsy.

2.4. Three-dimensional scaffolds construction

To create a three-dimensional scaffold for the chondrocytes, two types of compounds were used: fibrinogen, a protein present in the human plasma; and sodium alginate, a polysaccharide from brown algae. Both compounds are able to polymerise and thereby to form a mesh that encloses cells from cell suspension (Fig. 1).

2.4.1. Fibrin

Fibrin gels were prepared following previously described methods (Meana *et al.*, 1998). Fibrinogen from human plasma cryoprecipitates was used as the source of fibrin. Each unit of cryoprecipitate was heated at 37°C for 45–60 sec and, before complete thawing, it was centrifuged at 2204 g for 15 min at 4°C. The supernatant was discarded, the pellet dissolved in 10 ml of sterile saline, and finally, heated at 37°C to complete fibrinogen dissolution. Around 150–300 mg of fibrinogen were obtained from each unit of cryoprecipitate. The fibrinogen solution was stored at –80°C.

To produce 1 ml of fibrin gel, 200 μl of the fibrinogen solution were added to 800 μL of basal medium (supplemented with 10% HS) which contained 2.5×10^6 human chondrocytes and 333 U/K.I.U of bovine aprotinine. Immediately, the mixture was seeded in a plate of 1.9 cm^2 containing 66.6 μl of $CaCl_2$ 0.025 mM and 0.72 U of bovine thrombin and allowed to gelify at room temperature. The gels was covered with basal culture medium (10% HS) and incubated at 37°C, 5% CO_2 and 95% humidity. The culture medium was changed every other day.

2.4.2. Alginate

Cell entrapment was performed as described previously (Gou *et al.*, 1989; Häuselmann *et al.*, 1992). Chondrocytes were suspended at a density of 2.5×10^6 cells/ml in 1.2% sodium alginate solution in sterile water. Then, it was slowly passed through a 23-gauge needle to produce drops, which fell in 138 mM $CaCl_2$ solution. The beads were allowed polymerise further for 10 min in the calcium chloride solution. Then, after being washed in HBSS three times, the beads were embedded in basal culture medium (10% HS) and incubated at 37°C, 5% CO_2 and 95% humidity. The culture medium was changed every other day.

2.5. Histology

The cell-tissue constructs were controlled using an inverted light microscope. After three weeks in three-dimensional culture the gels were harvested for histological analysis.

Gels were removed from their culture dishes and fixed during 4 hours at 4°C in a solution of 2% glutaraldehyde and 2% formaldehyde. Afterwards, samples were washed abundantly in phosphate buffered saline (1 hour, changing frequently the washing solution), dehydrated with a series of alcohols of increasing concentration, cleared with two baths of xylene during 5 min each, and included in low melting point wax (35–40°C), taking care that time in wax did not excess 1 hour. Wax blocks were cut in slices of 5 to 8 μm deep with a MICROM HM340-E microtome.

To study at the optical microscopy level, slices were adhered to slides, deparafined and stained with hematoxiline-eosine and Mason trichromique techniques.

For the study at the Scanning Electron Microscopy (Philips XL30 ESEM), slices were adhered to metallic holders, immersed in xylene during 2 min to remove superficial wax, and shaded with a ion sputtering during 3 min.

2.6. Statistics

Statistical analysis for the different culture groups in the colorimetric WST-1 test was performed with SPSS for Windows. Kruskal-Wallis Test for independent samples was used to assess significant differences between the four assayed groups. *P*-values less than 0.005 were considered significant.

3. Results

Comparing the cell growth of tissue biopsies from the different donors it was observed the fact that those coming from the menisci provided an increased cellular proliferation that those from the condyles. Cell viability following isolation of chondrocytes from fresh articular cartilage by collagenase averaged 96%.

Depending on the supplement used for the basal culture medium (HS: human serum; FBS: foetal bovine serum; ULT: Ultroser; UCS: umbilical cord serum), different values of absorbance were obtained. The results of the proliferation assays showed a patent positive influence of the HS in front of the other supplements used for the basal medium. FBS and ULT reported similar results, while UCS turned out to be the supplement that less stimulated the proliferation of the chondrocytes (Fig. 2). There was significant differences (p-value < 0.0001) between the mean proliferation values of the groups assayed, except between the FBS and ULT groups.

Assaying different concentrations of HS for the basal culture medium, 10%, 15% and 20% HS yield the higher values, and they were similar.

In spite of serially subcultured chondrocytes growth as monolayer shown a de-differentiated fibroblast-like morphology, they were re-differentiated when transferred at three-dimensional construct, acquiring cartilaginous appearance (Figs. 2 to 7).

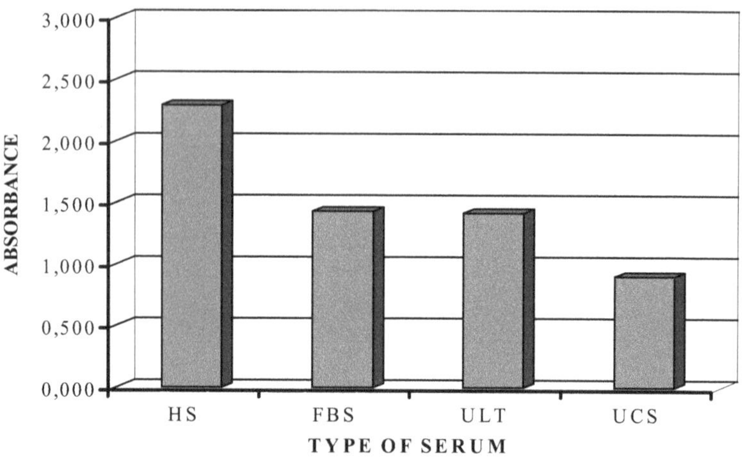

(a)

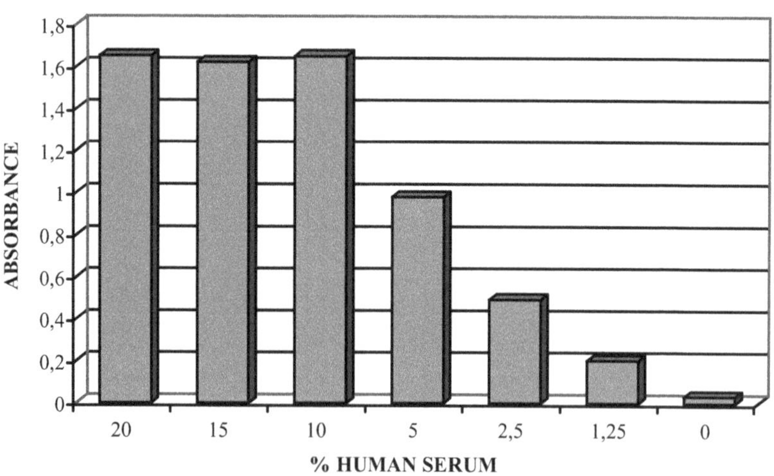

(b)

Fig. 2. Proliferation assays (a) depending on the supplement used for the basal culture medium (HS: human serum; FBS: foetal bovine serum; ULT: Ultroser; UCS: umbilical cord serum); and (b) concentration of HS for the basal culture medium.

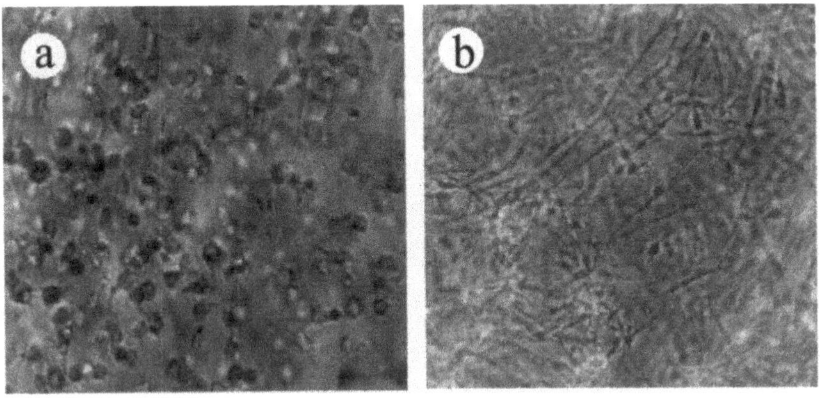

Fig. 3. The pictures show the appearance at optical microscope of : (a) fibrin gel containing chondrocytes of the cartilage biopsies, and (b) fibrin gel containing human dermis fibroblasts. It is clear that chondrocytes morphology (rounded) differs from that observed in fibroblasts (spindle-shaped).

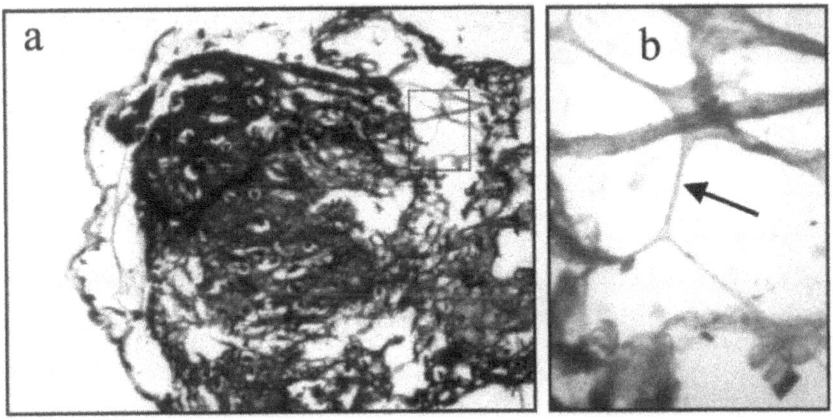

Fig. 4. (a) Histological section of a fibrin gel containing chondrocytes, in which it can be observed a resemblance with a nodule of cartilaginous aspect. (b) Augmented image corresponding to the marked zone in (a) with a discontinuous trace, it was identified as a collagenous band (arrow).

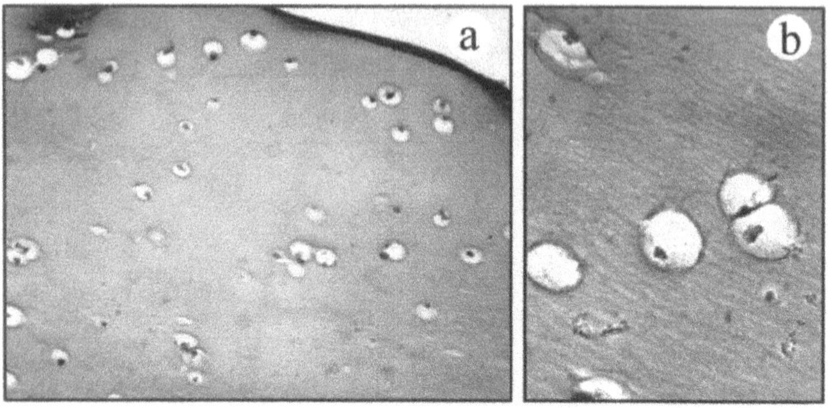

Fig. 5. (a) Histological section of an alginate gel containing chondrocytes. No spindle shaped cells were observed in the matrix suggesting that there was no change to a fibroblast cell type. (b) Augmented image showing a chondroid island formation into the gel.

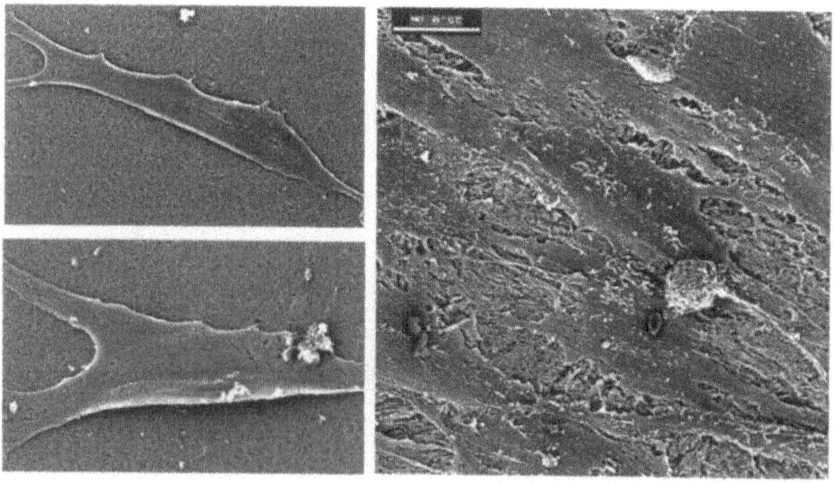

Fig. 6. Scanning electron microscopy images of the superficial zone of a fibrin gel containing cultured chondrocytes. At this zone, cells showed fibroblast-like morphology. These cells resemble those flattened chondrocytes observed in the superficial zone of articular cartilage *"in vivo"*, with their long axes parallel to the articular surface.

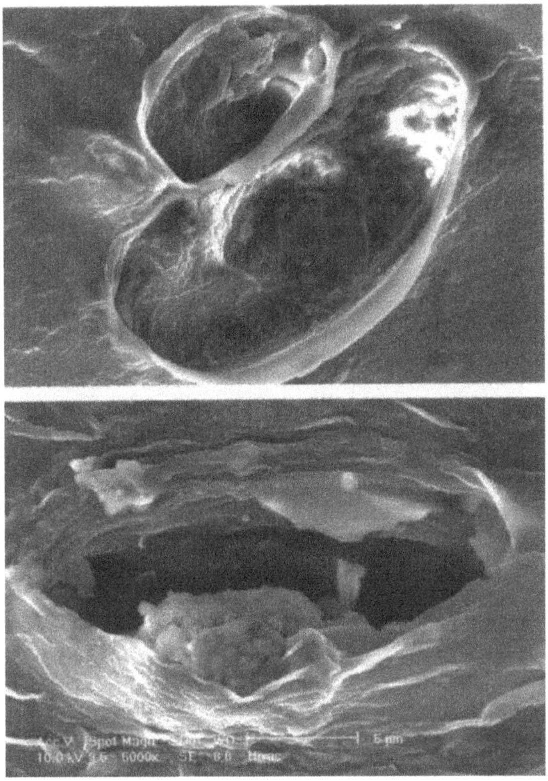

Fig. 7. Scanning electron microscopy images of cultured chondrocytes growth into the alginate gel. The upper image shows an isogenic group into the amorphus matrix. Lower image shows a rounded chondrocyte surrounded by its capsule.

4. Comment

Tissue bankers work to provide tissues or tissue substitutive to surgeons according to some criteria of quality and clinical effectiveness. Although appearing to be a simple, avascular matrix, hyaline cartilage possesses properties such as resistance to compression and the ability to distribute loads that cannot be fully replaced by any other tissue or device designed to date. Fresh cartilage allograft transplantation sensitises humoral and cell mediated immune systems (Globe *et al.*, 1999; Friedlander *et al.*, 1999; Shelton *et al.*, 1998), this sensitisation clearly does not preclude a satisfactory outcome. Freezing could reduce antigenicity of cryopreserved allografts but only a few cells survived the procedure (Stevenson, 1987; Csönge *et al.*, 1998). On the other hand, autografting implies an additional damage (donor site) and it is limited by the small amount of cartilage available in the body for transplantation to other sites. Transplantation of isolated autologous chondrocytes as a suspension presents additional problems because of the use of a liquid carrier, besides that the cells show a dedifferentiated phenotype at implantation.

Therefore, to create cartilage-like tissue "*in vitro*" which could be used to repair cartilage defects is a main goal that different groups of researches are investigating.

It is known that chondrocytes embedded in a gel delivery vehicle retains its phenotype and, in addition, it may provide a method which facilitates surgical handling. For cell culture in three-dimensional model, different types of substances to act as scaffold are being used: naturally derived polymers and synthetic ones. (Katsube *et al.*, 2000; Freed *et al.*, 1999; Grande *et al.*, 1997; Paige *et al.*, 1996; Vacanti and Upton, 1994; Ting *et al.*, 1998).

Chondrocytes of a cartilage biopsy can be expanded in culture, using autologous human serum as supplement of cultured medium (10% at least for optimal growth), until yielding cell population enough for tissue construct (Fig. 2). Moreover, fibrin of human origin can be used to obtain a gel containing cultured cells.

Therefore, it could be engineered a cartilage-like tissues starting from autologous materials (cells, serum and fibrin). Nevertheless, to achieve an enginneered tissue with the highly organised architecture of cartilage, its biochemical and biomechanical properties, and able to incorporate to surrounding cartilage after transplantation, there is still much work to be done (Temenoff and Mikos, 2000).

Currently, the investigation lines go toward the following objectives:

- Bioreactors present numerous advantages for growing large amounts of tissue quickly. NASA has designed a bioreactor that can be used with scaffolds growing in microgravity (Freed *et al.*, 1997).

- Culture of undifferentiated mesenchymal stem cells and the effect of bioactive factors such as platelet-derived growth factors and transforming growth factors inducing them to chondrogenic differentiation (Kieswetter *et al.*, 1997; Mackay *et al.*, 1998; Pittenger *et al.*, 1999).

Anyway, in addition to produce a therapy that could be clinically applicable, the development of "*in vitro*" culture systems that promotes chondrogenic growth presents an opportunity to study the complexities of chondrogenesis.

5. References

BUGBEE, W.D. and CONVERY, F.R. (1999). Osteochondral allograft transplantation, *Clin. Sports Med.* **18**(1), 67–75.

BRITTBERG, M., LINDAHL, A., NILSSON, A., OHLSSON, C., ISAKSSON, O. and PETERSON, L. (1994). Treatment of deep cartilage defects in the knee with autologous chondrocyte transplantation, *N. Engl. J. Med.* **331**(14), 889–895.

CSÖNGE, L., NEWMAN-GAGE, H., RIGLEY, T., BRAVO, D., CONRAD, E.U. and STRONG, D.M. (1998). Preservation of chondrocyte viability during long-term refrigerated storage of osteochondral allografts. In: *Advances in Tissue Banking*, G.O. Phillips, R. von Versen, D.M. Strong and A. Nather, eds., World Scientific, Singapore, Vol. 2, pp. 175–189.

FITZPATRICK, P.L. and MORGAN, D.A. (1998). Fresh osteochondral allografts: A 6-10-year review, *Aust. N. Z. J. Surg.* **68**(8), 573–579.

FREED, L.E., LANGER, R., MARTIN, I., PELLIS, N.R. and VUNJAK-NOVAKOVIC, G. (1997). Tissue engineering of cartilage in space, *Proc. Natl. Acad. Sci. USA* **94**(25), 13885–13890.

FREED, L.E., MARTIN, I. and VUNJAK-NOVAKOVIC, G. (1999). Frontiers in tissue engineering. *In vitro* modulation of chondrogenesis, *Clin. Orthop.* **367**(Suppl), 46–58.

FRIEDLAENDER, G.E., STRONG, D.M., TOMFORD, W.W. and MANKIN, HJ. (1999). Long-term follow-up of patients with osteochondral allografts. A correlation between immunologic responses and clinical outcome, *Orthop. Clin. North. Am.* **30**(4), 583–588.

GOBLE, E.M., KOHN, D., VERDONK. R. and KANE. S.M. (1999). Meniscal substitutes — Human experience, *Scand. J. Med. Sci. Sports* **9**(3), 146–157.

GOU, J., JOURDIAN, G.W. and MCCALLUM, D.K. (1989). Culture and growth characteristics of chondrocytes encapsulated in alginate beads, *Connect. Tissue Res.* **19**, 277–297.

GRANDE, D.A., HALBERSTADT, C., NAUGHTON, G., SCHWARTZ, R. and MANJI, R. (1997). Evaluation of matrix scaffolds for tissue engineering of articular cartilage grafts, *J. Biomed. Mater. Res.* **34**(2), 211–220.

HÄUSELMANN, H.J., AYDELOTTE, M.B., SCHUMACHER, B.L., KUETTNER, K.E., GITELIS, S.H. and THONAR, E.J. (1992). Synthesis and turnover of proteoglycans by human and bovine adult articular chondrocytes cultured in alginate beads, *Matrix* **12**, 116–129.

KATSUBE, K., OCHI, M., UCHIO, Y., MANIWA, S., MATSUSAKI, M., TOBITA, M. and IWASA, J. (2000). Repair of articular cartilage defects with cultured chondrocytes in Atelocollagen gel. Comparison with cultured chondrocytes in suspension, *Arch. Orthop. Trauma. Surg.* **120**(3–4), 121–127.

KIESWETTER, K., SCHWARTZ, Z., ALDERETE, M., DEAN, D.D. and BOYAN, B.D. (1997). Platelet derived growth factor stimulates chondrocyte proliferation but prevents endochondral maturation, *Endocrine* **6**(3), 257–264.

MACKAY, A.M., BECK, S.C., MURPHY, J.M., BARRY, F.P., CHICHESTER, C.O. and PITTENGER, M.F. (1998). Chondrogenic differentiation of cultured human mesenchymal stem cells from marrow, *Tissue Eng.* **4**(4), 415–428.

MEANA, A., IGLESIAS, J., DEL RIO, M., LARCHER, F., MADRIGAL, B., FRESNO, M.F., MARTIN, C., SAN ROMAN, F. and TEVAR, F. (1998). Large surface of cultured human epithelium obtained on a dermal matrix based on live fibroblast-containing fibrin gels, *Burns* **24**, 621–630.

PAIGE, K.T., CIMA, L.G., YAREMCHUK, M.J., SCHLOO, B.L., VACANTI, J.P. and VACANTI, C.A. (1996). De novo cartilage generation using calcium alginate-chondrocyte constructs, *Plast. Reconstr. Surg.* **97**(1), 168–178; discussion 179–180.

PITTENGER, M.F., MACKAY, A.M., BECK, S.C., JAISWAL, R.K., DOUGLAS, R., MOSCA, J.D., MOORMAN, M.A., SIMONETTI, D.W., CRAIG, S. and MARSHAK, D.R. (1999). Multilineage potential of adult human mesenchymal stem cells, *Science* **284**, 143–147.

RUDERT, M., WIRTH, C.J., SCHULZE, M. and REISS, G. (1998). Synthesis of articular cartilage-like tissue *in vitro, Arch. Orthop. Trauma. Surg.* **117**, 141–146.

SHELTON, W.R., TREACY, S.H., DUKES, A.D. and BOMBOY, A.L. (1998). Use of allografts in knee reconstruction: I. Basic science aspects and current status, *J. Am. Acad. Orthop. Surg.* **6**, 165–168.

SIMS, C.D., BUTLER, P.E., CAO, Y.L., CASANOVA, R., RANDOLPH, M.A., BLACK, A., VACANTI, CA. and YAREMCHUCK, M.J. (1998). Tissue engineered neocartilage using plasma derived polymer substrates and chondrocytes, *Plast. Reconstr. Surg.* **101**(6), 1580–1585.

SOLCHAGA, L.A., DENNIS, J.E., GOLDBERG, V.M. and CAPLAN, A.I. (1999). Hyaluronic acid-based polymers as cell carriers for tissue-engineered repair of bone and cartilage, *J. Orthop. Res.* **17**(2), 205–213.

STEVENSON, S. (1987). The immune response to osteochondral allografts in dogs, *J. Bone Joint Surg. Am.* **69**(4), 573–582.

TAVAKOL, K., MILLER, R.G., BAZETT-JONES, D.P., HWANG, W.S., MCGANN, L.E. and SCHACHAR, N.S. (1993). Ultrastructural changes of articular cartilage chondrocytes associated with freeze-thawing, *J. Orthop. Res.* **11**(1), 1–9.

TEMENOFF, J.S. and MIKOS, A.G. (2000). Review: Tissue engineering for regeneration of articular cartilage, *Biomaterials* **21**(5), 431–440.

TING, V., SIMS, C.D., BRECHT, L.E., McCARTHY, J.G., KASABIAN, A.K., CONNELLY, P.R., ELISSEEFF, J., GITTES, G.K. and LONGAKER, M.T. (1998). *In vitro* prefabrication of human cartilage shapes using fibrin glue and human chondrocytes, *Ann. Plast. Surg.* **40**(4), 413–420, discussion 420–421.

VACANTI, C.A. and UPTON, J. (1994). Tissue-engineered morphogenesis of cartilage and bone by means of cell transplantation using sinthetic biodegradable polymer matrices, *Clin. Plast. Surg.* **21**(3), 445–462.

WAKITANI, S., GOTO, T., YOUNG, R.G., MANSOUR, J.M., GOLDBERG, V.M. and CAPLAN, A.I. (1998). Repair of large full-thickness articular cartilage defects with allograft articular chondrocytes embedded in a collagen gel, *Tissue Eng.* **4**(4), 429–444.

Advances in Tissue Banking Vol. 6

11 PRESERVED ALLOGENIC RIB CARTILAGE IN RECONSTRUCTIVE SURGERY

D. ŚLADOWSKI, A. KOMENDER and J. KOMENDER
Department of Transplantology
Warsaw Medical University, Poland

H. MALCZEWSKA
Department of Histology & Embryology
Warsaw Medical University

1. Introduction

Despite the impressive developments in tissue engineering, rib cartilage is still frequently used for reconstructions in a face region (Komender *et al.*, 1986; Kryst, 1981; Meeuwsen and De Vries, 1996; Sailer, 1983; Thomassin, 2001) and for other clinical procedures (Tomford *et al.*, 1996). Transplants of allogenic cartilage are prepared in tissue banks from sterilised rib cartilage obtained from cadaveric donors (Komender and Komender, 1977). This material offers excellent physical properties enabling to obtain desired shape of implant in a relatively easy way. Graft might be carved from single cartilage, or might be composed from several elements glued together. In the opposite to living cartilage, preserved cartilage do not distort so often after transplantation. It offers long term support for soft tissues with slow degradation rate which usually does not occur within first four years after transplantation. The other advantage of preserved grafts is avoidance of additional trauma inflicted during autologus tissue retrieval.

Since 1963 (establishment of the Central Tissue Bank in Warsaw) more than 2500 grafts prepared in the Central Tissue Bank in Warsaw have been used in the reconstructive surgery. This vast clinical material allows clinical evaluation of the material in long time period after surgery. Our assessment was based on 440 complete, validated reports of reconstructive surgery performed in 11 surgical wards with the use of radiation-sterilised cartilage material. The average duration of illness before transplantation was 107.3 months with SD = 102.3. Time of hospitalisation was 23.5 days (SD = 15.2). Interval of time between surgery and estimation of the result of treatment was 98.2 months with SD = 46.0. The result of the treatment was estimated according to a four-grade scale: very good, satisfactory, difficult to estimate and unsatisfactory. It was found that 33.5% of all analysed patients achieved very good results of treatment, in 41.8% of the patients the result was satisfactory and in 19.9% failures were found. In approximately 4.1% of the patients, the result could not be classified in a conclusive way.

From the vast array of factors which might affect the results of the reconstructive surgery conducted with the use of cartilage material, several most important have been identified for analysis. The following variables were analysed: age, sex, diagnosis, location, duration of illness, the time of hospitalisation, general complications, local complications, reoperations, use of antibiotics, form of grafts (single fragment, several fragments, with patients own tissue), early estimated result after hospitalisation, interval between surgery and control examination, concomitant diseases, result of treatment, estimation of graft resorption and estimation of the role of the graft. Evaluation of the results was performed after one-year observation period. Obtained results indicate that in more than 75% of the cases, positive results of treatment may be achieved, however it must be remembered that the final outcome of the treatment depends on several factors which should be taken into consideration by the surgeon performing reconstructive surgery.

2. Material

The material prepared in The Central Tissue Bank in Warsaw is obtained from 18–55 years old donors, both sexes, excised 24 hours after death and preserved using the standard operating procedure. After removal of perichondrium, tissue is stabilised in 70% ethanol for 4 hours, then washed in 0.9% NaCl for 24 hours and immersed in a 0.9% NaCl solution. Closed vials with grafts are radiation-sterilised with a dose of 33 kGy, in gamma source. After sterility control, grafts are registered and sent to the surgical wards together with the appropriate questionnaire which should be returned back to the Tissue Bank for the final evaluation of graft performance.

Cartilage is usually transplanted mainly as a single fragment (56.4%), or two or three fragments (41.4%). Only occasionally preserved material is combined with autologus tissue (2.0%). Number of fragments has no effect on the final outcome of the surgery, but it must be remembered that there is a need for graft prepared and packed in a way facilitating the use of multiple fragments.

Our data indicate that cartilage is usually used in the post-traumatic surgery (47.5% of cases in our material) in the group of patients in their second (26.9%) and third (39.4%) decade of their life. Age distribution of patients from the other age groups are similar up to 50 years of age (7.7% of the patients were in the first decade, 10.9% in the fourth, 9.8% in the fifth, 5.3% of the patients were over fifty years of age) (Table 1). The sex distribution is similar in all age groups. The other areas of the use of cartilage are congenital deformations (28.9%), unspecific inflammations (16.8%), postoperative malformations (3.0%), specific inflammations (1.8%), malignant tumours (1.4%) and benign tumours (0.5%).

One of the most important factors affecting final results of the treatment is the age of the recipient. Successful outcome of the surgery increases with age, ranging from 70% in case of young patients (0–20 years) to more than 90% in case of patients older than 50 years (Fig. 1). This reflects age dependent decrease in resorption rate. In the most frequently reported age group (10–30 years) successful treatment can be expected in more than 75% of

Table 1. Results of cartilage transplantation in various age groups.

Age in years	Result of treatment									
	Very good		Satis-factory		Doubt-ful		Unsatis-factory		Total	
	n	%	n	%	n	%	n	%	n	%
0–10	9	27.3	15	45.5	0	0.0	9	27.3	33	8
11–20	36	30.5	46	39.0	4	3.4	32	27.1	118	27
21–30	58	33.5	77	44.5	7	4.0	31	17.9	173	40
31–40	17	35.4	19	39.6	4	8.3	8	16.7	48	11
41–50	23	54.8	12	28.6	1	2.4	6	14.3	42	10
> 51	5	21.7	15	65.2	2	8.7	1	4.3	23	5
Total	148	33.9	184	42.1	18	4.1	87	19.9	437	100

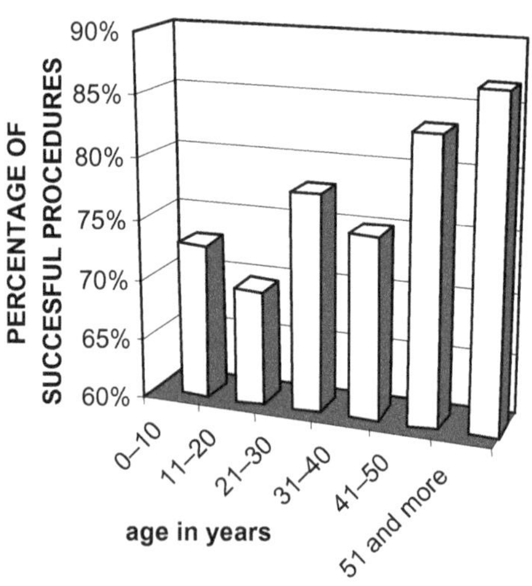

Fig. 1. Relation between age and success of the treatment with cartilage graft.

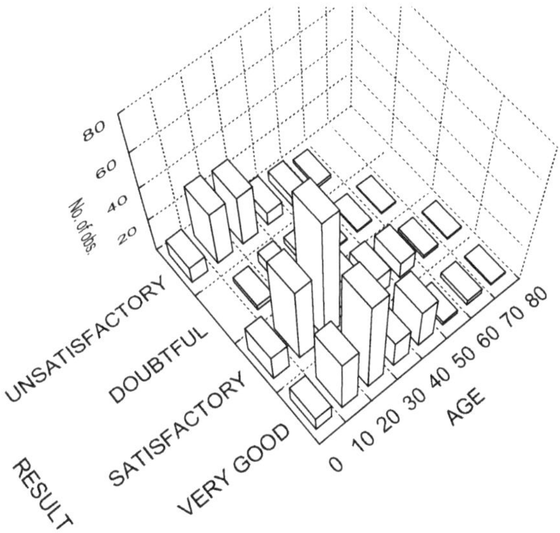

Fig. 2. Age dependent distribution of results.

cases. The fourth and fifth decades of life seem to be preferable for cartilage transplantation when unsatisfactory results are almost not existing (Fig. 2).

3. Observation

After surgery some local changes can be observed. The most common-oedema occurs in more than half cases (65.8%) and is not associated with the results of the surgery (Fig. 3). Other local changes are less frequent (purulence 5.9%, infiltration 2.5%, accelerated resorption of grafts 0.5%). It is interesting that in one fourth of all cases no oedema or other local changes after surgery can be observed (25.3%).

Preserved cartilage is predominantly used in nose surgery (more than a half of all reported cases 58.4%, reconstruction of ear (16.6%), and correction of mandible (11.1%) (Table 2). Unfortunately, in a long term observation (more than 7 years), the results of the nose reconstruction are usually hampered by the young age of the patients which usually undergo this kind of surgery. In this age

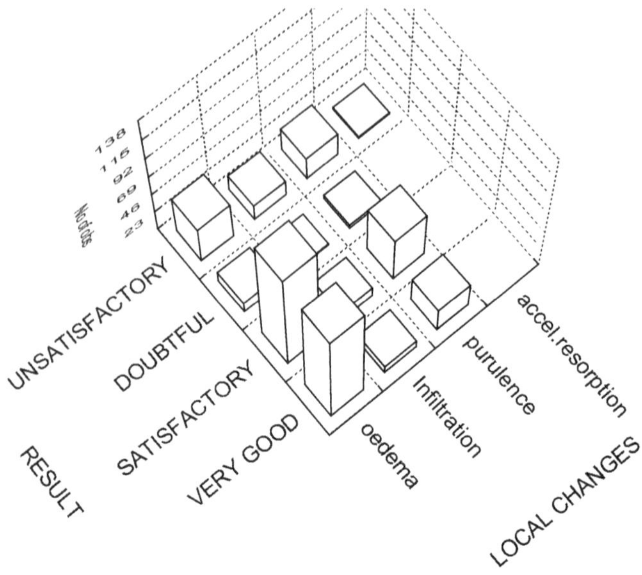

Fig. 3. Local changes observed after transplantation are not prognostic (except accelerated resorption).

Table 2. Location of transplanted cartilage and results of reconstructions.

Localisation	Result of Treatment									
	Very good		Satis-factory		Doubt-ful		Unsatis-factory		Total	
	n	%	*n*	%	*n*	%	*n*	%	*n*	%
Nose	75	29.2	119	46.3	10	3.9	53	20.6	257	59
Ear concha	3	17.8	33	45.2	1	1.4	26	35.6	73	17
Mandible	30	63.8	10	21.3	5	10.6	2	4.3	47	11
Maxilla	7	56.7	11	36.7	0	0.0	2	6.7	30	7
Orbit	8	50.0	6	37.5	1	6.3	1	6.3	16	4
Front	5	38.5	4	30.8	1	7.7	3	23.1	13	3
Total	48	33.9	183	42.0	18	4.1	87	20.0	436	100

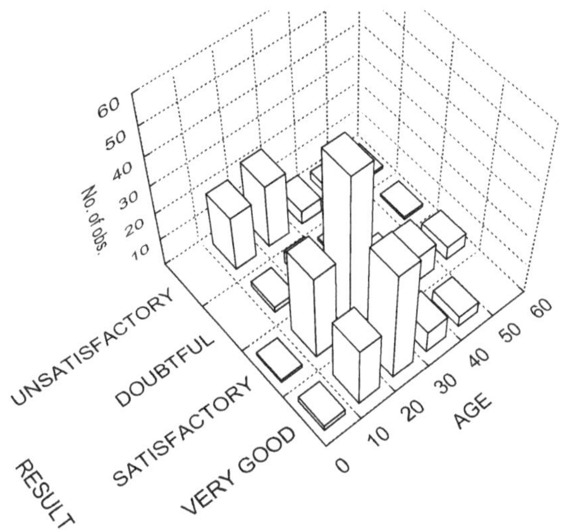

Fig. 4. Nose, age dependent distribution of results.

group (10–30 years) complete resorption of the transplanted material occurs in almost 30% of cases. There are only around 50% chances of obtaining positive results (58% in our material) (Fig. 4).

The final result of the treatment indicates that cartilage grafts are the most suitable for mandible corrections, where unsatisfactory results are observed very seldom (4.3%). The use of autologus bone obtained from iliac crest seems less attractive as failure can reach 24% (Bahat O, Fontanessi, 2001). In our material we observed over 63% of all "very good" results in case of patients with mandible reconstructions. Cartilage is transplanted to mandible not only to improve the shape of the bone but also to enhance the regeneration of the bone in the alveolar process. Similar distribution of the results of treatment was found in the group of patients with cartilage transplanted into maxilla. In case of nose surgery, the unsuccessful outcome of the treatment can be expected more frequently. Unsatisfactory results can be expected in almost 20% of all cases. In most cases, failure is caused by degradation and resorption of transplanted material. It must be stated however, that up to now, no other satisfactory treatment has been developed and even in case of graft resorption, the surgery can be performed once again. (In control examinations, full resorption of grafts was seen in 23.0% of the patients and limited resorption in 9.3% of the patients). It must be also remembered that resorption usually does not occur during first four years after transplantation.

Our research conducted on the problem of cartilage graft resorption resulted in creation of a new preservation procedure which slows down cartilage degradation (Pawlowski *et al.*, 1986) and we hope to observe more favourable results in the future. The most complicated surgery conducted with the use of preserved cartilage is reconstruction of the ear concha. Despite intensive resorption at this location, cartilage is quite frequently used (16.6% of all cases). Most of the "*unsatisfactory results*" can be expected after reconstruction in this location (35.6%) but still 63% of the results of treatment of patients are successful ("*very good*" and "*satisfactory*" together). The use of this material for ear

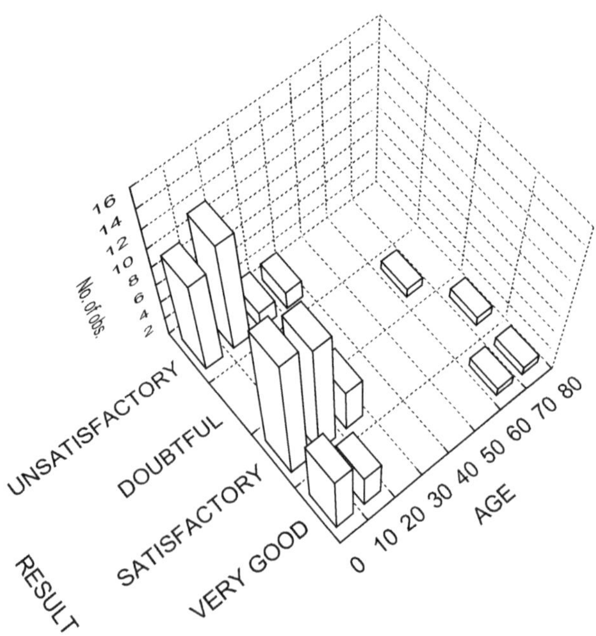

Fig. 5. Age and results of auricular concha reconstruction.

reconstruction seems to be very promising in case of older patients (> 60 years) when no unsatisfactory results have been observed (Fig. 5). From other locations less frequently reported, it must be stated that the use of preserved cartilage can be advocated as a material of choice in case of reconstructive surgery of maxilla and orbit where unsatisfactory results are very infrequent (6.7% and 6.3% respectively). The incidence of *"unsatisfactory results"* of treatment in children and young patients exceeded 27%. The tissues of children and young patients seem to react strongly to cartilage grafts, which is not observed in older age groups.

It is interesting that over 50% of all the *"unsatisfactory results"* can be expected in the group of patients that were operated for posttraumatic malformations. The results of treatment in this group were found significantly worse than that of the others.

Table 3. Results of cartilage transplantation in various diagnoses.

Diagnoses	Result of treatment									
	Very good		Satis-factory		Doubt-ful		Unsatis-factory		Total	
	n	%	*n*	%	*n*	%	*n*	%	*n*	%
Traumas	65	31.1	91	43.5	8	30.8	45	21.5	209	62
Congenital changes	36	28.3	52	40.9	4	30.1	35	27.6	127	8
Unspecific inflammations	34	46.6	29	39.7	5	60.8	5	6.8	73	22
Postoperative deformations	8	1.5	4	0.8	0	0.0	1	7.7	13	4
Specific inflammations	2	5.0	5	5.2	0	0.0	1	12.5	8	2
Tumours	3	2.9	2	8.6	1	4.3	1	14.3	7	2
Total	48	3.9	83	1.9	18	0.1	88	20.1	437	100

Transplantation of cartilage in "*congenital malformations*" gives even less favourable results: "*very good*" and "*satisfactory*" together 69.2% and "*unsatisfactory*" 27.6%. Transplantation of preserved cartilage in "*unspecific inflammations*" and "postoperative malformations" seems to be very effective (nearly 90% of positive results of surgery). The groups of "specific inflammations" and "tumours", however, not so numerous, presented a high percentage of positive results of treatment which indicates that the use of preserved cartilage should be advocated in this kind of treatment (Table 3).

Preserved costal, allogenic cartilage is the proper material for use in reconstructive surgery of the face. More than 33% of all operations were completed with full clinical success. In 42.0% of all operations, the results were found to be "*satisfactory*", which means that positive results should be expected in 75% of treated patients. It is also true that in nearly 20% of the patients, the result of treatment was "unsatisfactory" which means that the facial reconstructions were unsuccessful. This requires some additional explanation. If the successful transplantation of bone depends on the process of grafts substitution by regenerating the patient's own bone (creeping substitution), then the clinical of cartilage transplantation depends on a stable state of transplant in years (Komender, 1986; Kryst, 1981, Pawlowski, 1986). It often happens that for some unknown reasons, cartilage grafts are resorbed quickly. In our material the rapid resorption was found in 23% of all cases, while accelerated resorption was seen several days after transplantation in two patients (0.5% of the cases).

4. Conclusion

Costal, allogenic, preserved cartilage is often used for reconstruction of malformations in the region of the face. The examination of patients between 1 to 17 years after surgery, reveals positive results of treatment in 75% of cases. Unsatisfactory results of transplantation (19.9% in the whole group) are correlated mainly with younger patients, congenital or post-traumatic malformations and location in ear concha.

5. References

BAHAT, O. and FONTANESSI, R.V. (2001). Efficacy of implant placement after bone grafting for three-dimensional reconstruction of the posterior jaw, *Int. J. Periodontics Restorative Dent.* **21**, 220–231.

KOMENDER, J. and KOMENDER, A. (1977). Evaluation of radiation-sterilized tissue in clinical use. In: *Sterilization of Medical Products by Ionising Radiation*, E.R.L. Gaughran and A.J. Goudie, eds., Multisc Publ. Ltd., Montreal, p. 188.

KOMENDER, J., MALCZEWSKA, H. and KOMENDER, A. (1991). Therapeutic effects of transplantation of lyophilised and radiation-sterilised, allogeneic bone, *Clin. Orthop.* **272**, 38–49.

KOMENDER, J., MALCZEWSKA, H. and PAWLOWSKI, A. (1986). Preserved allogenic cartilage in reconstructive surgery, *Probl. Haematol. Transfusiol. Transpl.* **13**, 288–293.

KRYST, L. (1981). Przeszczepianie tkanek w chirurgii szczekowo-twarzowej. In: *Przeszczepy Biostatyczne*, J. Komender, ed., PZWL, Warsaw, Vol. II, pp. 151–161.

MEEUWSEN, F. and DE VRIES, PH.A. (1996). Preservation of human costal cartilage for transplants in nasal surgery. In: *4th International Conference European Association of Tissue Banks*, Byk Jr. Chr, A. Lechat and R. von Versen, eds., Monduzzi Editore Bologna, pp. 79–82.

PAWLOWSKI, A., MALEJCZYK, J., SLUBOWSKI, T., SLADOWSKI, D. and MOSKALEWSKI, S. (1986). Arrested resorption of costal cartilage grafts subjected to hydrochloric acid in rats, *Otolaryng. Pol.* **40**, 25.

SAILER, H.F. (1983). Transplantation of lyophilized cartilage in maxillo-facial surgery. In: *Experimental Foundations and Clinical Success*, Karger, Basel–New York, p. 178.

THOMASSIN, J.M., PARIS, J. and RICHARD-VITTON, T. (2001). Management and aesthetic results of support grafts in saddle nose surgery, *Aesthetic. Plast. Surg.* **25**, 332–337.

TOMFORD, W.W., OHLENDORF, C. and MANKIN, H.J. (1996). Articular cartilage cryopreservation and transplantation. In: *Orthopaedic Allograft Surgery*, A. Czitrom and H. Winkler, eds., Springer, Wien–New York, pp. 269–274.

Advances in Tissue Banking Vol. 6

12 PRESERVED BONE ALLOGRAFTS IN RECONSTRUCTIVE ORTHOPAEDICS

J. KOMENDER,[1] A. KOMENDER,[1] H. MALCZEWSKA[2] and W. MARCZYNSKI[3]

[1]Department of Transplantology, [2]Department of Histology & Embryology, Medical University in Warsaw, [3]Institute of Traumatology, Orthopaedics and Neurosurgery Central Clinical Hospital, Military Medical University in Warsaw, Poland

1. Introduction

Tissue banking experience was achieved over a long period (1963–2001) since the creation of tissue banks in Poland. Thanks to initiative of two outstanding Polish professors: Adam Gruca, director of Orthopaedic Clinic in Warsaw and Kazimierz Ostrowski, director of the Department of Histology & Embryology in Warsaw, organisation of tissue processing and co-operation with orthopaedic clinics, radiation-chemistry laboratories and microbiological units was started. Since 1963, a tissue bank has been operating in Warsaw and around 100 000 grafts of bone, cartilage, dura mater, skin and fascia have been prepared and used in various branches of reconstructive surgery. In 1967, the tissue bank in Katowice, connected to the Blood Transfusion Centre, was created. In 1978, a similar unit was organised in Kielce as the Cryobiological Unit in the local blood transfusion centre. These three units created the first network of multi-tissues banks in Poland. In the 1980s, in several medical institutions (Cardiosurgery Clinic in Zabrze, Institute — Centre of Health Child, Cardiosurgery Clinic in Kraków) the heart valves and blood vessels banks were organised. Later, two eye tissues banks (in Warsaw and in Lublin) were founded. All of these units are in contact, once a year within the frame of Polish Transplantation Society, when a session on tissue banks is organised. Most of the workers attend the meetings of EATB. Twice, in 1977 and in 1999, worldwide meetings of tissue bank specialists in Poland were organised. Implementation of radiation sterilisation as main sterilisation procedure has been adopted since the very beginning. Some studies performed in our tissue bank were fundamental for allografts radiation-sterilisation (Dziedzic-Goclawska, 1979; Dziedzic-Goclawska *et al.*, 1979; Komender, 1976; Ostrowski *et al.*, 1980; Ostrowski *et al.*, 1996). The Central Tissue Bank established good links of collaboration with orthopaedic units, and that enable us to perform interesting analyses of use of preserved allografts in clinics.

2. Transplantation of Lyophilised and Radiation-Sterilised Bone Grafts

In 1960s and 1970s, lyophilised bone most often was used for orthopaedic reconstructions. Thanks to the good co-operation with several orthopaedic units, records on the follow-up of patients after bone transplantation can be gathered. This analysis is based on 1014 cases of preserved (lyophilised and radiation-sterilised) allogenic bone transplantation, where 27 biological and clinical variables were taken into consideration and detailed results were published earlier (Komender *et al.*, 1991). The following general conclusions may be drawn: more than 91% of operated patients reached full restoration or significant improvement of their condition; after surgery 80.3% of patients reached their full physical fitness; in 5.1% of the patients the results of treatment were unsatisfactory; in cases of benign tumours and congenital changes, the highest number of "*very good*" results of treatment was achieved; unsatisfactory results of treatment most often appeared in post-traumatic deformation and degenerative diseases (Table 1).

Table 1. Transplantation of lyophilised, radiation-sterilised bone allografts. Results of treatment in various diagnoses.

Diagnoses	Result of treatment									
	Very good		Satis-factory		Difficult to estimate		Unsatis-factory		Total	
	n	%	*n*	%	*n*	%	*n*	%	*n*	%
Traumas	11	1.1	19	1.9	2	0.2	6	0.6	38	3.8
Benign tumours	107	10.6	68	6.7	6	0.6	5	0.5	186	18.4
Malignant tumours	0	0	2	0.2	0	0	0	0	2	0.2
Unspecific inflammations	6	0.6	30	3.0	1	0.1	4	0.4	41	4.1
Specific inflammations	42	4.2	86	8.5	5	0.5	7	0.7	140	13.9
Degenerative changes	34	3.4	38	3.8	3	0.3	7	0.7	82	8.1
Congenital change	141	14.0	213	21.1	11	1.1	17	1.7	382	37.8
Scolioses	25	2.5	73	7.2	7	0.7	4	0.4	109	10.8
Others	9	0.9	18	1.8	1	0.1	2	0.2	30	3.0
Total	375	37.1	547	54.2	36	3.6	52	5.1	1010	100

$Chi^2 = 75.479$; degrees of freedom, 24; $p < 0.01$

A comparative analysis of the distribution of clinical results obtained in this group was carried out using the Chi^2 test (Mainland, 1963). Some correlations and coincidences were presented. In order to find out the meaning of the correlations; the *"indices of coincidence"* were calculated according to Pearson (1904). Pearson's index permits a comparison of coincidence of different pairs of variables even with a various degree of freedoms, thus showing a strong relationship between the variables. As expected, numerous variables showed significant correlation with the estimated final result of treatment (Komender *et al.*, 2001).

Three sets of indices of coincidence for the *"final result of treatment"*, *"fitness for work"* and *"role of the graft"* are evaluated. A strong relationship was found between: *"rebuilding of grafts"*, *"early estimated result of treatment"* and *"re-operation"* with the *"final result of treatment"*. What concerns the rebuilding of transplanted allografts seems to be the most important and sensitive factor in the post-surgery observation. Less important factors seem to be *"diagnosis"* and *"anatomical localisation"*. When the patients with unsatisfactory results of treatment were analysed, the most often localisation is related to the vertebral column or crus. No significant coincidence was found with *"age"*, *"handicap level"* or other parameters. Numerous patients even in their 70s were operated with the use of preserved allografts and the age seemed having no influence on the final result of treatment.

The sequence of variables is different when they are compared with the *"fitness for work"* after surgery. This coincides with *"age"*, *"diagnosis"* and *"physical efficiency"*. Such characteristics as: *"complications intra- or post-surgery"*, *"wound healing"*, etc. do not show any significant relationship with *"fitness for work"*. The sequence of variables is different when they are compared with the *"role of the graft"*. A very strong coincidence is expressed with *"performed re-operation"* and with the *"rebuilding of graft"*. Reoperation is always an unfavourable incident in course of treatment and if surgeon is obliged to remove the allografts, a positive clinical effect can rarely be expected. Impaired graft substitution is also a rather bad prognostic factor. A comparison of coincidence indices shows how significant *"rebuilding of the grafts"* is in post-surgery estimation. Lyophilised bone allografts are not so readily used now.

3. Transplantation of Deep-Frozen and Radiation-Sterilised Allogenic Bone Grafts

In 1980s and 1990s, deep-frozen bone allografts are most often use for transplantation in humans. In the Institute of Traumatology,

Orthopaedics and Neurosurgery of the Military Medical Academy in Warsaw, during 1981–1995, biostatic bone allografts, frozen and radiation-sterilised were transplanted into numerous patients (Kwiatkowski and Ratynski, 1999; Marczynski *et al.*, 1999). The results achieved in group of 596 female (53%) and 529 male (47%) are presented here. The mean age of patients was 36 years and varied from 2 to 70 years. The grafts were most often used in the second decade of life for 336 patients (30%), or in the third decade for 247 (22%). The indications for bone transplantations varied, 32 diagnoses were categorised into five groups: bone union failure in 401 cases (36%), congenital anomalies in 307 cases (27%), benign tumours in 133 (12%), 68 (6%) of post inflammatory changes and arthroses, 216 (19%) of bone necrosis and others, with: hip joint hypoplasia in 168 cases (15%), scolioses in 146 cases (13%) and solitary cysts in 135 cases (12%). Various types of grafts were used, depending on indications: chips 663 (59%), bars 270 (24%), slices of bone 168 (15%) and solid, large grafts 22 (2%). The manner in which the graft was implanted depended on the diagnosis and location of the pathological change. The grafts were implanted as follows: without covering by periosteum in 517 cases (46%), in incomplete intraosseus apposition of grafts in 292 cases (26%), intraosseus in 213 cases (19%), in subperiosteal position 56 cases (5%) and by intramuscular implantation 45 grafts (4%). In spinal reconstructions, bars were often used — 198 cases (63%), while in femur surgery, chips were mainly used (184 — 78%). In hip surgery, slices of bone were most frequently transplanted (136 — 58%), while chips dominated in reconstructions of the crural bone (148 — 78%).

The result of treatment was analysed 2 to 11 years after surgery. The progress of rebuilding of bone grafts was evaluated in physical and X-ray examinations that were carried out at regular intervals after operation. Analysis shows that by the 3rd month, 517 (46%) grafts were completely rebuilt; the patients of this group were mainly in the second decade of life. Within 6 months after operation, a further 416 (37%) grafts were rebuilt, by the 9th month 101 (9%), and by the 12th month 22 (2%). Within next the 12 months, a further 54 (4.8%) were rebuilt. However, 10 grafts (0.8%) became sequestered with no sign of rebuilding. The number of grafts rebuilt

Table 2. Transplantation of frozen radiation-sterilised bone allografts. Rebuilding of grafts introduced in various position.

Position of graft	Rebuilt grafts		Total	
	n	%	*n*	%
No periosteal cover	466	90.1	517	100
Incomplete intraosseus	251	86.0	292	100
Intraosseus	187	87.8	213	100
Subperiosteal	56	100	56	100
Intramuscular*	0	0	46	100

*all grafts were resorbed

and those in the process of being rebuilt was 1022 (90%). All grafts implanted in muscles were partially or completely reabsorbed (Table 2).

The rebuilding of grafts varied, depending on numerous factors such as: diagnosis, site of grafting, age of the recipient, type, size and the way of the graft introduced. Grafts of spongy bone were rebuilt within 3 to 6 months, grafts of compact bone, in the form of bars were rebuilt within 6 to 24 months. A comparison of age and of graft substitution shows highly effective substitution in the first decade of life 96%. In the second decade, their effectiveness was also high (95%). In the subsequent decades of life, substitution of grafts diminishes. Satisfactory graft substitution was observed in 1022 cases (90.8%) of all patients. The number of transplantations that were estimated to be unsatisfactory was 103 (9.2%).

4. Conclusion

The application of allogenic, biostatic frozen grafts reduce the extent and duration of operations, and the *"creeping substitution"* of implanted bone lasts 3 to 8 months, thus progress of the graft substitution depends on its structure and the position of transplants. Observation of results of bone allografts application is still progressing and now the group of patients under observation is over two thousand. Development of surgical techniques, new generation of antibiotics and new technology of tissue banking certainly will influence the clinical results.

5. References

BLOEM, R.M., TOMFORD, W.W. and MANKIN, H.J. (1996). Histological observations on retrieved human allografts. In: *Orthopaedic Allografts Surgery*, A.A. Czitrom and H. Winkler, eds., Springer, Wien–New York, pp. 61–66.

BURCHARDT, H. and ENNENKING, W.P. (1978). Transplantation of bone, *Surg. Clin. North Am.* **58**, 403.

BURWELL, R.G. (1969). The fate of bone grafts. In: *Recent Advances in Orthopeadic*, A.G. Appley, ed., Churchill, London, p. 115.

DZIEDZIC-GOCLAWSKA, A. (1976). Effect of Radiation sterilisation on biostatic tissue grafts and their constituents. In: *Sterilisation by Ionizing Radiation*, E.R.L. Gaughran and A.J. Goudie, eds., Montreal, Multiscience, Vol. 2, p. 156.

DZIEDZIC-GOCLAWSKA, A., OSTROWSKI, K., STACHOWICZ, W. and MOUTIER, R. (1979). Decrease of crystallinity of bone mineral in osteopetrotic rats, *Metab. Bone Dis. Relat. Dis.* **2**: 33–37.

HEAD, W.C., MALININ, T.L. and BERKLACICH, F. (1987). Freeze-dried proximal femur allografts in revision total hip arthroplasty, *Clin. Orthop.* **215**, 109.

KOMENDER, A. (1976). Influence of preservation procedures on some mechanical properties of human haversian bone, *Mat. Med. Pol.* **10**, 13.

KOMENDER, A. (1977). Metody konserwacji tkanek. In: *Przeszczepy Biostatyczne*, J. Komender, ed., PZWL, Warsaw, Vol. I, pp. 33–48.

KOMENDER, J. and KOMENDER, A. (1977). Evaluation of radiation-sterilised tissue in clinical use. In: *Sterilisation of Medical Products by Ionizing Radiation*, E.R.L. Gaughran and A.J. Goudie, eds., Multisc Publ. Ltd., Montreal, p. 188.

KOMENDER, J., MALCZEWSKA, H. and KOMENDER, A. (1991). Therapeutic effects of transplantation of lyophilised and radiation-sterilised, allogeneic bone, *Clin. Orthop.* **272**, 38–49.

KOMENDER, J., MARCZYNSKI, W., TYLMAN, MALCZEWSKA, H., KOMENDER, A. and SLADOWSKI, D. (2001). Preserved tissue allografts in reconstructive surgery, *Cell. Tiss. Banking* **21**, 103–112.

KRYST, L. (1981). Przeszczepianie tkanek w chirurgii szcz_kowo-twarzowej. In: *Przeszczepy Biostatyczne*, J. Komender, ed., PZWL, Warsaw, Vol. II, pp. 151–161.

KWIATKOWSKI, K. and RATYNSKI, G. (1999). The usage of frozen allografts of spongy bone filling the loss of bone after hip prosthesis, *Ann Transpl.* **4**, 59–63.

Mainland, O. (1963). *Elementary Medical Statistics*, WB Saunders, London, Philadelphia, p. 231.

MARCZYNSKI, W., KOMENDER, J., BARANSKI, M. and KRAUZE, K. (1999). Frozen and radiation sterilised bone allografts in treatment of posttraumatic malformation of bone, *Ann. Transpl.* **4**, 41–45.

MARCZYNSKI, W., KOMENDER, J., STEPIEN, K. and BARANSKI, M. (1999). Fusion of spine in children scoliosis with frozen and radiation-sterilised bone allografts, *Ann. Transpl.* **4**, 30–34.

OSTROWSKI, K., DZIEDZIC GOCLAWSKA, A. and STACHOWICZ, W. (1980). Radiation induced paramagnetic entities in tissue mineral and their use in calcified tissue research. In: *Free Radicals in Biology*, W. Pryor, ed., Academic, New York, Vol. 4, p. 321.

OSTROWSKI, K., WLODARSKI, K., DZIEDZIC-GOCLAWSKA, A., MICHALIK, J. and STACHOWICZ, W. (1996). Allogeneic Bone grafts: Study of radiation sterilised bone tissue by electron paramagnetic resonance spectrometry and a new model of periosteal induction of osteogenesis. In: *Orthopaedic Allograft Surgery*, A.A. Czitrom and H. Winkler, eds., Springer, Wien–New York, pp. 45–52.

PEARSON, K. (1904). On the theory of contingency. In: *Memoirs, Biometric Series* No. 1. Draper's Co., London, p. 119.

SECTION III:

CELL PROCESSING AND PRESERVATION

Advances in Tissue Banking Vol. 6

13 ISOLATION AND CRYOPRESERVATION OF HAEMATOPOIETIC STEM AND PROGENITOR CELLS FROM PERIPHERAL BLOOD AND CORD BLOOD

WIESLAW WIKTOR-JEDRZEJCZAK
and ELZBIETA URBANOWSKA

Department of Haematology, Oncology, and
Internal Medicine, The Medical University of Warsaw
ul. Banacha 1a, 02-097 Warsaw, Poland

1. Introduction

There are at least two reasons for haematopoietic stem and progenitor cells to be present in the peripheral blood. The first goes back in ontogeny to the creation of the bone marrow, the second is related to the maintaining the unity of the bone marrow as an organ. Haematopoietic cells, in order to reach bones have to leave their earlier place of residence i.e. foetal liver and via peripheral blood home to spaces created by osteoclasts inside bones. They reach multiple bones and establish haematopoiesis there. If this process would be autonomous in each of these bones, then it is quite possible that over the years the haematopoiesis would look different in each of them. However, the only known variability in the composition of bone marrow between different bones is related to the quantitative relationships between red and yellow (i.e. fatty) bone marrow. Most likely, the unity of bone marrow is secondary to the constant exchange of stem and progenitor cells between various bones. This can only be achieved by their constant circulation between various bones via peripheral blood. While this was a theory, the first evidence confirming the presence of haematopoietic cells in the peripheral blood was published by Popp (1960) and Goodman and Hodgson (1962). The first evidence confirming this in man was published by McCredie *et al.* (1970), while first evidence demonstrating capacity of cryopreserved peripheral blood cells to restore haematopoiesis was published in 1975 by Debelak-Fehir and Epstein.

The first transplant of cells collected from peripheral blood in man was reported in 1979 by Goldman, who used it in accelerated phase of chronic myelocytic leukemia in an attempt to re-establish chronic phase. This was followed by several reports of successful establishment of haematopoiesis after transplantation of cells collected from peripheral blood (Korbling *et al.*, 1986; Kessinger *et al.*, 1986, 1988; Reiffers *et al.*, 1986).

However, normally the number of haematopoietic cells in peripheral blood was low and two other breakthroughs were essential for mastering the technique. The first was the identification of an antigen later named CD34 present almost exclusively on haematopoietic stem and progenitor cells (Civin *et al.*, 1984; Katz *et al.*, 1985). This has allowed later on for easy and fast enumeration of these cells and for their isolation. The second was the finding that the presence of these cells in peripheral blood was greatly increased after administration of cytotoxic drugs (Richman *et al.*, 1976) and of haematopoietic growth factors (Duhrsen *et al.*, 1988; Socinski *et al.*, 1988). This has allowed the collection of sufficient numbers of these cells for transplantation during a few or even one leukapheresis (Gianni *et al.*, 1989; Bensinger *et al.*, 1993).

However, there is one instance when peripheral blood naturally contains large number of haematopoietic stem and progenitor cells. This is peripheral blood of the foetus and consequently of the neonate. Besides, the neonate is so generous that it leaves approximately half of it in umbilical cord and placenta after being born. This cord and placenta are normally disposed as medical waste. However, from this cord and placenta, blood can be easily and safely collected, processed, stored and used for transplantation. The first evidence that cord blood may be a source of cells for transplantation in man was published in 1972 by Ende and Ende, while pioneers of modern techniques were Gluckman *et al.* (1989) and Broxmeyer *et al.* (1989) Therefore, cord blood is nothing else than this part of the peripheral blood of the neonate that is left after delivery in umbilical cord and placenta.

2. Biological Background for the Isolation and Cryopreservation of Haematopoietic Stem and Progenitor Cells

As of the beginning of 2001, there is still no direct assay available for the human haematopoietic stem cell, i.e. cell that has the full capacity for permanent (for the lifespan of given species) capacity to establish haematopoiesis. Such cells are present in murine marrow with frequency of 1/25 000 as determined using a limiting dilution assay in stem cell deficient W/W^v mouse (Wiktor-Jedrzejczak *et al.*, 1979; Boggs *et al.*, 1982). Their frequency in human marrow is most likely slightly lower as determined using one of surrogate stem cell assays available in human system: the assay for long-term culture initiating cell (Fraser *et al.*, 1992). Haematopoietic progenitor cells, i.e. cells with extensive proliferation and differentiation potential but unable to renew themselves, are also important from the transplantation perspective as they contribute significantly to the early and intermediate phases of engraftment. Haematopoietic engraftment has three phases that directly follow each other. The first two originate from late and early progenitor cells and are transient, while the third one is permanent and it originates from the haematopoietic stem cells.

Peripheral blood of the adult and cord blood (i.e. peripheral blood of the neonate) differ sharply in their content of progenitor cells, while the adult bone marrow falls in between. Adult peripheral blood particularly after procedure called "mobilisation", contains a very large number of progenitor cells which allow a very early engraftment. Cord blood has low number of progenitor cells but it has quite high content of stem cells (judging on reconstituting ability approximately 5 times of the bone marrow). This has profound consequences for the post-transplant care. Patients — transplanted with stem and progenitor cells obtained from adult peripheral blood — have early engraftment and short period of severe neutropenia and thrombocytopenia. At the opposite, cord blood recipients have very prolonged neutropenia and thrombocytopenia and require much more intensive supportive care prior reaching engraftment.

Similarly, as there is no method for direct identification of the stem cells there is also no simple method available for differentiating between stem and progenitor cells prior their transplantation. Their presence and relative activity can only be judged by analysing the effects of their transplantation. Permanent engraftment is a proof of stem cell activity, while transient engraftment is signifying transplantation of progenitor cells. Altogether, stem and progenitor cells compose less than 1% of the cell inoculum of any original transplant material being this peripheral blood, cord blood or bone marrow. Yet only these cells engraft and, hence, the transplantation process is limited to these cells with partial exclusion of immune cells. Differentiated immune cells (if present in the graft) engraft and are responsible for transferring to the recipient acquired immunity and to induce acute graft versus host reaction. Transplanted stem cells also eventually produce immune cells. Such cells generated in the new recipient may become more tolerant than the original donor immune cells, but they may induce chronic graft versus host disease.

While under ideal conditions a single haematopietic stem cell should be sufficient for the restoration of haematopoietic system (Abramson *et al.*, 1977), there are other factors that require transplantation of larger number of stem cells in human settings. These are: (i) multiple mismatches between donor and recipients in transplantation antigens, and (ii) necessity to maximally shorten the time of post-transplant life-threatening neutropenia and thrombocytopenia. Although donor and recipients for haematopoietic transplantation are always matched for at least a haplotype of major histocompatibility complex (MHC), there are always multiple and not identified mismatches in various minor histocompatibility antigens. From studies on mice, it is known that there exist at least 70 systems of such antigens. From the work of the late Gustavo Cudkovicz (1964) and Eva Lotzova (1977), it is known that there is a phenomenon of "allogeneic resistance" that requires increase in the number of stem cells for each such difference in order to allow for their engraftment, i.e. for breaking the allogeneic resistance. Even more, stem cells are needed to break resistance created by partial incompatibility in MHC. While even very severe neutropenia of duration below one week can be taken care of at the conditions of a single conventional hospital room, neutropenia of one month duration (like this observed after cord blood transplantation) usually requires extensive decontamination of the recipient and laminar air flow conditions and then is still life threatening. This problem is best taken care of by increasing the content of haematopoietic progenitors in the graft.

The first report of the possibility of cryopreservation of hematopoietic cells was published in 1955 by Barnes and Loutit, who used a technique originally developed in 1949 by Polge *et al.* for freezing spermatozoa. This technique, hinging on placing cells in hyperosmotic solution although refined is still used.

3. Manouvers to Increase the Yield of Haematopoietic Stem and Progenitor Cells in Peripheral and Cord Blood

As mentioned earlier, normally the frequency of haematopoietic stem and progenitor cells in the peripheral blood is low and requires multiple collections prior assuring sufficient number of cells for reasonably early engraftment. However, in pre-clinical experiments on haematopoietic growth factors, it was found that they cause shift of haematopoiesis to the periphery, i.e. to the spleen and the appearance of increased numbers of haematopoietic stem cells in the peripheral blood (Metcalf *et al.*, 1987). Similar phenomenon was earlier observed also during recovery phase after administration of cytostatic drugs. This has provided initial evidence for manouver called "mobilisation" that is aimed in recruiting haematopoietic stem cells to the peripheral blood and to facilitate their collection for subsequent use in transplantation. There are several possible mechanisms by which various factors may mobilise haematopoietic stem cells to the peripheral blood:

— reversal of the cytokine gradient between bone marrow and the peripheral blood: normally high cytokine concentration in the bone marrow attracts stem cells to the bone marrow; intravenous administration of haematopoietic growth factors may increase cytokine concentration in the peripheral blood over the bone marrow level and result in attracting this way haematopoietic cells to the peripheral blood.
— overgrowth of haematopoietic stem cells in the marrow and their "leakage" to the peripheral blood.
— modulation of surface adhesion molecules important for homing of haematopoietic cells to the bone marrow and for their residence there: this may include both action on adhesion molecules on haematopoietic cells and on adhesion molecules on haematopoietic microenvironment.
— depletion of haematopoietic cells in various locations in the bone marrow may prompt multiple regulatory mechanisms to induce regeneration and influx of stem cells from other locations.

The first two possibilities are considered unlikely, because it has been shown by Prosper *et al.* (1998) that during mobilisation, the number of CD34+ cells in the marrow remains stable. This is contrary to the expectations because the first postulated mechanism would imply decrease of CD34+ cells in the bone marrow due to their release to the peripheral blood, while the second mechanism would imply major increase of CD34+ cells in the marrow. However, if these mechanisms would act combined than the net result would be stable number of cells remaining in the marrow and their simultaneous increased presence in the peripheral blood (Zaucha *et al.*, 2000). Therefore, the participation of these mechanisms in mobilising stem cells to the peripheral blood cannot be excluded.

There is also good evidence for the third mechanism. Papayanopoulu *et al.* (1993) have shown that antibodies blocking interaction between fibronectin and its ligand adhesion molecule

VLA4 produce immediate egress of haematopoietic progenitors to the periphery. Prosper *et al.* (1998) have demonstrated a role in this process of $\alpha_4\beta_1$ integrin, while Pruijt *et al.* (1998) a role of β_2 integrin. Besides being a mechanism mediating the effects of cytokines, the modulation of adhesion molecules may evolve into direct clinical procedures employing molecules directly approaching these targets. The 4th mechanism is listed here for the explanation of the effects of cytostatic drugs. However, very little is known on the precise mechanisms by which these drugs mobilise stem and progenitor cells to the peripheral blood (Demirer *et al.*, 1999).

Currently, the above considerations have little impact on clinical practice in which standard protocols employing cytokines and cytostatic drugs are used. Collection of haematopoietic cells from peripheral blood for transplantation is a subject to additional limitations depending on whether the prospective donor is a healthy person (and in particular healthy person unrelated to the recipient) or whether the prospective donor is himself a planned recipient. Administration of medicines to healthy people is limited by ethical reasons (and in some countries also by legal reasons) and it is possible only when these medicines have no known long lasting or mutagenic effects. Consequently, the only drugs permitted in healthy donors are haematopoietic growth factors. On the other hand, in donors for autologous transplantation, the administration of cytostatic drugs may have additional beneficial effects besides causing mobilisation of haematopoietic cells, i.e. it is producing cytoreduction of the original neoplastic disease.

The standard growth factor mobilisation regimen is composed of five-day course of either G- or GM-CSF (recently, usually G-CSF) followed by leukapheresis (Gianni *et al.*, 1989; Bensinger *et al.*, 1993). The standard cytostatic regimen hinges on the administration of a single high dose of cyclophosphamide (4 or $7\,g/m^2$) and leukapheresis when the recovering neutrophil count would reach more than $1.5 \times 109/l$ (Demirer *et al.*, 1999). These standard regimens can be combined and used together as this is the case for majority of autologous transplantations performed worldwide (Jones *et al.*, 1994; To *et al.*, 1997; Dreger *et al.*, 2000). In the situation of unsuccessful mobilisation, it is first necessary to verify what time has elapsed between last growth factor injection and leukapheresis. This time should not be longer than a few hours and the cessation of growth factor administration on the morning of the day preceding collection is the most common reason for failed mobilisation. If this is not the case, there are several other routine manouvers that can increase yield of collected cells. They include increase of the growth factor dose (up to 4 times depending on patients tolerance) , combining G- and GM-CSF instead using either one of them, and trying other cytostatic drugs or their combinations. Several such combinations have been described known under eponyms ICE standing either for idarubicin, cytosine arabinoside, and etoposide (Carella *et al.*, 1997) or ifosfamide, carboplatin, and etoposide (Moskovitz *et al.*, 1999). We prefer to term this latter protocol ECI to avoid confusion. In allografting, G-CSF mobilised cells collected from peripheral blood seem to have similar therapeutic potential as bone marrow (Dlubek *et al.*, 1999; Dreger *et al.*, 2000).

Both in autologous and allogeneic setting, some donors would not mobilise. In autologous setting this includes mainly patients earlier heavily pretreated with cytostatic drugs, particularly those with myeloablative potential such as melphalan, busulfan, nitrosoureas and patients subjected to radiotherapy. The CD34 response to the single dose of G-CSF has been proposed as a test to identify patients with difficult mobilisation (Mijovic *et al.*, 1999). There are also some normal donors who do not mobilise in response to G-CSF. This may be regarded as a new abnormality of their haematopoietic system with unknown physiological significance.

There is a considerable need for a progress in this area. Current research includes testing novel (in this application) cytokines including Steel factor, Flt-3-ligand, IL-8 and other (Glaspy *et al.*, 1997, Pruijt *et al.*,1998).

There are no known manouvers to increase the presence of haematopoietic stem cells in the cord blood. Even if they exist, such manouvers would be extremely difficult to use because of ethical concerns arising from their possible effects on healthy newborn.

4. Collection of Haematopoietic Stem and Progenitor Cells from Peripheral Blood

Haematopoietic stem and progenitor cells are collected from peripheral blood using standard leukapheresis. Such possibility was first described by Korbling *et al.* (1980). As of 2001, there are three instruments commercially available to perform this procedure, including Cobe-Spectra, Baxter CS3000 and Fresenius AS104. Each of them (in older and newer versions) is supplied with a special programme optimising collection of mononuclear cells and detailed instructions to perform such a collection. The major limitation to the use of this technology from the side of the donor is a requirement of having two large diameter venous entries. Since donor is connected to the apparatus via outcoming line through which his blood is accessing the apparatus, and via incoming line through which his blood returns after being processed, these lines have to allow sufficient blood flow (usually not less than 30 ml/min). This is not possible with many utilised central venous catheters and ports. It is important to remember this restriction while selecting catheter or port type for a patient who eventually will become a candidate for stem cell collection. All separators use disposable closed system sets for collection. Usually, 10–16 l of blood are processed during one session at a flow rate of 30–80 ml/min, which means that a single session takes between 3 and 6 hours and that usually, no more than 2 patients can be served by 1 separator per day (Dreger *et al.*, 2000). Besides problems with venous access, the next important problem is calcium depletion resulting from the use of ACD-A solution for anticoagulation. This may require calcium supplementation at the end of separation. Additionally, depletion of platelets may create problems, especially in patients with low platelet counts initially. Usually, platelet count above $70 \times 10^9/l$ is required. For patients with platelet counts between 30 and $70 \times 10^9/l$, in order to carry leukapheresis (if necessary), it is required to transfuse them with platelets prior and immediately after the procedure. Larger problems occur when the donor is a child, because each separator has its external volume that normally is about 400 ml. For smaller children such volume may create a problem and some separators have attachments that may reduce external volume to 200 ml allowing to leukapherese smaller children. Another way to deal with this problem is to collect from the prospective donor 200 ml of blood 2 weeks in advance of stem cell collection and use this autologous blood to prime cell separator immediately prior leukapheresis (Diaz *et al.*, 2000).

Theoretically, in some cases, content of stem and progenitor cells in the peripheral blood may be high enough to allow for collection of sufficient cell number by simple donation of one unit (432 ml) of blood without leukapheresis. However, it is achieved rather rarely and therefore, it is difficult to suggest to use this as a standard. Recently, haematopoietic progenitors present in 750 ml of autologous blood collected prior chemotherapy but post G-CSF have been used to support recovery after administration of dose-dense (but not myeloablative) chemotherapy (Woll *et al.*, 2001).

5. Methods of Collection of Cord Blood for Transplantation

Originally, cord blood was collected by simple putting a bottle with medium and anticoagulant at the end of umbilical cord immediately after it was cut off the newborn and by allowing blood remaining in placenta and cord to freely flow to the bottle. Over time this procedure was completely replaced by closed systems, specially designed collection sets became commercially available and procedure underwent standardisation (Fraser *et al.*, 1998; Laughin, 2001; Linden *et al.*, 1997). At present, two major methods of collection are used: *In utero* and *ex utero*.

In *in utero* method, collection is carried out immediately after delivery of the neonate, when placenta still remains inside uterus. The needle of collecting set is introduced into the umbilical cord vein and the blood remaining in placenta and cord is allowed to freely flow to the collecting bag containing anticoagulant, usually CPD-A fluid. The reason for the use of CPD-A fluid instead of ACD-A fluid lies primarily in the difficulties in predicting the exact dose of cord blood collected. In contrast to CPD-A fluid, ACD-A fluid is not isoosmotic and requires mixing it with precise amount of blood. In a variation of this method, small amount of CPD-A fluid is first infused into the placenta through the same needle from the adjacent bag, and anticoagulated blood is then collected as before.

In *ex utero* method, blood is collected after delivery of placenta. This is also the only method used for collection after deliveries through caesarean section. Placenta is usually put on some support, and blood is collected as previously. Some teams use specially designed stations which could be driven to delivery room and contain all supplies necessary for standardised collection. Since the collected cell number is the most important factor determining the effect of cord blood transplantation a care is taken to collect as many cells as possible. This can be achieved by combining two methods (Wong *et al.*, 2001). Placental and cord blood is first collected using *in utero* method, and following delivery placenta is additionally lavaged with saline mixed with CPD fluid. Approximately 100 ml of saline-CPD fluid is introduced into placenta from the bag adjacent to collecting bag and then allowed to freely fill by gravitation the collecting bag. Such procedure may yield up to 50% of additional cells (Wiktor-Jedrzejczak and Urbanowska, unpublished).

After collection, labelled bags are stored in obstetric wards in specially assigned refrigerators till they are transferred (usually next morning) to the processing facility. Up to 24 hours storage at 4°C. has no significant effect on the number and viability of collected cells.

6. Methods of Quantification of Haematopoietic Stem and Progenitor Cells

As it has been mentioned, there is still no direct assay available for human haematopoietic stem cell. Consequently, all assays utilised are surrogate, measuring related cell populations with assumption that their numbers would be proportional to the number of stem cells. These assays include:

— determination of the number of mononuclear cells
— determination of the number of colony forming cells

— determination of the number of CD34+ cells
— determination of the number of long-term culture initiating cells (LTC-IC).
— transplantation into SCID mice.
— other assays under development or of exclusively research use.

Historically, determination of the number of mononuclear cells was the first index of haematopoietic potential of grafted preparations. Although, it has been widely criticised as basing judgement on enumeration of cell population distant to the evaluated cell population, there are reports confirming its reliability. For instance, for the cord blood most studies still refer mainly to the mononuclear cell count (Gluckman *et al.*, 1997).

The main problem with progenitor quantification using *in vitro* colony assays hinges on the time-lag between the day of testing (which is the day of the initiation of culture) and the day of result (which is the day of evaluating the culture). This time lag is between 7 and 14 days depending on which colony forming cell is being considered. This precludes of using these assays for these purposes, where immediate result is necessary (Verfaille *et al.*, 1999). In these assays the progenitor being tested or colony forming unit (CFU) is defined by the type(s) of cells that are generated. Progenitor forming both granulocytes and macrophages is termed GM-CFU, progenitor forming irregularly shaped colonies of red cells (or "bursts") is termed BFU-E, and the progenitor forming both granulocytes, erythrocytes, eosinophils, macrophages, and megakaryocytes is termed CFU-GEMM. Besides those, there are more restricted (e.g. G-CFU, MEG-CFU) or less restricted (Blast-CFU) progenitors that may be tested depending on the specific needs of given protocol. By far, the GM-CFU assay is the one the most frequently used (Spitzer *et al.*, 1980). Typically, these cultures are carried out by plating cells in semisolid or semiliquid medium, in recent years, most frequently in methylcellulose in the presence of growth factors such as G-CSF, GM-CSF, erythropoietin and others and scored using an inverted microscope after 14 days. Foci of variously defined numbers of cells (frequently more than 50) are counted as colonies and each of them is considering to represent one CFU. Although there is no common standard for the assay, commercially available kit (Methocult) purchased from Stem Cells (Vancouver, Canada) is playing this role.

In 1984, Civin *et al.* produced the first monoclonal antibody (My 10) against an antigen present on KG-1a cells, which subsequently received number 34 in CD. (cluster of differentiation) nomenclature and was found to be almost exclusively expressed on haematopoietic stem and progenitor cells (Krause *et al.*, 1996). In 1989, Siena *et al.* introduced determination of CD34 to the quantification of progenitor cells in the peripheral blood. The CD34 molecule was subsequently found to be composed of three major epitope classes characterised by different enzymatic sensitivity. Class I epitopes were found to be sensitive to neuramidase, chymopapain and glycoprotease, class II was insensitive to neuramidase, and class III was insensitive to all these enzymes (Krause *et al.*, 1996).

Many different monoclonal antibodies were then produced against of each of these epitopes. They include My 10, BI.3C5, ICH3, and 12.8 antibodies directed against class I epitopes, QBEnd-10 and 9C5 antibodies against class II epitopes, and 8G12, 581, BIRMA-K3 antibodies against class III epitopes (Arseniev *et al.*, 1999; Krause *et al.*, 1996). Class I antibodies were then restricted to use in procedures aimed in isolation of CD34+ cells, while class II and III antibodies are used for quantification of CD34+ cells. Initially, each laboratory was using its own slightly different protocol but the importance of the problem prompted attempts to standardise these assays.

Comparative studies have yielded markedly different results with different antibodies, and the main conclusion was to avoid QBEND10 antibody coupled with fluorescein (Arseniev *et al.*, 1999). Subsequently, several protocols have been adopted by various groups with two: Milan protocol (Siena *et al.*, 1994) and ISHAGE protocol (Gee and Lamb, 1994) being the most frequently used. Milan protocol is counting CD34+ cells as a fraction of leukocytes on the basis of forward-scatter and side-scatter characteristics. ISHAGE protocol determines CD34+ cells as a fraction of CD45+ cells (nucleated white cells). These protocols have been recently criticised because they involve erythrocyte lysis step prior analysis.

This lysis step may in a non-controlled way influence CD34+ cells. A new method was proposed that utilised fluorescent dye SYTO-13 that stains viable nucleated cells (Fornas *et al.*, 2000). This method allows for counting CD34+ cells in the whole blood. More recently, a new marker AC133 has been introduced to quantification of haematopoietic progenitor cells (Yin *et al.*, 1997; Miraglia *et al.*, 1997). At present, it has, at the most, accessory role to the CD34 determination.

Both quantification of LTC-IC and cells initiating human haematopoiesis in SCID mice as well as other methods have only research application (Verfaille *et al.*, 1999). By far assays employing transplantation of human haematopoietic stem cells into SCID mice are considered to be the best approximates of the actual repopulating capacity of human haematopoietic stem cells (Dick, 1996). These assays are used to verify major elements of techniques of stem cell isolation and preparation but not for testing individual patients in the course of therapeutic procedures. An example of attempts to develop other stem cells assays is a study of Young *et al.* (1999).

7. Methods of Isolation of Haematopoietic Stem and Progenitor Cells for Clinical Use

Finding that CD34 antigen is expressed on both haematopoietic stem and progenitor cells and availability of monoclonal antibodies against this antigen allowed for development of antibody-dependent methods of purification CD34+ cells. Subsequently, it was proven that these cells are capable of reconstituting haematopoiesis in man (Berenson *et al.*, 1991) and since then, they are finding increasing place in clinical transplantation (Brugger *et al.*, 1994; To *et al.*, 1997). As it has been mentioned, CD34 molecule has three different classes of epitopes that are recognised by different antibodies. Antibodies used in isolation procedures recognise Class I epitopes (Krause *et al.*, 1996). The first device that allowed for clinical scale purification of CD34+ cells was CEPRATE system representing more elaborated version of a method developed by Berenson *et al.* (1991). This system was using biotinylated monoclonal anti-CD34 antibody to mark the CD34+ cells with biotin. Subsequently, suspension of haematopoietic cells containing biotinylated CD34+ was passaged through avidin-immunoaffinity column. CD34+ cells have been trapped by avidin in the column while remaining cells passaged freely. Later on CD34+ cells were eluted from the column. However, over dispute on patent issues concerning the use of anti-CD34 antibodies the company (Cell-Pro) that was founded to manufacture this instrument ceased to operate. This has left on the market two later developed instruments: Isolex supplied by Nexell, and CliniMACS manufactured by Miltenyi-Biotec. Isolex is utilising mouse anti-CD34 antibody and anti-mouse IgG-coated paramagnetic beads that are trapped in a chamber (Croop *et al.*, 2000). CliniMACS represents extension of a system used earlier on laboratory scale. In this system, the cells are specifically labelled with superparamagnetic iron-dextran particles, which are covalently conjugated to mouse antihuman CD34 antibody. After labelling cells are passed through a separation column which is held in a field of strong permanent magnet. Labelled cells are retained in the column, while unlabelled cells passage through it. After removing column from the magnetic field cells are eluted, washed and available for further use (Schumm *et al.*, 1999).

Transplantation of purified autologous CD34+ cells is a novel treatment modality in autoimmune disorders (Van Laar, 2000), while in allogeneic setting it is used mainly for the so-called "megadose" grafts (Schwinger *et al.*, 2000).

8. Methods of Processing of Haematopoietic Stem and Progenitor Cells for Short- and Long-Term Storage

Collected haematopoietic stem and progenitor cells from either peripheral or cord blood can be stored in suspension for up to a few days in refrigerator or long term, frozen in –80° centigrade or in liquid nitrogen (–196° centigrade). Short-term refrigerator storage is possible only for the use in transplantation procedures that use conditioning of similarly short duration, for instance single agent melphalan. Recently, special solution has been developed for such storage (Burger *et al.*, 1999). Long-term storage is carried out in three main temperatures: in mechanical freezer −80° for usually not more than a few months, in mechanical freezer −135° or vapour phase of liquid nitrogen (−156°) for period of up to a few years, and in liquid phase of liquid nitrogen with the intention of indefinite storage. Storage in liquid phase of N_2 has disadvantage of facilitating viral contamination from external side of one bag to other bags, while it has advantage in very long-term storage as it precludes temperature variations during each tank opening. Although such temperature variations occur within deep minus range their effects on the long term storage are unknown and may be detrimental. The same problem concerns mechanical freezers and from that point of view horizontal freezers are preferred over vertical freezers. In vertical freezers, penetration of room temperature air to the freezer during its opening is negligible compared to horizontal freezers.

Usually, prior further processing, samples are taken from collected material for infectious agents testing, cell number and viability, *in vitro* culture assays, CD34 and other antigens determination by flow cytometry, and other assays (for instance histocompatibility antigens and blood group in the case of cord blood for use in unrelated recipients). Peripheral blood cells collected by leukapheresis usually do not require further preparation and may be directly forwarded toward freezing procedure. The same was the case originally with cord blood (Broxmeyer *et al.*, 1989). However, over time when cord blood banks became filled with thousands of frozen units a considerable interest arose in volume reduction, which would allow storage of significantly larger number of units in the same space. Currently, volume reduction prior freezing is a standard and it is achieved by first removing red cells and then reducing volume of plasma. Commonly, procedures originally developed by Rubinstein *et al.* (1995) are used for this purpose.

In this method, red cells are removed by hydroxyethyl starch sedimentation, and plasma volume is reduced by centrifugation. Similar results are obtained using gelatine sedimentation (Pahwa *et al.*, 1994), but this method is not used because of difficulties in its standardisation. Centrifugation over Ficoll-ditrizoate gradient has been shown to be much less effective (Regidor *et al.*, 1999) probably because cord blood progenitors have different density properties than adult progenitors.

For freezing, cell suspension is mixed with cryoprotectant solution of which major component is dimethyl sulfoxide (DMSO) in final concentration between 5 and 10% (Rowley, 1992) and autologous or heterologous plasma (final concentration 10–20%). For the cord blood Rubinstein *et al.* (1995) have developed special solution containing additionally hydroxyethyl starch that not only improves freezing but also improves thawing. In particular, it reduces cell clumping particularly, when cells are frozen in higher concentrations. Usually, the recommended cell concentration is between 1 and 4×10^8/ml. After addition of DMSO to cell suspension all manouvers leading to freezing have to be carried fastly because DMSO is considerably toxic to cells in solution. Cell suspension mixed with DMSO is usually in a closed system transferred to special freezing bags, that are commercially available. After filling, these bags are put in special metal cassettes and processed for freezing. Some transplant teams still use freezing vials instead of bags. This has disadvantage of making system open (although handling is carried out in laminar flow hood) and increasing workload (necessity to freeze hundreds of –5 ml vials per each cell suspension). However, it is considerably cheaper, splits the cell product into many vials making it less fragile to a single accident (like bag leakage), and also shortens the time of injection after thawing. In general, it cannot be used in the setting when transplant is collected in different centre that it is transplanted, because it precludes clear responsibility for a possible failure.

Freezing can be achieved by one of two ways: two-step freezing or controlled rate freezing. In two-step freezing bags or tubes containing cell suspension are first placed in mechanical –80° freezer (usually in a styrofoam box) and next day transferred to liquid nitrogen tank, except for the situation, when storage is in –80°C or –135°C mechanical freezers (Rowley, 1992; 1999). In controlled-rate freezing (usually 1°C/min) after reaching endpoint (usually –80°C or less) bags or tubes are transferred to place of final storage. There is no evidence that controlled-rate freezing improves the results when compared to two step freezing. However, controlled-rate apparatus provides a record of freezing procedure. Therefore, in the situation when transplantation procedure is performed in different facility than collection and storage (like the situation of transplantation of unrelated cord blood obtained from the bank located in different country) it is preferred to freeze using controlled-rate freezer and to provide a record of freezing procedure to the transplant centre (Dreger *et al.*, 2000).

9. Preparation of Stored Haematopoietic Stem and Progenitor Cells for Transplantation

The standard procedure is to thaw bags or tubes containing cell suspension in water-bath located adjacent to the patient bed and to inject or infuse thawed suspension immediately via the central vein without removing DMSO (Rowley, 1992, 1999). With freezing solution developed for cord blood by Rubinstein *et al.* (1995), it is now possible to centrifuge thawed solution, remove DMSO, resuspend it and finally infuse to the patient (Rowley *et al.*, 1999). However, particularly when admixture of neutrophils is large it is dangerous, because they are more fragile and when destroyed they release enzymes that are toxic to cells and DNA which makes suspension extremely viscous, and impossible to handle.

10. National and International Regulation, Standardisation and Cooperation in Utilisation of Stored Haematopoietic Stem and Progenitor Cells

Shortage of related hematopoietic graft donors has led to increasing use of unrelated donors. Because of extreme variability of HLA antigens, this required registration of millions of prospective donors and has led to the situations when quite frequently donor and recipient are located in different countries. To even greater extent this concerns cord blood, because it is usually stored long time before prospective recipients even got ill. All these problems have been successfully dealt with through the international cooperation and standardisation, and the existing standards are constantly improved (Fraser *et al.*, 1998; Hakenberg *et al.*, 1998). The organisations that are involved in setting up standards for haematopoietic cell processing and storage are International Society for Hematotherapy and Graft Engineering (ISHAGE) and European Group for Blood and Marrow Transplantation (EBMT) working additionally through the specialised bodies like FAHCT (Foundation for Accreditation of Hematopoietic Cell Therapy), JACIE (Joint Accreditation Committee ISHAGE/EBMT), and Netcord (organisation of cord blood banks having stored more than 1000 units each). These organisations have developed standards of collection, processing, testing, banking, selection and release of haematopoietic cells for transplantation. Institutions that are members of these organisations have direct access to constantly updated versions of these documents and are adjusting their procedures in a way fully complying with those standards. Some basic components of these standards have been independently published (Link *et al.*, 1995; Serke and Johnsen, 2001). Finally, members of such organisations voluntarily get access to their accreditation procedures that would eventually allow them to fully participate in world-wide exchange of haematopoietic transplants. Most of the communication is carried via Internet and can be accessed through BMDW server. However, for qualified entry into the system passwords are required that are restricted to the registered members.

Besides standardisation also legal status of haematopoietic progenitors procurement requires international cooperation. For instance, cord blood is variably regulated by either transfusion law or transplantation law (Annas, 1999). Sometimes, specific legal guidelines are developed for products such as cord blood (Linden *et al.*, 1997). Another regulatory approach to cellular products has been published by the American Food and Drug Administration (1997).

11. Concluding Remarks

During last years the isolation of haematopoietic stem and progenitor cells has become extremely well standardised procedure that is in routine use in hundreds facilities all over the world and supports many hundreds of hematopoietic transplant teams active world-wide. Despite enormous effort one aspect still remains a matter of frustration. It is the problem of *ex vivo* expansion (Srour *et al.*, 1999). Despite published claims that this and that laboratory developed a system solving this issue (Emerson, 1997; Lundell *et al.*, 1999) attempts to reproduce these data are usually not very successful (Shih *et al.*, 2000). Definitely, it is the most awaited development in the field. However, the largest expectations currently arise from the possible generation of haematopoietic stem cells from the embryonic stem cells (Anderson *et al.*, 2001) and not from the hope that finally someone will discover the magic trick to expand *in vitro* adult haematopoietic stem cells, that has not been discovered throughout a quarter of century of active pursuing of this issue.

12. Acknowledgements

This work was supported by the Polish State Committee for Scientific Research grant PBZ-029-12 to W. Wiktor-Jedrzejczak and Statutory Funds of the Medical University of Warsaw.

13. References

ABRAMSON, S., MILLER, R.G. and PHILLIPS, R.A. (1977). The identification in adult bone marrow of pluripotent and restricted stem cells of the myeloid and lymphoid systems, *J. Exp. Med.* **145**, 1567–1579.

ANDERSON, D.J., GAGE, F.H. and WEISSMAN, I.L. (2001). Can stem cells cross lineage boundaries? *Nature Med.* **7**, 393–395.

ANNAS, G.J. (1999). Waste and longing — The legal status of placental blood banking, *N. Engl. J. Med.* **340**, 1521–1524.

ARSENIEV, L., PICKERD, N., GOUDEVA, L., HERTENSTEIN, B. and GANSER, R. (1999). Comparative evaluation of commonly used clones and fluorochrome conjugates of monoclonal antibodies for CD34 antigen detection, *J. Hematother. Stem Cell Res.* **8**, 547–559.

BARNES, D.W.H. and LOUTIT, J.F. (1955). The radiation recovery factor. Preservation by the Polge-Smith-Parkes technique, *J. Natl. Cancer Inst.* **15**, 901–905

BENSINGER, W., SINGER, J., APPELBAUM, F., LILLEBY, K., ROWLEY, S., CLARKE, E., CLIFT, R., HANSEN, J., SHIELDS, T., STORB, R., WEAVER, C., WEIDEN P. and BUCKNER, C.D. (1993). Autologous transplantation with peripheral blood mononuclear cells collected after administration of recombinant granulocyte stimulating factor, *Blood* **81**, 3158–3163.

BERENSON, R.J., BENSINGER, W.I., HILL, R.S., ANDREWS, R.G., GARCIA-LOPEZ, J., KALAMASZ, D.F., STILL, B.J., SPITZER, G., BUCKNER, C.D., BERNSTEIN, I.D. and THOMAS, E.D. (1991). Engraftment after infusion of CD34+ marrow cells in patients with breast cancer or neuroblastoma, *Blood* **17**, 1717–1722.

BOGGS, D.R., BOGGS, S.S., SAXE, D.F., GRESS, L.A. and CANFIELD, D.R. (1982). Hematopoietic stem cells with high proliferative potential. Assay of their concentration in marrow by the frequency and duration of cure of W/W^v mice, *J. Clin. Invest.* **70**, 242–252

BROXMEYER, H.E., DOUGLAS, G.W., HANGOC, G., COOPER, S., BARD, J., ENGLISH, D., ARNY, M., THOMAS, L. and BOYSE, E.A. (1989). Human umbilical cord blood as a potential source of transplantable hematopoietic stem/progenitor cells, *Proc. Natl. Acad. Sci. USA* **86**, 3828–3832.

BRUGGER, W., HENSCHLER, R., HEIMFELD, S., BERENSON, R.J., MERTELSMAN, R. and KANZ, L. (1994). Positively selected autologous blood CD34+ cells and unseparated peripheral blood progenitor cells mediate identical hematopoietic engraftment after high-dose VP16, ifosfamide, carboplatin, and epirubicin, *Blood* **84**, 1421–1426.

BURGER, S.R., HUBEL, A. and McCULLOUGH, J. (1999). Development of an infusible-grade solution for non-cryopreserved hematopoietic cell storage, *Cytother.* **1**, 123–133.

CARELLA, A.M., CUNNIGHAM, I., LERMA, E., DEJANA, A., BENEVENUTO, F., PODESTA, M., CELESTI, L., CHIMIRRI, F., ABOTE, M., VASALLO, F., FIGARI, O., PARODI, C., SESSAREGO, M., VALBONESI, M., CARLIER, P., PRENCIPE, E., GATTI, A.M., Van den BERG, D., HOFFMAN, R. and FRASSONI, F. (1997). Mobilisation and transplantation of Philadelphia-negative peripheral-blood progenitor cells early in chronic myelogenous leukemia, *J. Clin. Oncol.* **15**, 1575–1582.

CIVIN, C.I., STRAUSS, L.C., BROVALL, C., FACKLER, M.J., SCHWARTZ, J.F. and SHAPER, J.H. (1984). Antigenic analysis of hematopoiesis. III. A hematopoietic progenitor cell surface antigen defined by a monoclonal antibody raised against KG-1a cells, *J. Immunol.* **133**, 157–165

CROOP, J.M., COOPER, R., SESHADRI, R., FERNANDEZ, C., GRAVES, V., KREISSMAN, S., SMITH, F.O., CORNETTA, K., WILLIAMS, D.A. and ABONOUR, R. (2000). Large-scale mobilization and isolation of CD34+ cells from normal donors, *Bone Marrow Transplant.* **26**, 1271–1279.

CUDKOWICZ, G. (1964). Hybrid resistance to parental marrow grafts. Association with the K region of H-2, *Science* **144**, 1339–1340, Correction of the data 147, 1056.

DEBELAK-FEHIR, K.M. and EPSTEIN, R.B. (1975). Restoration of hematopoiesis in dogs by infusion of cryopreserved autologous peripheral white cells following busulfan-cyclophosphamide treatment, *Transplantation* **20**, 63–67.

DEMIRER, T., BENSINGER, W.I. and BUCKNER, C.D. (1999). Peripheral blood stem cell mobization for high-dose chemotherapy, *J. Hematother* **8**, 103–113.

DIAZ, M.A., KANOLD, J., VINCENT, M.G., HALLE, P., MADERO, L. and DEMEOCQ, F. (2000). Using peripheral blood progenitor cells (PBPC) for transplantation in pediatric patients: a state-of-the-art review, *Bone Marrow Transplant.* **26**, 1291–1298.

DICK, J.E. (1996). Normal and leukemic human stem cells assayed in SCID mice, *Semin. Immunol.* **8**, 197–206.

DLUBEK, D., PACUSZKO, T., SUCHNICKI, K. and LANGE, A. (1999). G-CSF-mobilized leukapheresis products: Cellular characteristics and clinical performance in allografting, *J. Hematother.* **8**, 157–166.

DREGER, P., GLUCKMAN, E. and SCHMITZ, N. (2000). Source of haemopoietic cells. In: *Blood and Marrow Transplantation*, J.F. Apperley, E. Gluckman, A. Gratwohl and C. Craddock eds., ESH, EBMT, pp. 68–90.

DUHRSEN, U., VILLEVAL, J.-L., BOYD, J., KANNOURAKIS, G., MORSTYN, G. and METCALF, D. (1988). Effects of recombinant human granulocyte colony-stimulating factor on hematopoietic progenitor cells following chemotherapy in man, *Blood* **72**, 2074–2081.

EMERSON, S.G. (1997). *Ex vivo* expansion of hematopoietic precursorsors, progenitors, and stem cells: The next generation of cellular therapeutics, *Blood* **87**, 3082–3087.

ENDE, M. and ENDE, N. (1972). Hematopoietic transplantation by means of fetal (cord) blood. A new method, *Virginia Med.* **99**, 276–280.

FOOD AND DRUG ADMINISTRATION (1997). Proposed approach to regulation of cellular and tissue-based products, *J. Hematother.* **6**, 195–212.

FORNAS, O., GARCIA, J. and PETRIZ, J. (2000). Flow cytometry counting of CD34+ cells in whole blood, *Nature Med.* **6**, 833–836.

FRASER, C., SZILVASSY, S., EAVES, C. and HUMPHRIES, R. (1992). Proliferation of totipotent hematopoietic stem cells cultured at limiting dilution on supportive marrow stroma, *Proc. Natl. Acad. Sci. USA* **89**, 1968–1972.

FRASER, J.K., CAIRO, M.S., WAGNER, E.L., McCURDY, P.R., BAXTER-LOWE, L.A., CARTER, S.L., KERNAN, N.A., LILL, M.C., SLONE, V., WAGNER, J.E., WALLAS, C.H. and KURTZBERG, J. (1998). Cord blood transplantation study (COBLT): Cord blood bank standard operating procedures, *J. Hematother.* **7**, 521–561.

GEE, A.P. and LAMB, L.S. Jr. (1994). Preparation of samples for flow cytometric analysis: recommendations from the ISHAGE course in San Jose. In: *Hematopoietic Stem Cells. The Mulhouse Manual,* E. Wunder, H. Sovalat, P.R. Henon and S. Sarke, eds., AlphaMed Press, Dayton, pp. 67–77.

GIANNI, A.M., SIENA, S., BERGNI, M., TARELLA, C., STERN, A.C., PILERI A. and BONADONNA, G. (1989). Granulocyte-macrophage colony-stimulating factor to harvest circulating haemopoietic stem cells for autotransplantation, *Lancet* **9**, 580–583.

GLASPY, J.A., SHPALL, E.J., LeMAISTRE, C.F., BRIDDELL, R.A., MENCHACA, D.M, TURNER, S.A., LILL, M. and CHAP, L. (1997). Peripheral blood progenitor cell mobilization using stem cell factor in combination with filgrastim in breast cancer patients, *Blood* **90**, 2939–2951.

GLUCKMAN, E., BROXMEYER, H.E., AUERBACH, A.D., FRIEDMAN, H.S., DOUGLAS, G.W., DEVERGE, A., ESPEROO, H., THIERRY, D., SOCIE, G., LEHN, P., COOPER, S., ENGLISH, D., KURTZBERG, J., BARD, J. and BOYSE, E.A. (1989). Hematopoietic reconstitution in a patient with Fanconi's anemia by means of umbilical cord blood from an HLA-identical sibling, *N. Engl. J. Med.* **321**, 1174–1178.

GLUCKMAN, E., ROCHA, V., BOYER CHAMMARD, A., LOCATELLI, F., ARCESE, W., PASQUINI, R., ORTEGA, J., SOVILLAT, G., FERREIRA, E., LAPORTE, J-P., FERNANDEZ, M. and CHASTANG, C. (1997). Outcome of cord transplantation from related and unrelated donors, *N. Engl. J. Med.* **337**, 373–381.

GOLDMAN, J.M. (1979). Autografting cryopreserved buffy coat cells for chronic granulocytic leukaemia in transformation, *Exp. Hematol.* **7**(Suppl. 5), 389–397.

GOODMAN, J.W. and HODGSON, G.S. (1962). Evidence for stem cells in the peripheral blood of mice, *Blood* **19**, 702–708.

HAKENBERG, P., KOGLER, G. and WERNET, P. (1998). NETCORD: A cord blood allocation network, *Bone Marrow Transplant.* **22**(Suppl. 1), S17–S18.

JONES, H.M., JONES, S.A., WATTS, M.J., KHWAJA, A., MILLS, W., FIELDING, A., GOLDSTONE, A.H. and LINCH, D.C. (1994). Development of a simplified single-apheresis approach for peripheral-blood progenitor-cell transplantation in previously treated patients with lymphoma, *J. Clin. Oncol.* **12**, 1693–1702.

KATZ, F., TINDLE, R.W., SUTHERLAND, D.R. and GREAVES, M.F. (1985). Identification of a membrane glycoprotein associated with hemopoietic progenitor cells, *Leuk. Res.* **9**, 191–197.

KESSINGER, A., ARMITAGE, J.O., LANDMARK, J.D., SMITH, D.M. and WEISENBURGER, D. (1988). Autologous peripheral hematopoietic stem cell transplantation restores hematopoietic function following marrow ablative therapy, *Blood* **71**, 723–727.

KESSINGER, A., ARMITAGE, J.O., LANDMARK, J.D. and WEISENBURGER, D. (1986). Reconstitution of human hematopoietic function with autologous cryopreserved circulating stem cells, *Exp. Hematol.* **14**, 192–197

KORBLING, M., DORKEN, B., HO, A.D., PEZZUTTO, A., HUNSTEIN, W. and FLIEDNER, T.M. (1986). Autologous transplantation of blood-derived hematopoietic stem cells after myeloablative therapy in a patient with Burkitt's lymphoma, *Blood* **67**, 529–535.

KORBLING, M., FLIEDNER, T.M. and PFLIEGER, H. (1980). Collection of large quantities of granulocyte macrophage progenitor cells (CFUc) in man by means of continuous-flow leukapheresis, *Scand. J. Haematol.* **24**, 22–27.

KRAUSE, D.S., FACKLER, M.J., CIVIN, C.I. and MAY, W.S. (1996). CD34: Structure, biology, and clinical utility, *Blood* **87**, 1–13.

LAUGHLIN, M.J. (2001). Umbilical cord blood for allogeneic transplantation in children and adults, *Bone Marrow Transplant.* **27**, 1–6.

LINDEN, J.V., PRETI, R.A. and DRACKER, R. (1997). New York State guidelines for cord blood banking, *J. Hematother.* **6**, 535–541.

LINK, H., SCHMITZ, N., GRATWOHL, A. and GOLDMAN, J. (1995). Standards for specialist units undertaking blood and marrow stem cell transplants - recommendations from the EBMT, *Bone Marrow Transplant.* **16**, 733–736.

LOTZOVA, E. (1977). Resistance to parental, allogeneic, and xenogeneic hemopoietic grafts in irradiated mice, *Exp. Hematol.* **5**, 215–235.

LUNDELL, B.I., MANDALAM, R.K. and SMITH A.K. (1999). Clinical scale expansion of cryopreserved small volume whole bone marrow aspirates produces sufficient cells for clinical use, *J. Hematother.* **8**, 115–128.

McCREDIE, K.B., HERSH, E.M. and FREIREICH, E.J. (1971). Cells capable of colony formation in the peripheral blood of man, *Science* **71**, 293–294.

METCALF, D., BEGLEY, C.G., WILLIAMSON, D.J., NICE, E.C., DELAMARTER, J., MERMOD, J.-J., THATCHER, D. and SCHMIDT, A. (1987). Hemopoietic responses in mice injected with purified recombinant murine GM-CSF, *Exp. Hematol.* **15**, 1–9.

MIJOVIC, A., PAGLIUCA, A. and MUFTI, G.J. (1999). The "G-CSF test": The response to a single dose of granulocyte colony-stimulating factor predicts mobilization of hemopoietic progenitors in patients with hematologic malignancies, *Exp. Hematol.* **27**, 1204–1209.

MIRAGLIA, S., GODFREY, W., YIN, A.H., ATKINS, K., WARNKE, R., HOLDEN, J.T., BRAY, R.A., WALLER, E.K. and BUCK, D.W. (1997). A novel five-transmembrane hematopoietic stem cell antigen: isolation, characterization, and molecular cloning, *Blood* **90**, 5013–5021.

MOSKOWITZ, C.H. BERTINO, J.R., GLASSMAN, J.R., HEDRICK, E.E., HUNTE, S., COADY-LYONS, N., AGUS, D.B., GOY, A., JURCIC, J., NOY, A., O'BRIEN, J., PORTLOCK, C.S., STRAUS, D.S., CHILDS, B., FRANK, R., YAHALOM, J., FILIPPA, D., LOUIE, D., NIMER, S.D. and ZELENETZ, A.D. (1999). Ifosfamide, carboplatin, and etoposide: A highly effective cytoreduction and peripheral-blood progenitor-cell mobilization regimen for transplant-eligible patients with non-Hodgkin's lymphoma, *J. Clin. Oncol.* **17**, 3776–3785.

PAHWA, R.N., FLEISCHER, A., THAN, S. and GOOD, R.A. (1994). Successful hematopoietic reconstitution with transplantation of erythrocyte-depleted allogeneic human umbilical cord blood cells in a child with leukemia, *Proc. Natl. Acad. Sci. USA* **91**, 4485–4488.

PAPAYANNOPOULOU, T. and NAKAMATO, B. (1993). Systemic treatment of primates with anti-VLA4 leads to an immediate egress of hemopietic progenitors to periphery, *Proc. Natl. Acad. Sci. USA* **90**, 9374–9378.

POLGE, C., SMITH, A.U. and PARKES, A.S. (1949). Revival of spermatozoa after vitrification and dehydration at low temperatures, *Nature (Lond.)* **164**, 666.

POPP, R.A. (1960). Erythrocyte repopulation in X-irradiated recipients of nucleated peripheral blood cells of normal mice, *Proc. Soc. Exptl. Biol.* **103**, 232–238.

PROSPER, F., STRONCEK, D., McCARTHY, J.D. and VERFAILLE, C.M. (1998). Mobilization and homing of peripheral blood progenitors is related to reversible downregulation of alpha4beta1 integrin expression and function, *J. Clin. Invest.* **101**, 2456–2467.

PRUIJT, J.F.M., vanKOOYK, Y., FIGDOR, C.G., LINDLEY, I.J.D., WILLEMZE, R. and FIBBE, W.E. (1998). Anti-LFA-1 blocking antibodies prevent mobilization of hematopoietic progenitor cells induced by interleukin-8, *Blood* **91**, 4099–4105.

REGIDOR, C., POSADA, M., MONTEAGUD, M., GARAULET, C., SOMOLINOS, N., FORES, R., BRIZ, M. and FERNANDEZ, M.-N. (1999). Umbilical cord blood for unrelated transplantation: Evaluation of cell separation and storage methods, *Exp. Hematol.* **27**, 380–385.

REIFFERS, J., BERNARD, P., DAVID, B., VEZON, G., SARRAT, A. and MARIT, G. (1986). Successful autologous transplantation with peripheral blood hemopoietic cells in a patient with acute leukemia, *Exp. Hematol.* **14**, 312–315.

RICHMAN, C.M., WEINER, R.S. and YANKEE, R.A. (1976). Increase in circulating stem cells following chemotherapy in man, *Blood* **47**, 1031–1039.

ROWLEY, S.D. (1992). Hematopoietic stem cell cryopreservation: A review of current techniques, *J. Hematother.* **1**, 233–250.

ROWLEY, S.D. (1999). Hematopoietic stem cell transplantation. In: *Hematopoietic Cell Transplantation*, E.D. Thomas, K.G. Blume and S.J. Forman, eds., Blackwell Sci., Malden, pp. 481–492.

ROWLEY, S.D., FENG, Z., YADOCK, D., HOLMBERG, L., MacLEOD, B. and HEIMFELD, S. (1999). Post-thaw removal of DMSO does not completely abrogate infusional toxicity or the need for pre-infusion histamine blockade, *Cytother.* **1**, 439–446.

RUBINSTEIN, P., DOBRILA, L., ROSENFIELD, R.E., ADAMSON, J.W., MIGLIACCO, G., MIGLIACCIO, A.R., TAYLOR, P.E. and STEVENS, C.E. (1995). Processing and cryopreservation of placental/umbilical cord blood for unrelated bone marrow reconstitution, *Proc. Natl. Acad. Sci. USA* **92**, 10 119–10 122.

SCHUMM, M., LANG, P., TAYLOR, G., KUCI, S., KLINGEBIEL, T., BUHRING, H.-J., GEISELHART, A., NIETHAMMER, D. and HANDGRETINGER, R. (1999). Isolation of highly purified autologous and allogeneic peripheral CD34+ cells using the CliniMACS device, *J. Hematother.* **8**, 209–218.

SCHWINGER, W., URBAN, Ch., LACKNER, H., KERBL, R., BENESCH, M., DORNBUSCH, H.J., SOVINZ, P., SCHAUENSTEIN, K., SCHUMM, M. and HANDGRETINGER, R. (2000). Unrelated peripheral blood stem cell transplantation with "megadoses" of purified CD34+ cells in three children with refractory severe aplastic anemia, *Bone Marrow Transplant.* **25**, 513–517.

SERKE, S. and JOHNSEN, H.E. (2001). A European reference protocol for quality assessment and clinical validation of autologous haematopoietic blood progenitor and stem cell grafts, *Bone Marrow Transplant.* **27**, 463–470.

SHIH, C.-C., DiGIUSTO, D. and FORMAN, S.J. (2000). *Ex vivo* expansion of transplantable human hematopoietic stem cells: Where do we stand in the year 2000? *J. Hematother. Stem Cell Res.* **9**, 621–628.

SIENA, S., BREGNI, M., BRANDO, B., RAVAGNANI, F., BONADONNA, G. and GIANNI, A.M. (1989). Circulation of CD34+ hematopoietic cells in the peripheral blood of high-dose cyclophosphamide treated patients: Enhancement by intravenous recombinant human granulocyte-macrophage colony-stimulating factor, *Blood* **74**, 1905–1914.

SIENA, S., BREGNI, M., DINICOLA, M., PECCATORI, F., MAGNI, M., BRANDO, B., RAVAGNANI, F. and GIANNI, A.M. (1994). Milan protocol for clinical CD34+ cell estimation in peripheral blood for autografting in patients with cancer. In: *Hematopoietic stem cells. The Mulhouse Manual,* E. Wunder, H. Sovalat, P.R. Henon and S. Sarke, eds., AlphaMed Press, Dayton, pp. 23–30.

SOCINSKI, M.A., ELIAS, A., SCHNIPPER, L., CANNISTRA, S.A., ANTMAN, K.H. and GRIFFIN, J.D. (1988). Granulocyte-macrophage colony stimulating factor expands the circulating haemopoietic progenitor cell compartment in man, *Lancet* **2**, 1194–1198.

SPITZER, G., VERMA, D.S., FISHER, R., ZANDER, A., VELLEKOOP, L., LITAM J., McCREDIE, K.B. and DICKE, K.A. (1980). The myeloid progenitor cell — Its value in prediciting hematopoietic recovery after autologous bone marrow transplantation, *Blood* **55**, 317–323.

SROUR, E.F., ABONOUR, R., CORNETTA, K. and TRAYCOFF, C.M. (1999). *Ex vivo* expansion of hematopoietic stem and progenitor cells: Are we there yet? *J. Hematother.* **8**, 93–102.

TO, L.B., HAYLOCK, D.N., SIMMONS, P.J. and JUTTNER, C.A. (1997). The biology and clinical uses of blood stem cells, *Blood* **89**, 2233–2258.

VAN LAAR, J.M. (2000). Immune ablation and stem-cell therapy in autoimmune disease: Immunological reconstitution after high dose immunosuppression and haematopoietic stem-cell transplantation, *Arthritis Res.* **2**, 270–275.

VERFAILLIE, C.M., PLOEMACHER, R., DI PERSIO, J., SUTHERLAND, R., SERKE, S., JOHNSEN, H., NOGA, S. and NEGRIN, R. (1999). Assays to determine hematopoietic stem cell content in blood or marrow grafts, *Cytother.* **1**, 41–49.

WIKTOR-JEDRZEJCZAK, W., SZCZYLIK, C., GORNAS, P., SHARKIS, S.J. and AHMED, A. (1979). Different marrow cell numbers requirements for the haematopoietic colony formation and the cure of the W/W^v anemia, *Experientia* **35**, 546–547.

WOLL, P.J., THATCHER, N., LOMAX, L., HODGETTS, J., LEE, S.M., BURT, P.A., STOUT, R., SIMMS, T., DAVIES, R. and PETTENGELL, R. (2001). Use of hematopoietic progenitors in whole blood to support dose-dense chemotherapy: A randomized phase II trial in small-cell lung cancer patients, *J. Clin. Oncol.* **19**, 712–719.

WONG, A., YUEN, P.M.P., LI, K., YU, A.L.M. and TSOI, W.C. (2001). Cord blood collection before and after placental delivery: Levels of nucleated cells, haematopoietic progenitor cells, leukocyte subpopulations and macroscopic clots, *Bone Marrow Transplant.* **27**, 133–138.

YIN, A.H., MIRAGLIA, S., ZANJANI, E.D., ALMEIDA-PORADA, G., OGAWA, M., LEARY, A.G., OLWEUS, J., KEARNEY, J. and BUCK, D.W. (1997). AC133, a novel marker for human hematopietic stem and progenitor cells, *Blood* **90**, 5002–5012.

YOUNG, J.C., WU, S., TRAVIS, M., LUENS, K.M., OSBORNE, L., SCOLLAY, R. and HILL, B. (1999). Quantification of the proliferative potential of highly enriched human primitive hematopoietic progenitor cells using a stroma-free limiting dilution assay with automated sorting, *Cytother.* **1**, 147–156.

ZAUCHA, J.M., KNOPINSKA-POSLUSZNY, W., BIENIASZEWSKA, M., MYSLIWSKI, A. and HELLMAN, A. (2000). The effect of short G-CSF administration on the numbers and clonogenic efficiency of hematopoietic progenitor cells in bone marrow and peripheral blood of normal donors, *Ann. Transplant.* **5**, 20–26.

Advances in Tissue Banking Vol. 6

14 PRE-CLINICAL CHARACTERISATION OF FRESHLY ISOLATED AND CRYOPRESERVED HUMAN HEPATOCYTES

ALINA OSTROWSKA, XUN HU and DONALD C. BODE

Tissue Transformation Technologies
Edison, New Jersey 08837, USA

1. Introduction

There is a pressing need for alternative therapeutic methods effective in the management of patients with inherited and acquired liver insufficiency. Improvements in hepatocyte isolation have generated enthusiasm for allogeneic hepatocyte transplantation as a potential treatment to provide temporary liver function in such patients. In addition, pilot studies have demonstrated that isolated human hepatocytes can be cryopreserved and banked to be available for future clinical and commercial use on demand (Adams *et al.*, 1995; Hengstler *et al.*, 2000a). Data from reported experiments indicate that properly prepared human hepatocytes retain their liver-specific characteristics and may be successfully used in such cell-based therapies (Strom *et al.*, 1997). However, the successful use of hepatocytes in clinical trials depends on reliable demonstration of functional and morphological integrity of freshly isolated and cryopreserved cells. Pre-clinical evaluation of hepatocytes involves microscopic examination and precise *in vitro* analysis of cell viability and multiple metabolic activities, which identifies the capacity of these cells to remain functional *in vivo* after transplantation. This article is intended as an overview of the state of the art in human hepatocyte isolation, cryopreservation and characterisation from a pre-clinical perspective.

2. Criteria for Acceptance of Human Liver Tissue for Hepatocyte Isolation

In many cases, viable human liver tissue that could be a valuable resource for biomedical research is, instead, discarded. The purpose of our hepatocyte procurement program is to recover that material, and distribute it in forms that are useful for either clinical (cell transplantation, artificial liver device) or research (biotransformation and toxicity of drugs, the study of liver diseases) applications. We are currently processing transplant-grade hepatocytes from non-diseased human livers that are recovered for transplantation but cannot be used for that purpose. All livers are collected from brain-dead donors with informed consent from the next-of-kin and obtained through a network consisting of organ procurement organisations (OPOs). At the time of referral, all specimens are deemed unsuitable for transplantation due to donor age, high fat content, anatomical anomalies or damage, biopsy results, liver enzyme levels, etc. The donated livers are protected from ischaemia injury by flushing with ice-cold University of Wisconsin solution immediately after vascular clamping. The interval from initial flush to hepatocyte isolation is never more than 30 hours and is often much less. Accurate assessment of histopathological changes is always required before accepting a liver for hepatocyte isolation. Establishment of the fat content (level of micro- or macrovesicular steatosis) and the stage of development of the most common abnormalities (fibrosis, cirrhosis) is critical for further determination of liver perfusion and enzymatic digestion parameters. Generally, the fat content must be lower than 50% and there should be no signs of severe perivenular fibrosis, necrosis or fully established cirrhosis. Although liver microbiopsies provide valuable insights, a complete biochemical evaluation of the donor enzyme profile is also required to complement the biopsy findings.

Liver enzyme values, particularly transaminases (SGOT, SGPT) must be within normal limits or nearly so.

Quality controls in the liver procurement process include serology test results, which must be consistent with FDA (Food and Drug Administration), AATB (American Association of Tissue Banks) and/or MTF (Musculoskeletal Transplant Foundation) standards. Donors of all livers accepted for hepatocyte isolation are negative for anti-HIV 1&2, hepatitis B surface antigen, hepatitis C antibody and syphilis. Hepatitis B core antibody, if tested, must be negative. In the case of liver donors for hepatocyte transplantation, a negative HIV PCR test is also required. An additional FDA requirement is to calculate the plasma dilution factor in order to rule out the possibility of false-negative test results in cases where the liver donor received a significant volume of intravenous fluids due to blood loss. However, the use of a plasma dilution algorithm to determine tissue and organ donor suitability is a regulatory requirement with little scientific basis (Eastlund, 2000).

3. Isolation, Cryopreservation and Assessment of Hepatocyte Viability and Morphology

The development of human hepatocyte transplantation has been hampered by the limited availability of adequate numbers of fresh, viable cells due to the ongoing shortage of liver donors. Each year there is an increase in the number of individuals in need of a liver transplant. Data indicate, however, that the number of organs donated during the last five years has reached a plateau (Wicks, 2000). Thus, there is no surplus of healthy organs from which freshly isolated hepatocytes can be prepared when needed. In most cases, the yield of primary hepatocytes from a single lobe of a fresh human liver is greater than five billion cells, sufficient for cell transplantation and well in excess of the capacity of large-scale *in vitro* assays. A practical solution to this problem is cryopreservation and banking of isolated hepatocytes. Thus, there is a continuing need for further improvement in the methods of long-term preservation and recovery of human hepatocytes to maximise the efficient use and minimise the waste of this valuable resource.

Pre-clinical evaluation of human hepatocytes requires an isolation technique that provides a sufficient number of cells with a high degree of purity and viability. The successful isolation of highly viable human hepatocytes remains a significant challenge for many laboratories. Our procedures have been developed over the past seven years, based on hundreds of isolations. Our laboratory has been routinely isolating, cryopreserving and banking human hepatocytes that can be used for clinical cell transplantation (Bilir *et al.*, 1995). Each year, approximately 50 human livers that could not be used for whole organ transplantation are processed. A novel method has been developed and enables the large-scale isolation of human hepatocytes with high yield, viability and functional integrity.

Hepatocytes are isolated from an encapsulated liver sample (preferably the left lateral segment) by a modified two-step perfusion technique introduced by Seglen (1976). The dissected encapsulated liver fragment (150–300 g) with one cut surface is placed in the custom-made perfusion apparatus and three to five hepatic vessels are cannulated with tubing attached to a multi-channel manifold. The sample is perfused initially (recirculation technique) with calcium-free Hanks' balanced salt solution (HBSS) supplemented with 0.5 mM EGTA (ethylene glycol-bis(β-aminoethyl ether)-N,N,N',N'-tetraacetic acid) for 30 minutes and then with approximately 0.05% collagenase dissolved in Leibowitz's L-15 medium at 37°C until the tissue is fully digested. The digested liver is removed, immediately cooled with ice-cold L-15 medium and the cell suspension is strained serially through progressively smaller stainless steel sieves, with a final filtration through 100-μm and 60-μm nylon mesh.

Special attention must be given to the quality and concentration of collagenase used for hepatocyte isolation. There appears to be a significant relationship between the quality and yield of hepatocytes and the type of commercially available collagenase (Alpini *et al.*, 1994; Strom *et al.*, 1997; Berry and Phillips, 2000).

However, due to the large degree of inter-individual variability, it is difficult to standardise the digestion conditions through the use of human liver tissue. It is recommended that the activity of a given collagenase preparation be tested via the isolation of hepatocytes from rat liver. Thus far, we have obtained the best results using collagenase type I (Worthington), Liver Digestion Medium (Gibco BRL), and collagenase P (Roche). Our recent studies indicate that the enzyme blend Liberase (Roche), commonly used for isolation of pancreatic islets, generally does not enhance hepatocyte yield and viability. In many cases, Liberase causes rapid and aggressive digestion of human liver tissue, which increases the risk of overdigestion and deterioration of cell integrity.

Initial, crude preparations of freshly isolated hepatocytes are contaminated by dead and damaged cells as well as by non-parenchymal cells, including Kupffer cells, endothelial cells, and lymphocytes. The focus of standard hepatocyte isolation methods is the elimination of all of these potential contaminants. The rationale for eliminating dead or damaged hepatocytes is self-evident, whereas for some applications the retention of viable nonparenchymal cells may be desirable. Modifications of animal hepatocyte preparations, such as depletion of antigen-presenting nonparenchymal cells may result in ablation of the immune response, which is important for reducing *in vivo* alloreactivity have been reported (Karrer *et al.*, 1997; Ostrowska *et al.*, 1999). For this reason, all cell preparations (by centrifugation in Percoll) are further purified when the initial viability is lower than 80%. Assessment of cell purity requires morphological and/or immunohistochemical techniques. In our experiments, the quantity and viability of freshly isolated (or cryopreserved/thawed) hepatocytes are estimated by trypan blue exclusion, according to standard procedures described for isolated hepatocytes (Berry *et al.*, 1991). To better qualify the viability of the prepared cells, we use another vital stain for hepatocytes, 5,6-carboxyfluorescein diacetate succinimidyl ester (CFSE). Staining with CFSE correlates with hepatocyte function and permits the visualisation of the cells by fluorescence microscopy (Soriano *et al.*, 1992; Ostrowska *et al.*, 1999).

It was observed that trypan blue exclusion tests and CFSE labelling were also inversely correlated with LDH (lactate dehydrogenase) measurements. LDH is a large enzyme, located almost entirely in the cytoplasm, and its leakage is a good marker of cell damage in the suspension (Berry *et al.*, 1991).

Another useful means of evaluating the purity of hepatocyte preparations is staining with an antibody to CD45 (leukocyte common antigen), which is expressed on the surface of T and B lymphocytes, mononuclear phagocytes and polymorphonuclear leukocytes, but not on hepatocytes. By using this simple method, it is demonstrated that Percoll purification significantly reduces the contamination of hepatocyte preparations by CD45+ nonparenchymal cells. Any hepatocyte preparation characteristics that are detectable with a fluorescent label can be evaluated by fluorescence-activated cell sorting (FACS; Afford *et al.*, 1999).

Although viable human hepatocytes can be successfully isolated in large yields, they have very low growth capacity that does not permit subcultivating. Thus, cryopreserved hepatocytes have become very attractive as a model system for both commercial and clinical use. However, there are no standard, readily available protocols for cryopreservation of human hepatocytes. Only a few studies have dealt with human hepatocytes after cryopreservation and variable results were obtained. But for the most part, it has been well documented that cryopreservation maintains the morphological and functional integrity of isolated hepatocytes *in vitro* and that prolonged storage does not lead to loss of cell viability and function (Chesne *et al.*, 1993). Various factors such as composition of freezing mixture, concentration of cryoprotectant, cooling rate and conditions of cell thawing can significantly influence the recovery of cryopreserved hepatocytes. The consensus is that L-15 medium or University of Wisconsin solution supplemented with 10% DMSO (dimethyl sulfoxide), and 20% FBS (fetal bovine serum), together with controlled-rate freezing (–3°C/min) provided maximal hepatocyte viability after thawing. Most often, the human hepatocyte suspensions have been cryopreserved in 1- to 5-ml vials ml for *in vitro* testing. However,

Table 1. Viability of freshly isolated and cryopreserved/thawed hepa-tocytes as determined by trypan blue exclusion and CFSE labelling.

	Hepatocyte preparations			
Liver #	Fresh Trypan blue	CFSE	Cryopreserved Trypan blue	CFSE
1	91% ± 1.2	93%	72% ± 2.3*	75%
2	89% ± 2.3	91%	74% ± 2.3*	78%
3	94% ± 1.7	95%	85% ± 2.9	82%
4	74% ± 1.2	76%	68% ± 2.3	71%
5	74% ± 1.6	78%	66% ± 2.3*	63%
6	77% ± 1.2	81%	48% ± 1.2*	53%
7	86% ± 1.2	84%	66% ± 0.6*	69%

*$p < 0.05$ vs. fresh hepatocytes

in hepatocyte preparations for transplantation, blood cryogenic bags of larger capacity (100 to 150 ml) are used. Cryopreserved hepatocytes are maintained in the vapor phase above liquid nitrogen (approximately –175°C) for months or years until needed. For cell thawing, the bags or vials are removed from liquid nitrogen and thawed rapidly in a 37°C water bath.

It is essential to compare the morphological and functional parameters of cryopreserved/thawed hepatocytes with those determined for freshly isolated cells. Although the viability of fresh cell preparations (based on trypan blue exclusion and CFSE staining) in our studies is generally high, it always decreases somewhat following cryopreservation (Table 1). Electron microscopic examination reveals that the appearance of cryopreserved hepatocytes is similar to that of normal cells (not shown). However, moderate signs of cellular derangement can be found in some cells, including slightly swollen mitochondria, large lipid droplets, and development of vesicular forms in endoplasmic reticulum. In primary cultures, the plating efficiency and survival of isolated hepatocytes depends on the experimental conditions such as extracellular matrix.

Plating efficiency is a good marker of the overall integrity of the isolated cells. Freshly prepared hepatocytes attach to collagen-coated dishes within 5 hours after seeding, and form confluent cultures by twelve hours, at which point the cells are flattened and form a monolayer across the dish. In our studies, the plating efficiency of cells seeded immediately after isolation averages from 75% to 98%. In contrast, at least eight hours are required for initial attachment of cryopreserved/thawed hepatocytes and the final plating efficiency is significantly lower as well (22% to 72%). There is a dramatic decline in hepatocyte plating efficiency in most, but not all, cryopreserved/thawed preparations, even in those cases where cells showed high viability in suspension and acceptable levels of integrity. Chesne *et al.* (1993) and Adams *et al.* (1995), on the other hand, have reported that the plating efficiency of cryopreserved/thawed hepatocytes is similar to that observed using fresh hepatocytes. Another study reported that cryopreserved hepatocyte cultures are in a state allowing practical application, but should be further optimised (Hengstler *et al.*, 2000b).

A variety of tests are available for the detection of hepatocellular damage in culture. In our laboratory, the functional integrity of fresh and cryopreserved hepatocytes is routinely evaluated by the MTT assay. The product of the 3-(4,5-dimethylthiazol-2-yl)-2,5-diphenyl tetrazolium bromide (MTT) reaction is always expressed more intensely in cultures of freshly isolated hepatocytes, and is generally correlated with plating efficiency.

4. Assessment of Hepatocyte Function

Confidence in the clinical use of human hepatocytes would be increased by demonstrating that both freshly prepared and cryopreserved cells retain various liver-specific functions. Our laboratory has established a variety of characterisation assays to evaluate the functional integrity of isolated human hepatocytes. To determine whether these cells maintain their differentiated phenotype, several specific properties including energy status, secretion of urea, and the activities of major phase I (cytochrome

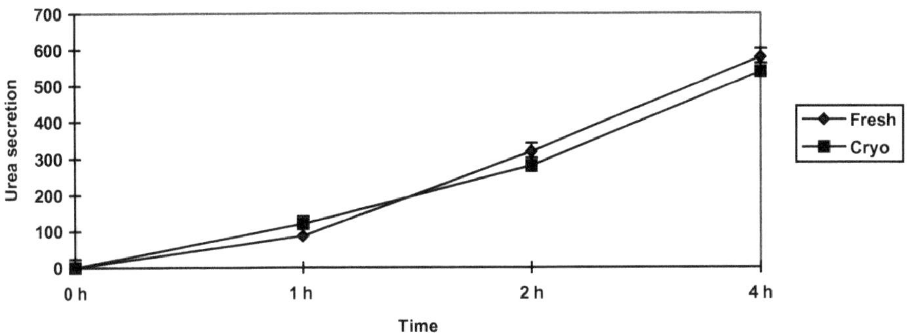

Fig. 1. Average basal rates of urea synthesis for fresh and cryopreserved/thawed hepatocytes from all processed livers (nmol/10^6 cells/4 hours).

P450 [CYP] 1A1/2, CYP2A6, CYP2C9, CYP2C19, CYP2D6, CYP2E1, CYP3A4) and phase II (uridine diphosphate glucuronyl transferase [UGT], sulfotransferase [ST], glutathione S-transferase [GST]) drug-metabolising enzymes are examined.

To measure ureagenesis, one million cells are suspended in serum-free culture medium and incubated with 5 mmol/L ammonium chloride for 0, 1, 2, and 4 h. After incubation, the cells are centrifuged and the urea concentration in the medium is measured using a spectrophotometric urea nitrogen test kit. The process of ureagenesis is found to be well maintained in fresh and cryopreserved hepatocyte preparations, consistent with other reports (Pang *et al.*, 1997, Matsuda *et al.*, 1999). Average basal rates of urea synthesis for fresh and cryopreserved/thawed hepatocyte preparations are presented graphically in Fig. 1.

Energy status of hepatocytes is evaluated by measurement of intracellular ATP content. Briefly, firefly luciferin/luciferase is added to 1×10^5 cells and the light output is measured using a luminometer. Based on our results, we conclude that when cryopreserved hepatocytes are thawed, purified and cultured briefly under optimised conditions, there is only a slight decrease in intracellular ATP content. Figure 2 shows data from eight human hepatocyte preparations.

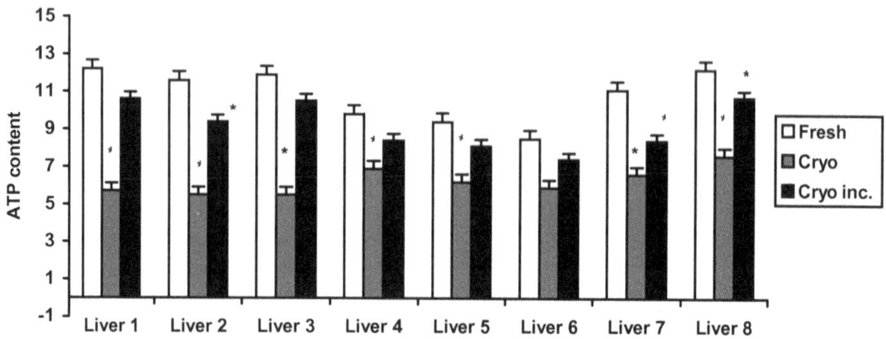

Fig. 2. Energy status of freshly isolated hepatocytes and cryopreserved hepatocytes (immediately after thawing and after incubation for 30 min at 37°C) as evaluated by measurement of intracellular ATP content (nmol/10^6 cells): $*p < 0.05$ vs. fresh hepatocytes.

To determine CYP-dependent metabolic function, 0.2×10^6 cells are mixed with a specific probe of desired concentration (Table 2) and incubated at 37°C for one hour. Metabolites are extracted by 0.2 ml methanol. The cell debris and protein are removed by high-speed centrifugation and metabolites in supernatant are identified by reversed-phase HPLC analysis. Although drug metabolism capacity varies from one liver to another, cryopreserved hepatocytes retain most of the metabolising activity expressed by freshly isolated cells from the same preparation. Table 3 shows values representing enzyme activities (ethoxycoumarin O-deethylase [ECOD]), which represents several individual CYP forms; and CYP3A4) in both preparations of cells incubated with 25 μM of the probe substrates. The small difference in mean activities between fresh and cryopreserved/thawed preparations is not statistically significant. The large standard deviations are expected and reflect the fact that the expression and activities of these liver enzymes vary dramatically between individuals. Similar results have been reported by several authors (e.g. Pearce *et al.*, 1996; Li *et al.*, 1999; Hengstler *et al.*, 2000a). Since enzyme activities in human hepatocytes are expressed at very different levels, it is believed that a more critical comparison, rather than routine statistical

Table 2. Summary of assays applied for enzyme activities.

Enzyme	Probe	Metabolite	Monitoring wavelength (nm)
CYP2A6	Coumarin	7-Hydroxycoumarin	324
CYP3A4	Testosterone	6b-Hydroxytestosterone	254
CYP2C9	Tolbutamide	4-Hydroxytolbutamide	240
CYP2C19	Mephenytoin	4-Hydroxymephenytoin	240
CYP2D6	Dextromethorphan	Levorphanol	280
CYP2E1	Chlorzoxazone	6-Hydroxychlorzoxazone	287
ECOD (several CYPs)	7-Ethoxycoumarin	7-Hydroxycoumarin	324
CYP1A	7-Ethoxyresorufin	Resorufin	535
UGT	Umbelliferone glucuronide	Umbelliferone	324
ST	Umbelliferone	Umbelliferone sulfate	450
GST	1-Chloro, 2,4-dinitrobenzene (CDNB)	CDNB conjugate of glutathione	340

Table 3. Enzymatic activities of freshly isolated and cryopreserved/thawed human hepatocytes.

	n	Enzyme activities (nmole/hour/million cells)	
		CYP3A4	ECOD
Freshly isolated	19	9.27 ± 6.59	4.40 ± 2.67
Cryopreserved/thawed	22	8.24 ± 6.72	4.00 ± 2.73

calculation, should be performed using the same batch of cells before and after cryopreservation. As illustrated in Fig. 3 for six representative preparations, in almost all cases cryopreservation does not substantially alter the activities of major cytochrome P450 isozymes. Likewise, for the most part cryopreservation has little effect on the major phase II enzymes, as shown in Fig. 4. Thus, there is a good correlation between fresh and cryopreserved cells in terms of the overall activities of the major xenobiotic transformation enzymes, as reported by others (Ruegg *et al.*, 1997).

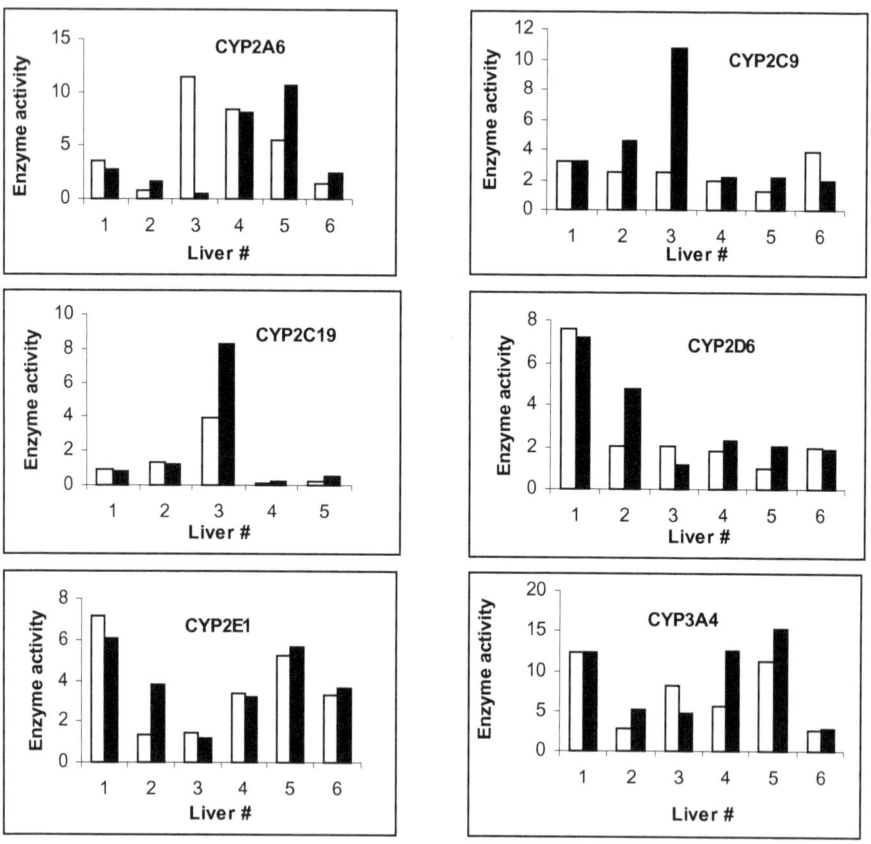

Fig. 3. Comparison of the major cytochrome P450 activities of freshly isolated and cryopreserved/thawed hepatocytes (nmol/hour/million cells). Open bar, fresh cells; solid bar, cryopreserved/thawed cells.

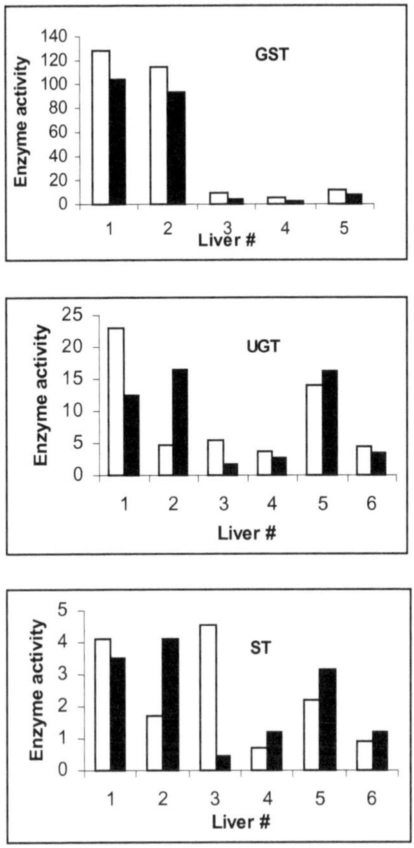

Fig. 4. Comparison of UGT, ST, and GST activities of freshly isolated and cryopreserved/thawed hepatocytes (nmol/hour/million cells). Open bar, fresh cells; solid bar, cryopreserved/thawed cells.

5. Conclusions

Several attempts have been made to use isolated human hepatocytes in clinical cell transplantation and artificial liver support to treat a variety of conditions, including liver-based inborn errors of metabolism, chronically diseased liver, and acute liver failure. Some recent clinical trials have shown that both freshly prepared and cryopreserved hepatocytes may be useful for this kind of hepatic support (Strom *et al.*, 1997; Fox *et al.*, 1998, Bilir *et al.*,

2000; Fisher *et al.*, 2000). Both cell preparations maintain typical and intact morphology at the light and electron microscopic levels. In addition, our characterisations of freshly isolated and cryopreserved hepatocytes show that after cryopreservation they maintain energy status and various liver-specific functions such as urea synthesis and drug metabolising enzyme activities. It is demonstrated that billions of viable human hepatocytes can be isolated and cryopreserved on a routine basis. However, it is clear that further refinements in the methods of hepatocyte isolation, cryopreservation and banking are needed to utilise more fully these valuable cells in the clinic.

6. References

ADAMS, R.M., WANG, M., CRANE, A.M., BROWN, B., DARLINGTON, G.J. and LEDLEY, F.D. (1995). Effective cryopreservation and long-term storage of primary human hepatocytes with recovery of viability, differentiation, and replicative potential, *Cell Transplant.* **4**, 579–586.

AFFORD, S.C., RANDHAWA, S., ELIOPOULOS, A.G., HUBSCHER, S.G., YOUNG, L.S. and ADAMS, D.H. (1999). CD40 activation induces apoptosis in cultured human hepatocytes via induction of cell surface Fas ligand expression and amplifies Fas-mediated hepatocyte death during allograft rejection, *J. Exp. Med.* **189**, 441–446.

ALPINI, G., PHILLIPS, J.O., VROMAN, B. and LARUSSO, N.F. (1994). Recent advances in the isolation of liver cells, *Hepatology* **20**, 494–514.

BERRY, M.N. EDWARDS, A.M., BARRITT, G.J. (1991). Isolated hepatocytes: Preparation, properties and applications. In: *Lab. Tech. Biochem. Mol. Biol.*, Elsevier, Amsterdam, Vol. 21, pp. 1–460.

BERRY, M.N. and PHILLIPS, J.W. (2000). The isolated hepatocyte preparation: 30 years on, *Biochem. Soc. Trans.* **28**, 131–135.

BILIR, B., OSTROWSKA, A., HANKS, T. and KARRER, F.M. (1995). Controlled rate freezing of hepatocytes: A step closer to human hepatocyte banking, *Hepatology* **22**(Pt. 2), 211A.

BILIR, B.M., GUINETTE, D., KARRER, F., KUMPE, D.A., KRYSL, J., STEPHENS, J., McGAVRAN, L., OSTROWSKA, A. and DURHAM, J. (2000). Hepatocyte transplantation in acute liver failure, *Liver Transpl.* **6**, 32–40.

CHESNE, C., GUYOMARD, C., FAUTREL, A., POULLAIN, M.G. FREMOND, B., DE JONG, H. and GUILLOUZO, A. (1993). Viability and function in primary culture of adult hepatocytes from various animal species and human beings after cryopreservation, *Hepatology* **18**, 406–414.

EASTLUND, T. *et al.* (2000). Hemodilution due to blood loss and transfusion and reliability of cadaver tissue donor infectious diease testing, *Cell. Tissue Bank* **1**, 121–127.

FOX, I.J., CHOWDHURY, J.R., KAUFMANN, S.S., GOERTZEN, T.C., CHOWDHURY, N.R., WARKENTIN, P.I., DORKO, K., SAUTER, B.V. and STROM, S.C. (1998). Treatment of the Criggler-Najjar syndrome type I with hepatocyte transplantation, *N. Engl. J. Med.* **14**, 1422–1426.

FISHER, R.A., BU, D., THOMPSON, M., TISNADO, J., PRASAD, U., STERLING, R., POSNER, M. and STROM, S. (2000). Defining hepatocellular chimerism in a liver failure patient bridged with hepatocyte infusion, *Transplantation* **69**, 303–307.

HENGSTLER, J.G., UTESCH, D., STEINBERG, P., PLATT K.L., DIENER, B., RINGEL, M., SWALES, N., FISCHER, T., BIEFANG, K., GERL, M., BÖTTGER, T. and OESCH, F. (2000). Cryopreserved primary hepatocytes as a constantly available *in vitro* model for the evaluation of human and animal drug metabolism and enzyme induction, *Drug Metab. Disp.* **4**, 81–118.

HENGSTLER, J.G., RINGEL, M., BIEFANG, K., HAMMEL, S., MILBERT, U., GERL, M., KLEBACH, M., DIENER, B., PLATT,

K.L., BÖTTGER, T., STEINBERG, P. and OESCH, F. (2000). Cultures with cryopreserved hepatocytes: Applicability for studies of enzyme induction, *Chem. Biol. Int.* **125**, 51–73.

KARRER, F.M., BILIR, B., GILL, R., OSTROWSKA, A., STEGALL, M., WACHS, M. and KAM, I. (1997). Cryopreserved hepatocytes are poor stimulators of *in vitro* cytotoxicity, *Transplant. Proc.* **29**, 1111–1112.

LI, A.P., CHUANG, L., BRENT, J.A., PHAM, C., FACHETT, A., RUEGG, C.E. and SILBER, P.M. (1999). Cryopreserved human hepatocytes: Characterisation of drug-meatabolizing enzyme activities and applications in higher throughput screening assays for hepatotoxixity, metabolic stability, and drug-drug interactions potential, *Chem. Biol. Int.* **121**, 17–35.

MATSUDA, H., YAGI, T., MATSUOKA, J., YAMAMURA, H. and TANAKA, N. (1999). Subzero nonfreeszing storage of isolated rat hepatocytes in University of Wisconsin solution, *Transplantation* **67**, 186–191.

OSTROWSKA, A., KARRER, F.M. and BILIR, B. (1999). Histological identification of purified and cryopreserved allogeneic hepatocytes following transplantation in a murine model without host immunosuppression, *Transplant Int.* **12**, 188–194.

PANG, J.M., ZALESKI, J. and KAUFFMAN, F.C. (1997). Toxicity of allyl alcohol in primary cultures of freshly isolated and cryopreserved allogeneic hepatocytes maintained on hydrated collagen gels, *Toxicol. Appl. Pharmacol.* **142**, 87–94.

PEARCE, R.E., McINTYRE, C.J., MADAN, A., SANZGIRI, U., DRAPER, A.J., BULLOCK, P.L., COOK, D.C., BURTON, L., NEVINS, C. and PARKINSON, A. (1996). Effects of freezing, thawing, and storing human liver microsomes on cytochrome P450 activity, *Arch. Biochem. Biophys.* **15**, 145–169.

RUEGG, C.E., SILBER, P.M., MUGHAL, R.A., ISMAIL, J., LU, C., BODE, D.C. and LI, A.P. (1997). Cytochrome-P450 induction and conjugated metabolism in primary human hepatocytes after cryopreservation, *In Vitro Toxicol.* **10**, 217–222.

SEGLEN, P.O. (1976). Preparation of isolated rat liver cells, *Method Cell Biol.* **13**, 29–83.

SORIANO, H.E., LEWIS, D., LEGNER, M., BRANDT, M., BALEY, P., DARLINGTON, G., FINEGOLD, M. and LEDLEY, F.D. (1992). The use of Dil-marked hepatocytes to demonstrate orthotopic, intrahepatic engraftment following hepatocellular transplantation, *Transplantation* **54**, 717–723.

STROM, S.C., FISHER, R.A., THOMPSON, M.T., SANYAL, A.J., COLE, P.E., HAM, J.M. and POSNER, M.P. (1997). Hepatocyte transplantation as a bridge to orthotopic liver transplantation in terminal liver failure, *Transplantation* **63**, 559–569.

WICKS, M.N. (2000). Biotechnological solutions to the organ crisis. In: *Medscape Transplantation*, Medscape, Inc.

Advances in Tissue Banking Vol. 6

15 PURIFICATION AND CHARACTERISATION OF HAEMATOPOIETIC STEM CELLS

ANDRZEJ LANGE, DOROTA DLUBEK, TOMASZ PACUSZKO, JAROSLAW DYBKO, BARBARA WYSOCZANSKA, ANNA LABA and JOLANTA BOCHENSKA

BMT Unit, K. Dluski Hospital/Institute of Immunology and Experimental Therapy, Wroclaw, Poland

1. Introduction

Haematopoietic stem cell transplantation is a curative procedure in some haematological malignancies. Allogeneic transplantation supports haematological recovery after high dose chemo/radiotherapy and in addition serves as an immunotherapeutic tool. The latter activity is exerted by transplanted lymphocytes, which recognise neoplastic cells in the course of alloreactivity. This is known as graft versus leukaemia (GvL) effect (Barret, 1997; Porter and Antin, 1997; Shlomchik and Emerson, 1996). Alloreactivity, however, can cause a life threatening complication clinically seen as acute graft versus host disease (aGvHD) (Klingebiel and Schlegel, 1998; Hill *et al.*, 1997). Both GvL and aGvHD originate from a lymphocyte pool. There are suggestive data showing that both reactivities can involve different subset of cells (Murphy and Longo, 1997; Jiang *et al.*, 1997; Fowler *et al.*, 1996). Acute GvHD is associated with toxicity of conditioning regimen, which enhances local pro-inflammatory cytokine production and with reactivity directed against virus infected cells (Ferrara, 2000; Holler *et al.*, 1997). GvL is exerted by T lymphocytes but likely expanding specifically against neoplastic cells antigens still persisting after myeloablative conditioning regimen as minimal residual disease. Both reactivities can be abolished by removal of T cells (Fehse *et al.*, 2000; Mavroudis *et al.*, 1998). Traditionally bone marrow (BM) is used as a source of transplant material and a routine prevention of aGvHD is based on Cyclosporine-A. Patients receiving a graft from unrelated or family alternative donors have a higher risk of aGvHD (Mickelson *et al.*, 2000). In the presented in this paper approach we describe the use of peripheral blood stem cells (PBSC) instead of BM and the use of counterflow centrifugal elutriation (CCE) to remove an excess T cell in G-CSF stimulated leukapheresis product obtained from healthy donors (G-CSF LP) for alternative transplantation.

2. Methods

2.1. Leukapheresis

Seventy-six G-CSF LPs obtained from 41 matched family donors and from 8 matched unrelated donors was used for 50 allotransplantation in 49 patients (12 children and 47 adults). Donors received G-CSF (Neupogen, filgrastim, Roche, Basel, Switzerland) usually at a dose of 5 μg/kg b.w. twice daily for 5 days prior to leukapheresis. Here 68 leukapheresis were performed with the use of continuous flow separator COBE Spectra (COBE BCT, Inc., Blood Component Technology, Lakewood, CO) and 8 G-CSF LPs for 7 patients was sent from foreign places. Volume of processed blood depended on the recipient demand and weight of the donor and varied from 2.3 L to 12.5 L (Mean ± SEM: 8.1 L ± 0.4) of blood. Donors for unrelated recipients underwent leukapheresis in our on in co-operative international units. G-CSF administration and blood volume was similar to that used in leukapheresis performed in family donor.

BM obtained by iliac crest aspiration under a general anaesthesia was used for 56 allotransplantation. All G-CSF LPs and BM were analysed for the number of nucleated cells with the use of CELL-DYN 1600 Blood Cell Analyser.

2.2. Flow cytometry

Profile of cells in a whole population of G-CSF LPs and in harvested BM was analysed on FACStar^{PLUS} or FACS Calibur flow cytometer (Becton Dickinson, Mountain View, CA) with an argon-ion laser tuned at 488 nm. Data acquisition and analysis were performed using vendorsupplied software. Cells were prepared for immunophenotyping as previously described (Dlubek *et al.*, 1999). For description of haematopoietic progenitors double-colour staining with anti-CD34 and anti-HLADR specific antibodies were used. Additionally all G-CSF LPs were analysed for T cells (CD3), monocytes (CD14) and granulocytes (CD15) profile. The results were compared with those describing the BM.

2.3. Counterflow centrifugal elutriation (CCE)

G-CSF LPs obtained from 9 donors, using for 10 alternative allotransplantation was subfractionated by CCE with the use of J6-MC centrifuge equipped with a JE-5.0 rotor and single large (40 ml) chamber (Beckman Instruments, Palo Alto, CA). Masterflex peristaltic pump (Cole-Palmer Instruments, Chicago, IL) provided continuous flow of elutriation media. Elutriation medium contains physiological saline (0.9%, Polfa, Kutno, Poland) supplemented with 1% D-glucose (Pliva, Kraków, Poland), disodium ethylenediaminetetraacetic acid (EDTA 0.3mM) and 0.5% human serum albumin (HSA, ZLB, Berno, Switzerland). pH was adjusted to 7.2 and osmolarity was 285 ± 5 mOsm. Steady state of the elutriation system was achieved with a rotor speed of 3000 rpm (kept constant throughout elutriation) and 70 ml/min of medium flow at 20°C. Then the elutriation chamber was loaded with 1.0×10^{10} cells. The loaded cells initiate elutriation and 700 ml of first fraction was collected at flow rate 70 ml/min to exhaustion platelets, red blood cells and very small lymphocytes. Then following fractions were collected by increasing step-wise flow rate (110, 140 and 170 ml/min) and with the same volume of elutriation medium (700 ml). At each new flow rate cells of incrementally larger density and size were eluted. The last fraction was obtained by elution of all remaining cells when rotor was off (R/O).

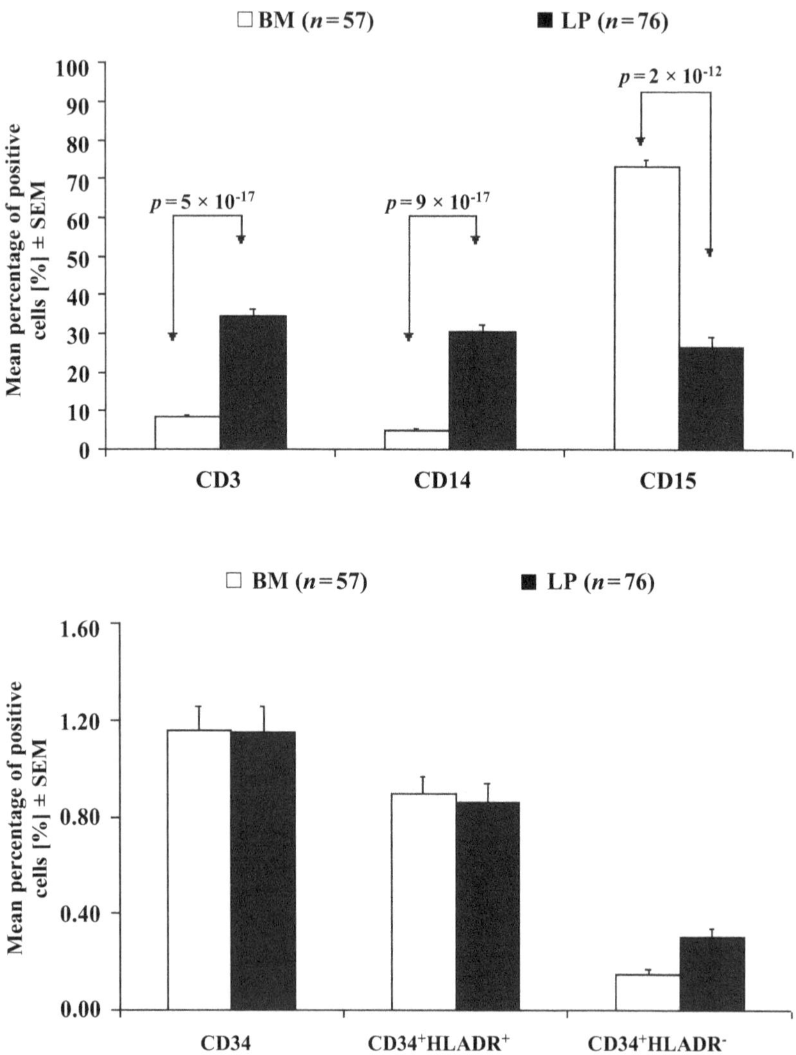

Fig. 1. Percentage (mean ± SEM) of cells in BM (white columns, $n = 57$) and in G-CSF mobilised LP (dark columns, $n = 76$) from healthy donors.

3. Results

Percentage of haematopoietic progenitor cells ($CD34^+$) was similar in G-CSF LPs and BM inoculi (Mean ± SEM: 1.15% ± 0.11 vs. 1.16% ± 0.10, Fig. 1) but G-CSF LPs $CD34^+$ cells were more frequently $HLADR^-$ than those in BM (Fig. 2). G-CSF LPs had significantly higher proportion of $CD3^+$ cells (34.6% ± 1.6 vs. 8.2% ± 0.7, $p = 5 \times 10^{-17}$) and $CD14^+$ cells (30.5% ± 1.6 vs. 4.9% ± 0.3, $p = 9 \times 10^{-17}$) but lower of $CD15^+$ cells (26.7% ± 2.6 vs. 73.2% ± 2.0, $p = 2 \times 10^{-12}$) than BM inoculi (Fig. 1).

A number of nucleated cells/kg b.w. of recipient was higher in G-CSF LPs than in BM recipients ($7.71 \times 10^8 \pm 0.85$ vs. $3.07 \times 10^8 \pm 0.23$, $p = 1 \times 10^{-7}$, Fig. 3). This was also valid for all cell subsets except of granulocytes (Fig. 3).

The leukocyte separation was performed with the use of CCE method in 9 patients (Table 1). Patient UPN 229 was transplanted two times due to the first graft rejection. For children recipients one or two CCE runs of G-CSF LP cells from single leukapheresis were used for transplantation. The target number of $CD34^+$ cells

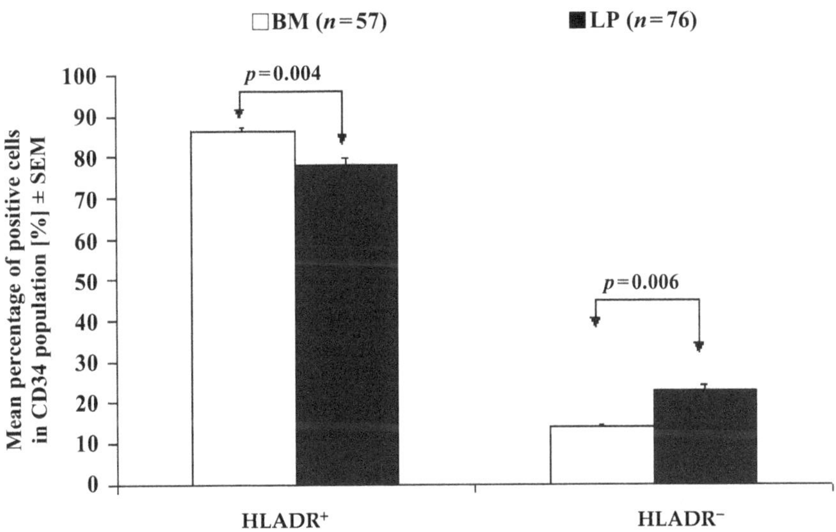

Fig. 2. HLADR positivity of $CD34^+$ cells (mean ± SEM) in BM (white columns, $n = 57$) and in G-CSF mobilised LP (dark columns, $n = 76$) from healthy donors.

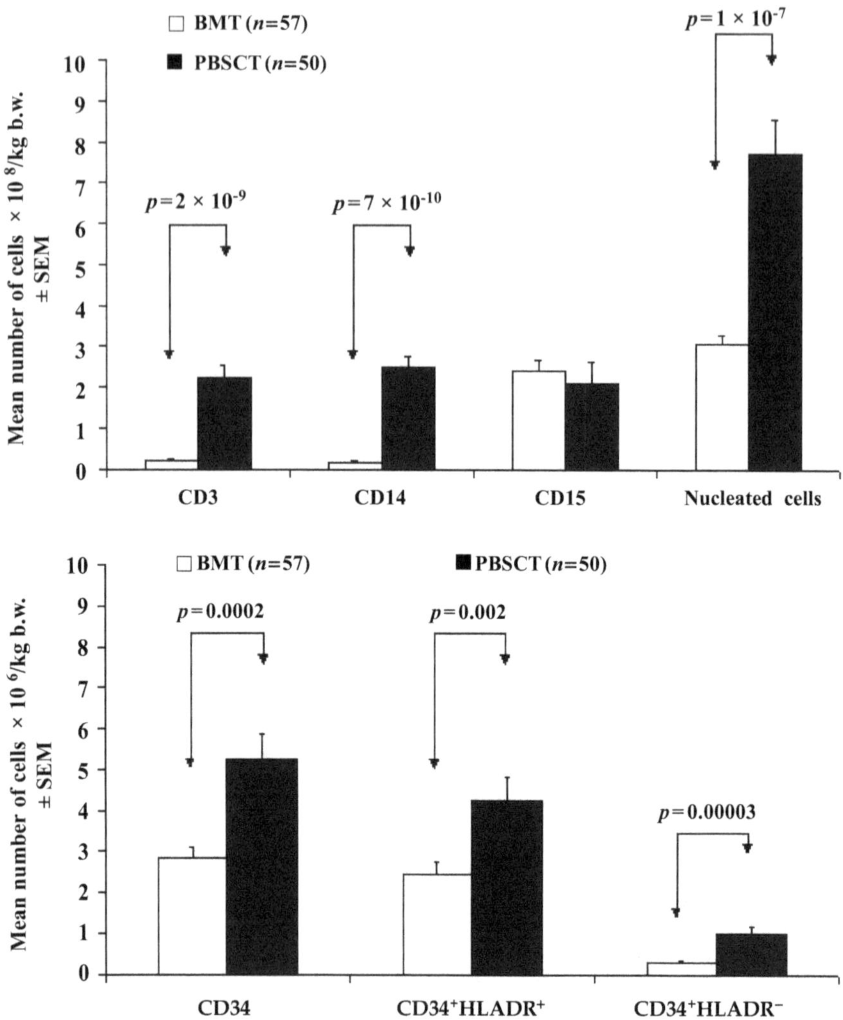

Fig. 3. Total number of cells per kg of body weight of recipients (mean ± SEM) in BM (white columns, $n = 57$) and in G-CSF mobilised LP (dark columns, $n = 50$) inoculi from healthy donors.

(3×10^6 cells/kg b.w. of recipient) was achieved in 6 out of 7 CCE procedures performed for children recipients. For adults three consecutive day leukapheresis procedures followed by 8–10 CCE runs were performed. In spite of this effort we failed to reach the

Table 1. Characteristics of donor-recipient pairs and counterflow centrifugal elutriation (CCE) products used for transplantation.

	Children < 16 years	Adults > 16 years
Number of patients (number of transplantation)	6 (7)	3 (3)
Donors age (years: range, median)	25–34 (30)	22–44 (40)
Donors weight (kg: range, median)	60–90 (78)	54–64 (62)
Recipients age (years: range, median)	1–5 (1.5)	17–47 (18)
Recipients weight (kg: range, median)	4.2–15 (12)	70–80 (70)
Number of leukapheresis (range)	1–2	3
Number of CCE processing cells (range, median)	10×10^9–50×10^9 (2×10^9)	80×10^9–100×10^9 (90×10^9)
Number of CCE runs (range, median)	1–5 (2)	8–10 (9)
Number of fractions selected for transplantation originated from one CCE run (range, median)	1–2 (1)	1–3 (1)
Number of transplanted $CD34^+$ cells/kg b.w. of recipient (range, median)	1.4×10^6–10.1×10^6 (4.1×10^6)	0.4×10^6–1.9×10^6 (0.8×10^6)
Number of transplanted $CD3^+$ cells/kg b.w. of recipient (range, median)	0.5×10^6–25.4×10^6 (1.7×10^6)	1.4×10^6–7.9×10^6 (1.6×10^6)

target number of $CD34^+$ cells for all adult patients transplanted. The main target of the CCE procedure is depletion of $CD3^+$ cells. To meet the requirements of a low number of $CD3^+$ and safe number of $CD34^+$ cells, fractions to be chosen for transplantation should not have a higher proportion of $CD3^+$ than $CD34^+$ cells. This approach was successfully implemented in six out of seven children transplantations but only in one out of three adult transplantations. In all transplanted CCE fractions a proportion of

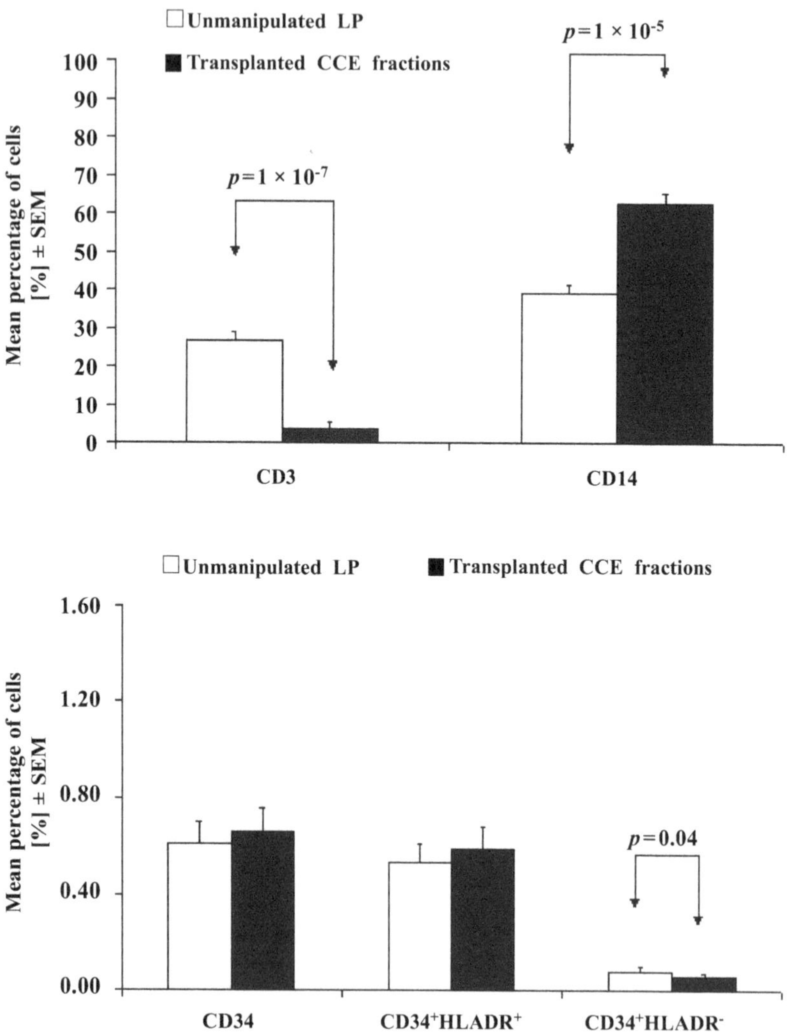

Fig. 4. Percentage (mean ± SEM) of cells in unmanipulated G-CSF mobilised LP from healthy donors (white columns) and in transplanted CCE fractions (dark columns).

$CD3^+$ cells was significantly lower than in unmanipulated G-CSF LP. CCE fractions had a similar proportion of total $CD34^+$ cells but lower $CD34^+HLADR^-$ cells as compared to primary G-CSF LP material (Fig. 4). Finally, transplanted CCE fractions (Fig. 5) had $CD34^+$ cells in a range from 0.4×10^6 cells/kg b.w. to $10.1 \times$

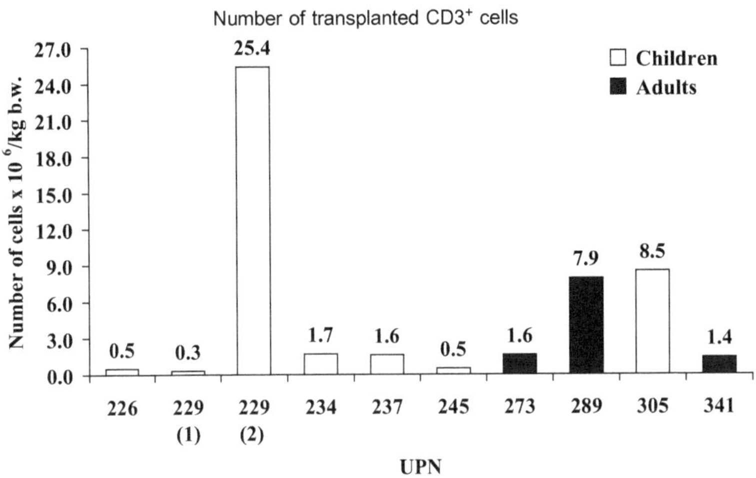

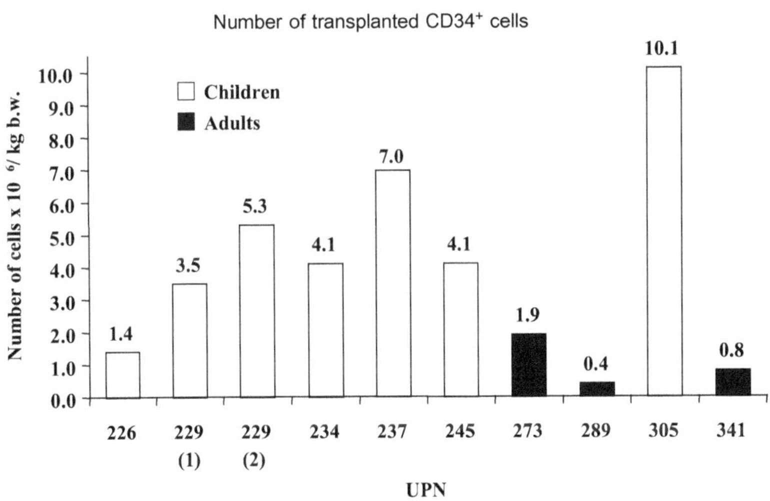

Fig. 5. Number of T cells (CD3) and progenitor cells (CD34) per kg of body weight of recipients in transplanted CCE fractions in children (white columns) and in adults (dark columns) transplantations.

10^6 cells/kg b.w. (median = 3.8×10^6) and $CD3^+$ cells ranged from 0.5×10^6 cells/kg b.w. to 25.4×10^6 cells/kg b.w. of recipients (median = 1.7×10^6). Depletion of T cells was about one to two logs as compared to primary G-CSF LPs (median = 2.9×10^8 $CD3^+$/kg b.w. of recipients in primary G-CSF LPs vs. 1.7×10^6 $CD3^+$/kg b.w. of recipients in CCE transplanted fractions).

4. Discussion

This study compared a large number of G-CSF LPs and BM used for allogeneic transplantation. Both transplant materials were obtained from healthy donors. Collection of LP, however, was preceded by G-CSF stimulation. Administration of G-CSF induces the presence of $CD34^+$ cells and an increase of $CD15^+$ cells and $CD14^+$ cells in blood (Molineux and Dexter, 1994). This enabled harvesting of an appreciable number of $CD34^+$ cells during leukapheresis and caused a high percentage of monocytes in G-CSF LP. Granulocytes were not seen in excess in G-CSF LP (Fig. 1). These characteristics of the output of G-CSF LP finally resulted with a similar proportion of $CD34^+$, higher of $CD14^+$ and $CD3^+$ and lower of $CD15^+$ cells in G-CSF LP as compared to BM inoculi (Fig. 1). This showed that with the use of continuous flow blood cell separator (COBE Spectra) equipped with the MNC-WBC or Auto-WBC programmes we collected mostly mononuclear cells independent whether they were large with low density ($CD14^+$ cells) or small with higher density ($CD3^+$ cells). This was an important feature of the leukapheresis procedure as similar differences in size and density can be seen analysing a profile of $CD34^+$ cells. Early progenitors ($CD34^+CD38^-HLADR^-$) are rather small but when they proceed consecutive differentiation steps they may become bigger (Herbein *et al.*, 1994; Chang *et al.*, 1997). G-CSF LPs contained indeed a substantial proportion of both $CD34^+HLADR^+$ and $CD34^+HLADR^-$ (Fig. 1). This is translated into clinical practice. PBSC (G-CSF stimulated LP) secure a prompt and life-long haematological reconstitution (Schmitz *et al.*, 1996) due to the presence of primitive and lineage committed precursors (Dlubek *et al.*, 1999). G-CSF LP provides an opportunity to collect more cells than when transplant material is obtained from iliac crest (Dlubek *et al.*, 1999). The other distinct feature of G-CSF LPs is a high proportion of CD3$^+$ cells (Fig. 1). Finally, G-CSF LP (PBSC) recipients receive more T cells than those transplanted with the marrow (Fig. 3). In family matched donors it is not translated into a higher risk of aGvHD (Dlubek *et al.*, 1999; Miflin *et al.* 1997; Schmitz *et al.*, 1996; Tanaka *et al.*, 1995). In alternative donor transplantation, which is associated with a higher risk of aGvHD this large number of CD3$^+$ cells in PBSC may play a role. Therefore, T cell depletion is used in some centres with positive results (Elhasid *et al.*, 2000; Buckley *et al.*, 1999; Dickinson *et al.*, 1997; Martin-Hernandez *et al.*, 1997). In this study, the efficacy of CCE in obtaining a transplant material poorer in T cells is presented. This task was however difficult due to the co-purification of CD3$^+$ cells with primitive haematopoietic precursors. Therefore, we had to accept the presence in transplant material a number of CD3$^+$ cells exceeding advocated $0.5 - 1 \times 10^6$ cells/kg b.w. of recipient (Elashid *et al.*, 2000; O'Donnell *et al.*, 1998). The use of CCE helped to deplete T cells in a range of one to two logs but only in three out of 10 allotransplantations this target could be reached (Fig. 5).

The number of CD34$^+$ cells was used as a main denominator of fractions used for transplantation. Target value for CD34$^+$ cells was 3×10^6 CD34$^+$ cells/kg b.w. of the recipient (Dlubek *et al.*, 1999), achieved more frequently in children than in adult transplantations. Acceptance a higher number of T cells for the sake of the presence of CD34$^+$HLADR$^-$ cells resulted in the presence of early progenitors in CCE transplanted fractions. The proportion of these cells was still lower than that in primary G-CSF LPs and in BM (Figs. 1 and 4). It seems that the number of CD34$^+$HLADR$^-$ cells in CCE transplanted fractions was sufficient what was suggested by a satisfactory haematological reconstitution seen in all patients followed longer than one year post-transplant (not shown).

It appears that G-CSF stimulated LP is advantageous for transplantation as compared to BM with respect to the number of CD34$^+$ cells including early progenitors. This is of a special value as the number of CD34$^+$ cells in transplanted cell population constitute the main predictive factor for the outcome of allogeneic haematopoietic stem cell transplantation (Sharp *et al.*, 2000; Singhal *et al.*, 200; Vogel *et al.*, 2000). CCE can be used to decrease the number of T cells, which is high in G-CSF LPs. This technique is more suitable for children transplantation having adult donors as it is easier to obtain a sufficient number of $CD34^+$ cells per kg of body weight of minor recipients. CCE removal of T cells is not as vigorous as in other techniques i.e. immunomagnetic T cell depletion, CD34 positive selection (Feugier *et al.*, 1997) but depletion is substantial and clinically effective. Among 9 patients none suffered from aGvHD life threatening complications (not shown).

5. References

BARRETT, A.J. (1997). Mechanisms of the graft-versus-leukemia reaction, *Stem Cells* **15**(4), 248–258.

BUCKLEY, R.H., SCHIFF, S.E., SCHIFF, R.I., MARKERT, L., WILLIAMS, L.W., ROBERTS, J.L., MYERS, L.A. and WARD, F.E. (1999). Hematopoietic stem-cell transplantation for the treatment of severe combined immunodeficiency, *N. Engl. J. Med.* **340**(7), 508–516.

CHANG, Q., HARVEY, K., AKARD, L., THOMPSON, J., DUGAN, M., ENGLISH, D. and JANSEN, J. (1997). Counterflow centrifugal elutriation as a method of T cell depletion may cause loss of immature $CD34^+$ cells, *Bone Marrow Transplant.* **19**(11), 1145–1150.

DICKINSON, A.M., REID, M.M., ABINUN, M., PEAK, J., BRIGHAM, K., DUNN, J. and CANT, A.J. (1997). *In vitro* T cell depletion using Campath 1M for mismatched BMT for severe combined immunodeficiency (SCID), *Bone Marrow Transplant.* **19**(4), 323–329.

DLUBEK, D., PACUSZKO, T., SUCHNICKI, K. and LANGE, A. (1999). G-CSF mobilized leukapheresis products: Cellular characteristics and clinical performance in allografting, *J. Hematother.* **8**, 157–166.

ELHASID, R., BEN ARUSH, M.W., KATZ, T., GAN, Y., SHECHTER, Y., SAMI, I., POSTOVSKY, S., REISNER, Y. and ROWE, J.M. (2000). Successful haploidentical bone marrow transplantation in Fanconi anemia, *Bone Marrow Transplant.* **26**(11), 1221–1223.

FEHSE, B., GOLDMANN, M., FRERK, O., BULDUK, M. and ZANDER, A.R. (2000). Depletion of alloreactive donor T cells using immunomagnetic cell selection, *Bone Marrow Transplant.* **25**(Suppl. 2), 39–42.

FERRARA, J.L. (2000). Pathogenesis of acute graft-versus-host disease: Cytokines and cellular effectors, *J. Hematother. Stem. Cell. Res.* **9**(3), 299–306.

FEUGIER, P., THEVENIN, S., BILLOT, M., SERRURIER, B., LOIGEROT, C., SCHOONEMAN, F., JANOT, C., HERVE, P. and RACADOT, E. (1997). Comparison of T cell depletion strategies from bone marrow, umbilical cord and peripheral blood using five separation systems, *Hematol. Cell. Ther.* **39**(2), 67-73.

FOWLER, D.H., BREGLIO, J., NAGEL, G., ECKHAUS, M.A. and GRESS, R.E. (1996). Allospecific $CD8^+$ Tc1 and Tc2 populations in graft-versus-leukemia effect and graft-versus-host disease, *J. Immunol.* **157**(11), 4811–4821.

HERBEIN, G., SOVALAT, H., WUNDER, E., BAERENZUNG, M., BACHORZ, J., LEWANDOWSKI, H., SCHWEITZER C., SCHMITT, C., KIRN, A. and HENON, P. (1994). Isolation and identification of two $CD34^+$ cell subpopulation from normal human peripheral blood, *Stem. Cells.* **12**, 187–197.

HILL, G.R., KRENGER, W. and FERRARA, J.L. (1997). The role of cytokines in acute graft-versus-host disease, *Cytokines Cell Mol. Ther.* **3**(4), 257–266.

HOLLER, E., ERTL, B., HINTERMEIER-KNABE, R., RONCAROLO, M.G., EISSNER, G., MAYER, F., FRAUNBERGER, P., BEHRENDS, U., PFANNES, W., KOLB, H.J. and WILMANNS, W. (1997). Inflammatory reactions induced by pretransplant conditioning — An alternative target for modulation of acute GvHD and complications following allogeneic bone marrow transplantation? *Leuk. Lymphoma* **25**(3–4), 217–224.

JIANG, Y.Z., BARRETT, A.J., GOLDMAN, J.M. and MAVROUDIS, D.A. (1997). Association of natural killer cell immune recovery with a graft-versus-leukemia effect independent of graft-versus-host disease following allogeneic bone marrow transplantation, *Ann. Hematol.* **74**(1), 1–6.

KLINGEBIEL, T. and SCHLEGEL, P.G. (1998). GVHD: Overview on pathophysiology, incidence, clinical and biological features, *Bone Marrow Transplant.* **21**(Suppl. 2), 45–49.

MARTIN-HERNANDEZ, M.P., ARRIETA, R., MARTINEZ, A., GARCIA, P., JIMENEZ-YUSTE, V. and HERNANDEZ-NAVARRO, F. (1997). Haploidentical peripheral blood stem cell transplantation with a combination of CD34 selection and T cell depletion as graft-versus-host disease prophylaxis in a patient with severe combined immunodeficiency, *Bone Marrow Transplant.* **20**(9), 797–799.

MAVROUDIS, D.A., DERMIME, S., MOLLDREM, J., JIANG, Y.Z., RAPTIS, A., VAN RHEE, F., HENSEL, N., FELLOWES, V., ELIOPOULOS, G. and BARRETT, A.J. (1998). Specific depletion of alloreactive T cells in HLA-identical siblings: A method for separating graft-versus-host and graft-versus-leukaemia reactions, *Br. J. Haematol.* **101**(3), 565–570.

MICKELSON, E.M., PETERSDORF, E., ANASETTI, C., MARTIN, P., WOOLFREY, A. and HANSEN, J.A. (2000). HLA matching in hematopoietic cell transplantation, *Hum. Immunol.* **61**(2), 92–100.

MIFLIN, G., RUSSELL, N.H., HUTCHINSON, R.M., MORGAN, G., POTTER, M., PAGLIUCA, A., MARSH, J., BELL, A., MILLIGAN, D., LUMLEY, M., COOK, G. and FRANKLIN, I. (1997). Allogeneic peripheral blood stem cell transplantation for haematological malignancies — An analysis of kinetics of engraftment and GVHD risk, *Bone Marrow Transplant.* **19**(1), 9–13.

MOLINEUX, G. and DEXTER, T.M. (1994). Biology of G-CSF, In: *Filgrastim (r-metHuG-CSF) in Clinical Practice*, G. Morstyn and T.M. Dexter, eds., Marcel Dekker INC, New York, pp. 1–21.

MURPHY, W.J. and LONGO, D.L. (1997). The potential role of NK cells in the separation of graft-versus-tumor effects from graft-versus-host disease after allogeneic bone marrow transplantation, *Immunol. Rev.* **157**, 167–176.

O'DONNELL, P.V., JONES, R.J., VOGELSANG, G.B., SEBER, A., AMBINDER, R.F., FLINN, I., MILLER, C., MARCELLUS, D.C., GRIFFIN, C., ABRAMS, R., BRAINE, H.G., GREVER, M., HESS, A.D., PIANTADOSI, S. and NOGA, S.J. (1998). $CD34^+$ stem cell augmentation of elutriated allogeneic bone marrow grafts: Results of a phase II clinical trial of engraftment and graft-versus-host disease prophylaxis in high-risk hematologic malignancies, *Bone Marrow Transplant.* **22**(10), 947–955.

PORTER, D.L. and ANTIN, J.H. (1997). Graft-versus-leukemia effect of allogeneic bone marrow transplantation and donor mononuclear cell infusions, *Cancer Treat Res.* **77**, 57–85.

SCHMITZ, N., BACIGALUPO, A., LABOPIN, M., MAJOLINO, I., LAPORTE, J.P., BRINCH, L., COOK, G., DELILIERS, G.L., LANGE, A., ROZMAN, C., GARCIA-CONDE, J., FINKE, J., DOMINGO-ALBOS, A. and GRATWOHL, A. (1996). Transplantation of peripheral blood progenitor cells from HLA-identical sibling donors. European Group for Blood and Marrow Transplantation (EBMT), *Br. J. Haematol.* **95**(4), 715–723.

SHARP, J.G., KESSINGER, A., LYNCH, J.C., PAVLETIC, Z.S. and JOSHI, S.S. (2000). Blood stem cell transplantation: Factors influencing cellular immunological reconstitution, *J. Hematother. Stem. Cell. Res.* **9**(6), 971–981.

SHLOMCHIK, W.D. and EMERSON, S.G. (1996). The immunobiology of T cell therapies for leukemias, *Acta Haematol.* **96**(4), 189–213.

SINGHAL, S., POWLES, R., TRELEAVEN, J., KULKARNI, S., SIROHI, B., HORTON, C., MILLAR, B., SHEPHERD, V., TAIT, D., SASO, R., ROWLAND, A., LONG, S. and MEHTA, J. (2000). A low $CD34^+$ cell dose results in higher mortality and poorer survival after blood or marrow stem cell transplantation from HLA-identical siblings: Should $2 \times 10(6)$ $CD34^+$ cells/kg be considered the minimum threshold? *Bone Marrow Transplant.* **26**(5), 489–496.

TANAKA, J., KASAI, M., IMAMURA, M. and ASAKA, M. (1995). Clinical application of allogeneic peripheral blood stem cells transplantation, *Ann. Hematol.* **71**(6), 265–269.

VOGEL, W., SCHEDING, S., KANZ, L. and BRUGGER, W. (2000). Clinical applications of $CD34(^+)$ peripheral blood progenitor cells (PBPC), *Stem Cells* **18**(2), 87–92.

Advances in Tissue Banking Vol. 6

16 PANCREATIC ISLET CELL PROCESSING FOR TRANSPLANTATION

SHINICHI MATSUMOTO,[1,2] R. BRIAN STEVENS[2] and D. MICHAEL STRONG[1]

[1]Puget Sound Blood Center and Program, Seattle, WA 98104, USA,
[2]University of Washington Medical Center, Department of Surgery, Division of Transplantation, Seattle, WA 98195, USA

1. Introduction

The results and understanding of pancreatic islet allotransplantation as a cure for insulin dependent diabetes mellitus have shown significant improvement over the last few years (International Islet Transplant Registry News Letter #8, 1999; Keymeulen *et al.*, 1998; Jaeger *et al.*, 1996). Recently, the Edmonton (Canada) group reported 100% cure of type 1 diabetes after two or more islet transplants (Shapiro *et al.*, 2000). In order to confirm their results, the NIH proposed to repeat the Edmonton protocol in other islet transplantation centres through the Immune Tolerance Network (ITN). The JDFI Center for Human Islet Transplantation in Seattle was selected as one of these islet isolation centres and the isolation protocol was adjusted to make it similar to Edmonton's protocol. In this chapter the islet isolation strategies and the adjusted protocol used are described.

2. Organ Procurement and Preservation of Pancreas Prior to Islet Isolation

Although many factors can improve the outcome of human islet isolation, using a high quality organ is the most important factor. Procuring the pancreas properly and minimising preservation injury are important in achieving a high quality level. Human pancreata were obtained from multi-organ donors after in situ perfusion of 3–4 l of the University of Wisconsin (UW) solution (Kneteman *et al.*, 1994). Keeping the pancreata cold after the UW flush is critical because warm ischemic injury deteriorates the quality of pancreata quickly. Pancreata are usually immersed in UW solution (600–800 ml per pancreas); however, the two-layer (University of Wisconsin solution (UW)/perfluorocarbon) preservation method has proven to be superior for pancreas preservation prior to islet isolation (Matsumoto *et al.*, 2000b). Current results concerning two-layer preservation will be discussed later in this chapter.

3. Principles of Pancreatic Islet Cell Processing

The automated method for isolation of human pancreatic islets described by Camillo Ricordi was followed (Ricordi *et al.*, 1988). Pancreatic islet cell processing was performed in a class 100 clean room; however, the entire procedure (isolation and purification) can be performed in a standard tissue culture laboratory, equipped with two biological safety cabinets (class 100), two refrigerated centrifuges, and a COBE 2991 cell processor (Ricordi *et al.*, 1994). Initially the pancreas was distended by intraductal perfusion of HBSS solution including collagenase. After distention the pancreas was placed into a continuous digestion device (CDD) and the pancreas was digested chemically with collagenase and mechanically with shaking CDD. Once islets were liberated from the pancreas, islets were rescued from the collagenase solution by diluting with albumin containing solution and cooling. Islets were purified from exocrine tissues using a COBE 2991 cell processor with density gradients.

4. Pancreas Distention with Collagenase Solution

Selection of an adequate collagenase was important in successful islet isolation. Human islet recovery was improved by using a purified enzyme, Liberase (Linetsky *et al.*, 1997; Olack *et al.*, 1999). Five hundred mg of Liberase was diluted with 350 ml of Hank's balanced salt solution (HBSS). As collagenase requires calcium, it was supplemented with 3.5 mM calcium dichloride. Then, 24 mM HEPES was added for buffering pH. Phenol red should be avoided because it causes clumping during isolation (Matsumoto, unpublished data).

Collagenase solution was loaded into the pancreatic duct using a controlled perfusion technique (Lakey *et al.*, 1999). After trimming and washing, the pancreas was immersed in antibiotic solution (Gentamicin, Vancomycin, Fungizone) for 5 min. The pancreas was split to find the main pancreatic duct. Next, 16 or 18 gauge needles were inserted into both the head and tail part of main pancreatic ducts. Cold collagenase solution was injected at 60–80 mmHg for the first 5 min and 160–180 mmHg for the next 5 min (Lakey *et al.*, 1999). Any leakage was clamped using mosquito clamps.

5. Pancreatic Digestion Using the Continuous Digestion Device (CDD)

The continuous digestion device (CDD) was developed by Ricordi (Ricordi *et al.*, 1988) and this system was modified according to the Edmonton group protocol (Lakey *et al.*, 1999) (Fig. 1). A distended pancreas was cut into 7 to 9 pieces; these pieces and all collagenase solution that had leaked out of pancreas during loading were transferred to CDD chamber. The recirculating collagenase solution in the CDD was warmed to 37°C within 5 min and held at 37 ± 0.5°C throughout dissociation of the pancreas (Lakey *et al.*, 1999). During the digestion phase, the CDD chamber was rotated and shaken vertically 3–4 times every 10 to 15 seconds, instead of vigorous shaking. The CDD chamber should be shaken gently for a young donor (age 15 to 35) and more vigorously for an older donor (36 to 70 years old) (Ricordi *et al.*, 1994). Samples were first

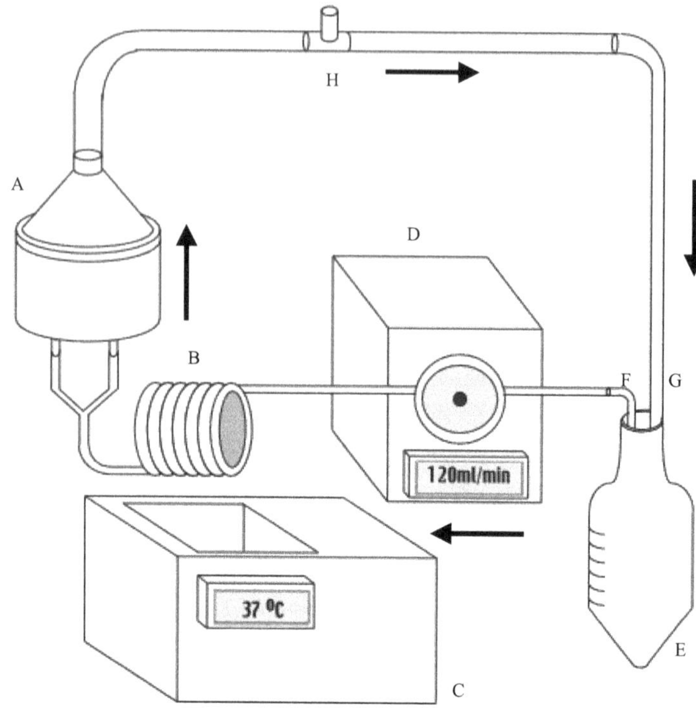

Fig. 1. During digestion phase, the pancreas was put into the continuous digestion device (CDD) chamber (A) and collagenase solution was warmed by a heat exchanger (B) and water bath (C). Collagenase solution was recirculated by pump (D) from 250 ml conical reservoir (E) through afferent glass pipette (F) to efferent glass pipette (G). Samples were taken to check the status of digestion through the sampling port (H). Arrows indicate flow direction of collagenase solution. During dilution phase culture media was sucked using afferent glass pipette (F) and digestate was collected into 250 ml conical tubes through efferent glass pipette (G). At this phase solution was not recirculated.

removed at 8 min and then every 2 min thereafter to monitor digestion (Ricordi *et al.*, 1994). Samples were stained with freshly prepared and filtrated dithizone solution (50 mg dithizone powder, 5 ml DMSO, 45 ml HBSS) for the immediate identification of islets (Latif *et al.*, 1988; Ricordi *et al.*, 1990). When the majority of islets were free from exocrine tissue, the digest was diluted with warm culture medium (Dilution Phase). Culture media was prepared at

37°C and 4 l at room temperature for dilution. Initially, culture media was used at 37°C and then switched to culture media at room temperature. Diluted digest was collected into 250 ml conical tubes pre-filled with 15 ml 20% human albumin in ice slush. Digestate was washed with cold culture media supplemented with 30 ml/l 20% human albumin and antibiotics.

6. Islet Purification Using Continuous Osmolality and Density Gradient

The cleaved islets were purified with continuous osmolality and density gradient using histidine lactobionate nicotinamide (HLN) solution and iodixanol (Matsumoto *et al.*, 1999). Histidine lactobionate (HL) solution was proven to be superior to UW solution to preserve human and rat islets (Deflino *et al.*, 1993). The HL solution was adjusted for density gradient and developed HLN solution (Table 1). Iodixanol was originally made for X-ray contrast and does not include endotoxin. Iodixanol combined with UW solution was shown to be an excellent density gradient medium for pig islet purification (van der Berg *et al.*, 1998). The density

Table 1. The composition of histidine-lactobionate-nicotinamide (HLN) solution.

Lactobionate	100 mmol/l
NaOH	115 mmol/l
L-histidine	30 mmol/l
KOH	7 mmol/l
Raffinose	30 mmol/l
$CaCl_2$	0.5 mmol/l
$MgSO_4$	5 mmol/l
NaH_2PO_4	25 mmol/l
Nicotinamide	5 mmol/l
Osmolality	335 mOsm/kg
pH	7.4

range of the continuous gradient was adjusted specifically to the density of each pancreatic digestate because the density of digestive from human pancreata is not constant. Density of digestive was determined by centrifuging small volumes (150 μl) of digestive in 5 ml aliquots of eight different density solutions (1.085, 1.090, 1.095, 1.100, 1.105, 1.110, 1.115, 1.120 g/cm^3), and observing whether the disaggregated particles pelleted, floated, or distributed evenly throughout the solution. Once the optimum density parameters of the density-gradient solution had been determined, the washed, pelleted digestate was resuspended in 200 ml of gradient "recombination" solution that had been adjusted to the highest density that still allowed pelleting. The range of density of recombination solution was usually between 1.094 and 1.114 g/cm^3. In general, a high quality pancreas for islet isolation has a high density of digestate.

For the lightest density layer (capping layer), HLN solution was used alone (density = 1.030 g/cm^3, osmolality = 330 mOsm, 60 ml). For the gradient layer, a light density solution and heavy density solution was used. The light density solution consisted of HLN solution and iodixanol (density = 1.060 g/cm^3, osmolality = 330 mOsm, 125 ml) and the heavy density solution consisted of high osmolality HLN solution (500 mOsm) and iodixanol (density = 1.090–1.110 g/cm^3, osmolality = 500 mOsm, 125 ml). To produce a high osmolality HLN solution, 210 mmol/l histidine and 10 mmol/l lactobionic acid was added to regular HLN solution. For the heaviest layer (recombination layer), a high osmolality HLN solution and iodixanol was used (density = 1.094–1.114 g/cm^3, osmolality = 500 mOsm, 200 ml). The ratio of HLN solution to iodixanol was changed to adjust the density of heavy and recombination solution but the density of light or capping solution was not adjusted. After washing the pancreatic digest, the heaviest solution (recombination solution) was mixed with the pancreatic digest and loaded on a COBE 2991 cell processor (Cobe Laboratories, Denver, CO). The digestate was gravity fed into the COBE 2991 cell processor and centrifugation was initiated at 1000 rpm (200 G). The system was bled of air, and the bottom solution and digested tissue were overlaid with 250 ml of a continuous density-gradient solution from a two-reservoir gradient former. The solution was pumped into the spinning cell processor at 60 ml/min, beginning with high density and ending with low. These layers were then finally "capped" by the addition of 60 ml of low density solution lacking iodixanol. When stratification was complete, the contents of the spinning COBE bag were expressed and collected into twelve 250 ml conical centrifuge tubes prefilled with 200 ml cold culture media and 15 ml 20% human albumin. Purified islets were collected and washed using culture media. Washed islets were cultured until transplantation. Next, 15 000 IE purified islets were cultured in 10 ml CMRL1066 culture media and 30 ml of 20% human albumin at room temperature with 5% CO_2 in a T75 flask (Sarstedt, Inc., Newton, NC).

7. Islet Evaluation

The islet mass was quantified, and the purity was assessed by two independent observers according to the criteria established at the 1989 International Workshop on Islet Assessment (Ricordi *et al.*,

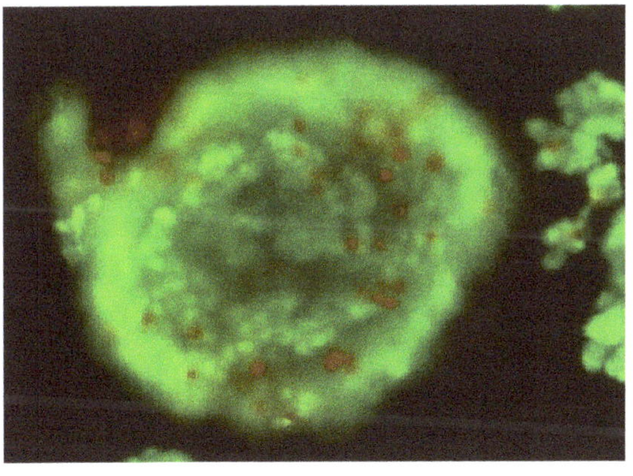

Fig. 2. Human islet stained by acridine orange/propidium iodine (AO/PI) staining. Viable cells were stained by acridine orange and become green. Non-viable cells were stained by propidium iodine and become red. The viability of this islet was more than 90%.

1990). The viability of pure islets was evaluated using acridine orange/propidium iodide (AO/PI) staining (Bank, 1988). Low concentrations of acridine orange (10 μmol/l) and propidium iodide (15 μmol/l) (AO/PI) were used to visualise living and dead islet cells simultaneously (Fig. 2). Fifty islets were counted and the viability of each islet was visually determined; the average viability was then calculated. Cultured islet function was evaluated using insulin release by glucose stimulation. Aliquots of cultured 150 IE islets were incubated with 2.8 mM or 20 mM glucose for 2 hours at 37°C with 5% CO_2. Insulin concentration was analysed with a human insulin ELISA kit (ALPCO Insulin ELISA kit, Windham, NH). Insulin concentrations for the peak stimulation portion (20 mM) of testing were divided by the insulin levels obtained for the basal time period (2.8 mM) to determine the stimulation index. Viability staining should be more than 70% and the stimulation index should be more than 1.5 for transplantable islets.

8. Two-Layer Preservation of Human Pancreata Prior to Islet Isolation.

One of the most critical factors for the success of islet isolation is to obtain high quality of human pancreas. Therefore, the two-layer preservation method was applied prior to islet isolation. Two-layer preservation was originally developed for the whole pancreas preservation and consists of UW solution and perfluorocarbon with continuous oxygenation. The two-layer method has proven to be superior to UW solution storage (Kawamura *et al.*, 1989). With the two-layer method, pancreas grafts were directly oxygenated and maintain high tissue adenosine triphosphate (ATP) concentration (Matsumoto *et al.*, 1996). This method was applied for pancreas preservation prior to clinical whole-pancreas transplantation without related complication (Matsumoto *et al.*, 2000a). Before applying the two-layer method prior to islet isolation, this method was studied (Matsumoto *et al.*, 2000b). Eight human pancreata from multi-organ donors were used in the study. Preparation of the pancreata at organ recovery included flushing with chilled UW solution (I.E. Du

Pont de Nemours Co., Waukegan, IL) and pancreata were packaged for transport by immersion in chilled UW solution in a sealed container. For transportation, discontinuous oxygenation of the two-layer preservation was used to avoid carrying an oxygen tank. Perfluorocarbon covered by UW solution was oxygenated using 100% oxygen for at least 15 min with a flow rate of 100 ml/min. After putting the pancreas between the two solutions, the container was closed and surrounded by ice and shipped.

Upon arrival at the Northwest Tissue Center each organ was aseptically transected into approximately equal halves (head and tail). The pancreas halves were placed in either UW solution at 4°C or in the two-layer method at 4°C for 14 to 27 hours, with a mean of 20 hours. Each group included four heads and four tails. The mean total preservation time (the interval between cross clamping and processing for islet recovery) was 30.2 ± 3.2 hours for the two-layer method group and 30.2 ± 3.0 hours for the UW group. The islets were then isolated and the islet yields were compared, viability was assessed by AO/PI staining and stimulation index. Approximately a double yield was obtained when the two-layer preservation method was used compared to UW storage in both digestate and post-purification (Fig. 3). Viability was also significantly better with the two-layer method and the stimulation index was better (Fig. 4).

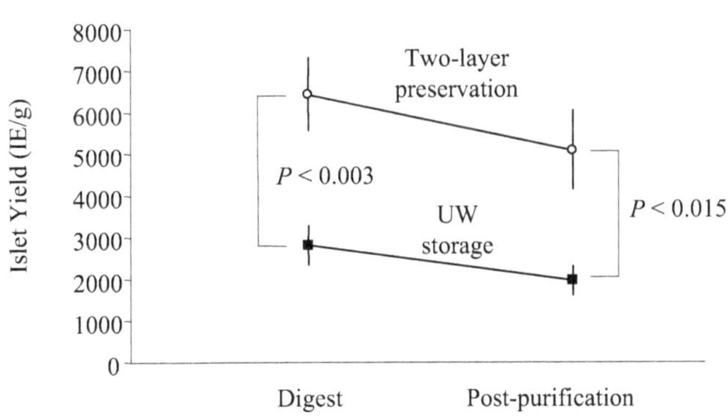

Fig. 3. Islet yields from human pancreata preserved by the two-layer method (○) or UW storage (■). Islet yields were about double when pancreas was preserved by the two-layer method both in digest and post-purification.

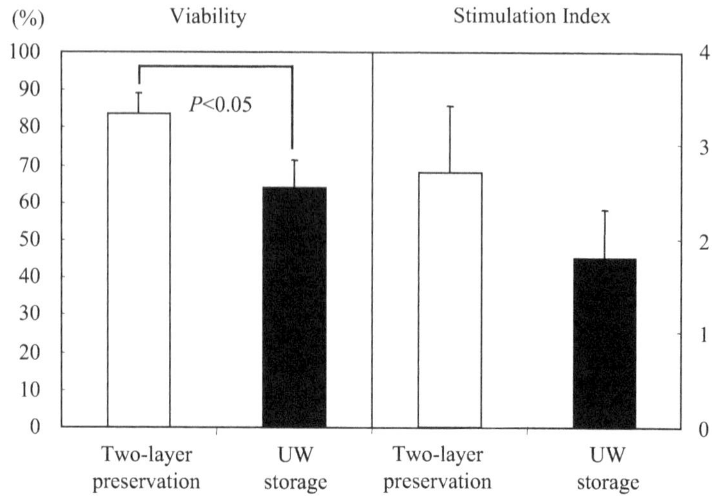

Fig. 4. Viability and stimulation index of glucose challenge test of islets from human pancreata preserved by the two-layer method (□) or UW storage (■). Islets from the pancreas preserved by the two-layer method showed better viability and stimulation index.

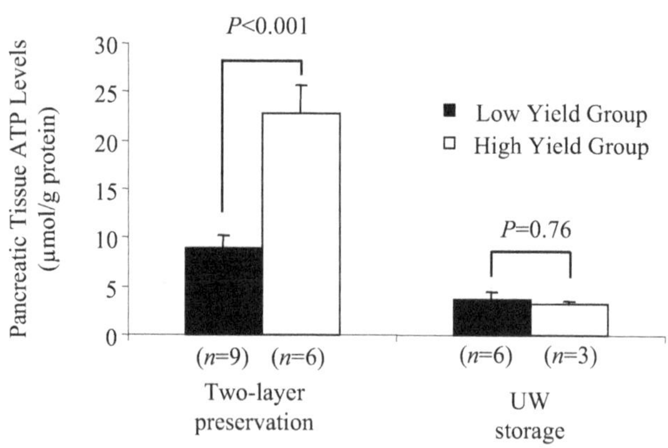

Fig. 5. Pancreatic tissue ATP levels after two-layer preservation or UW storage in low yield group (■) (islet yield below average) and in high yield group (□) (islet yield above average). High yield is associated with high tissue ATP levels after preservation by the two-layer method.

It was discovered that tissue ATP levels after two-layer preservation correlated to islet yields, thereby permitting evaluation of the donor pancreata before islet isolation. In this study, at the end of preservation and before islet isolation a piece of the pancreas was rapidly frozen until analysis. ATP measurements on each biopsy were performed using the ATP measurement kit (Sigma Diagnostics, Inc., St. Louis, MO) and adjusted by protein content. Pancreatic tissue ATP levels after two-layer preservation was substantially high in high yield group (Islet yield more than average) however, tissue ATP levels after UW storage was not different between high yield group and low yield group (Fig. 5). These data suggested that expensive isolation attempts could be avoided that resulted in poor islet yields by measuring tissue ATP after preservation by the two-layer method.

9. Discussion and Conclusion

The current success of Edmonton group takes islet allotransplantation from experimental events to clinical treatment. The differences between their protocol and previous protocols included (i) minimising the time length between procurement and transplantation (isolating islets as soon as possible and avoiding culture prior to transplantation), (ii) avoiding steroids and using islet friendly immunosuppressant (sirolimus, daclizumab and low dose FK506), and (iii) transplanting a large number of islets even if this required two or more pancreata. Seven out of seven type 1 diabetic patients became insulin free and these patients also had no more hypoglycemic episodes (Shapiro *et al.*, 2000). In order for islet transplantation to supersede pancreas transplantation, it will be essential for islet transplantation to succeed using one donor pancreas. To achieve this, the two-layer method of preservation should be applied prior to islet isolation, utilising an endotoxin free purification protocol. These modifications should be able to improve islet isolation substantially and allow islet transplantation to replace pancreas transplantation to cure type 1 diabetes without major surgery.

10. Acknowledgements

The Human Islet Transplantation in Seattle (HITS) Consortium is a joint programme of the Pacific Northwest Research Institute, the Puget Sound Blood Center, Fred Hutchinson Cancer Research Center, Swedish Hospital, the University of Washington, and the Virginia Mason Research Center. This work was funded in part by Human Islet Isolation Facility Development Grand funds sponsored by the Virginia Mason Research Center, Puget Sound Blood Center, and by the Juvenile Diabetes Foundation Grant Number 4-1999847.

We thank Ms. Mary West for skillful manuscript preparation and Mr. Theodore Rigley and Ms. Sabrina Qualley for their technical help.

11. References

BANK, H.L. (1988). Rapid assessment of islet viability with acridine orange and propidium iodide, *In Vitro Cell. Dev. Biol.* **24**(4), 266–273.

DEFLINO, V.D., GRAY, D.W., LEOW, C.K., SHIMIZU, S., FERGUSON, D.J. and MORRIS, P.J. (1993). A comparison of four solutions for cold storage of pancreatic islets, *Transplantation* **56**(6), 1325–1330.

INTERNATIONAL ISLET TRANSPLANT REGISTRY NEWS LETTER #**8**: (1999).

JAEGER, C., HERING, B.J., DYRBERG, T., FEDERLIN, K. and BRETZEL, R.G. (1996). Islet cell antibodies and glutamic acid decarboxylase antibodies in patients with insulin-dependent diabetes mellitus undergoing kidney and islet-after-kidney transplantation, *Transplantation* **62**(3), 424–426.

KAWAMURA, T., KURODA, Y., SUZUKI, Y., FUJIWARA, H., FUJINO, Y., YAMAMOTO, K. and SAITOH, Y. (1989). Seventy-two-hour preservation of the canine pancreas by the two-layer (Euro-Collins' solution/perfluorochemical) cold storage method, *Transplantation* **47**(5), 776–778.

KEYMEULEN, B., LING, Z., GORUS, F.K., DELVAUX, G., BOUWENS, L., GRUPPING, A., HENDRIECKX, C., PIPELEERS, M.-M., VAN, S.-C., SALMELA, K. and PIPELEERS, D.G. (1998). Implantation of standardized beta-cell grafts in a liver segment of IDDM patients: Graft and recipients characteristics in two cases of insulin-independence under maintenance immunosuppression for prior kidney graft, *Diabetologia* **41**(4), 452–459.

KNETEMAN, N.M., LAKEY, J.R.T., KIZILISIK, L.T., AO, Z., WARNOCK, G.L. and RAJOTTE, R.V. (1994). Cadaver pancreas recovery technique. Impact on islet recovery and *in vitro* function, *Transplantation* **58**(10), 1114–1119.

LAKEY, J.R.T., WARNOCK, G.L., SHAPIRO, A.M.J., KORBUTT, G.S., AO, Z., KNETEMAN, N.M. and RAJOTTE, R.V. (1999). Intraductal collagenase delivery into the human pancreas using syringe loading or controlled perfusion, *Cell Transplant.* **8**(3), 285–292.

LATIF, Z.A., NOEL, J. and ALEJANDRO, R. (1988). A simple method of staining fresh and cultured islets, *Transplantation* **45**(4), 827–830.

LINETSKY, E., BOTTINO, R., LEHMANN, R., ALEJANDRO, R., INVERARDI, L. and RICORDI C. (1997). Improved human islet isolation using a new enzyme blend, Liberase, *Diabetes* **46**(7), 1120–1123.

MATSUMOTO, S., KANDASWAMY, R., SUTHERLAND, D.E., HASSOUN, A.A., HIRAOKA, K., SAGESHIMA, J., SHIBATA, S., TANIOKA, Y. and KURODA, Y. (2000a). Clinical application of the two-layer [University of Wisconsin solution (UW)/perfluorochemical (PFC) plus O2] method of pancreas preservation before transplantation, *Transplantation* **70**(5), 771–774.

MATSUMOTO, S., KURODA, Y., HAMANO, M., KIM, Y., SUZUKI, Y., KU, Y. and SAITOH, Y. (1996). Direct evidence of pancreatic tissue oxygenation during preservation by the two-layer method, *Transplantation* **62**(11), 1667–1670.

MATSUMOTO, S., QUALLEY, S., RIGLEY, T., MARSH, C. and STEVENS, R.B. (2000b). Prolonged preservation of the human pancreas prior to islet isolation using the two-layer (University of Wisconsin solution [UW]/Perfluorocarbon) method, *Transplantation* **69**(8), S213.

MATSUMOTO, S., SHIBATA, S., KIRCHHOF, N., HIRAOKA, K., SAGESHIMA, J., ZHANG, X.W., GILMORE, R.T., ANSITE, D.J., ZHANG, H.J., SUTHERLAND, D.E. and HERING, B.J. (1999). Immediate reversal of diabetes in primates following intraportal transplantation of porcine islets purified on a new histidine-lactobionate-iodixanol gradient, *Transplantation* **67**(7), S220.

OLACK, B.J., SWANSON, C.J., HOWARD, T.K. and MOHANAKUMAR, T. (1999). Improved method for the isolation and purification of human islets of Langerhans using Liberase™ enzyme blend, *Hum. Immunol.* **60**(12), 1303–1309.

RICORDI, C., DIGON, B.J. III, MINTZ, D.H. and ALEJANDRO, R. (1994). Human islet separation. In: *Pancreatic Islet Transplantation.* Vol. 1: Procurement of Pancreatic Islets. R.P. Lanza and W.L. Chick, eds., R.G. Landes Company, Austin, pp. 97–105.

RICORDI, C., GRAY, D.W., HERING, B.J., KAUFMAN, D.B., WARNOCK, G.L., KNETEMAN, N.M., LAKE, S.P., LONDON, N.J., SOCCI, C., ALEJANDRO, R. *et al.* (1990). Islet isolation assessment in man and large animals, *Acta Diabetol. Lat.* **27**(3), 185–195.

RICORDI, C., LACY, P.E., FINKE, E.H., OLACK, B.J. and SCHARP, D.W. (1988). Automated method for isolation of human pancreatic islets, *Diabetes* **37**(4), 413–420.

SHAPIRO, A.M., LAKEY, J.R., RYAN, E.A., KORBUTT, G.S., TOTH, E., WARNOCK, G.L., KNETEMAN, N.M. and RAJOTTE, R.V. (2000). Islet transplantation in seven patients with type 1 diabetes mellitus using a glucocorticoid-free immunosuppressive regimen, *N. Engl. J. Med.* **27**, **343**(4), 230–238.

VAN DER BURG, M.P., BASIR, I. and BOUWMAN, E. (1998). No porcine islet loss during density gradient purification in a novel Iodixanol in University of Wisconsin solution, *Transplant. Proc.* **3**(2), 362–363.

Advances in Tissue Banking Vol. 6

17 TYPE I COLLAGEN AND A RECOMBINANT TGF-β1 SERVE AS A SCAFFOLD FOR BONE MARROW MESENCHYMAL STEM CELLS

J.A. ANDRADES and J. BECERRA

Department of Cell Biology, Genetics and Physiology
Faculty of Sciences, University of Málaga. Campus
Universitario de Teatinos, 29071-Málaga, Spain

1. Introduction

Chemotaxis, proliferation and differentiation are central events in bone induction cascade. Transforming growth factor-betas (TGF-βs) are a superfamily of molecules closely related in structure and function. Their presence during cell growth, differentiation and synthesis-degradation of extracellular matrix (ECM), have been widely reported in the last two decades (Roberts and Sporn, 1989; Nimni, 1997). Although widely distributed, bone was the tissue where TGF-βs were firstly reported to be closely related with regulation of cell function. Besides, bone represents the body compartment with the most abundant deposit of this factor (Centrella *et al.*, 1988; Bostrom and Asnis, 1998). Belonging to this gene family, bone morphogenetic proteins (BMPs) represent a large group of factors of the TGF-β family, first isolated from bone matrix and currently related to a wide range of activities in several tissues and organs (Urist, 1965; Reddi, 1992; Welch, 1998).

Some carriers are effective in releasing the TGF-β family factors at the required rate and do not interfere with the normal process of osteogenesis (Gao *et al.*, 1997; Bax *et al.*, 1999; Hecht *et al.*, 1999), whereas other do. In general, only a few approaches provide practical modalities for delivery in a clinical setting (Groeneveld and Burger, 2000). We will focus on collagen matrices which have received high acceptance both in experimental animal models and in ongoing clinical trials (Rao, 1995; Khor, 1997). Type I collagen is the most abundant extracellular proteins of bone, and is essential for it strength. The most widely used carrier for BMPs is demineralised bone matrix (DBM) from cortical bone of various animal sources (Groeneveld *et al.*, 1999), which consists of more than 99% of type I collagen. This protein has been shown to stimulate osteoblastic differentiation of cells in culture (Shi *et al.*, 1996). *In vivo*, the addition of type I collagen was shown to aid bone formation in grafts containing osteogenin (Muthukumaran *et al.*, 1988), although collagen gels have been shown by others as an obstacle for callus formation in fracture healing (Bax *et al.*, 1999).

A mesenchymal stem cell (MSC) can be defined by its ability to maintain its population at an existing state or level, and to give rise to a number of differentiated cell types, such as chondrocytes and osteocytes. They are positively or negatively affected by biological agents, including hormones, growth factors (GFs), and cytokines, derived from cells of the same or different lineages (Roberts and Sporn, 1992). Osteogenic cell lines are believed to be derived from a common MSC, probably resident of the bone marrow (BM). It has been known for some time that different bones arise from different embryonic lineages, but knowledge of the nature of commitment steps from early stem cells to committed, but primitive, osteoprogenitors through later stage progenitors, preosteoblasts, and functional osteoblast is still relatively weak. MSCs are capable of self-renewal and undergo expansion in the presence of TGF-β1, a pleiotropic cytokine with autocrine and paracrine functions (Roberts and Sporn, 1992). There is more and more evidence that there may be a common progenitor cell of the haematopoietic and mesenchymal lineages. Such as cell families can reside within the marrow stroma (Horwitz *et al.*, 1999; Pittinger *et al.*, 1999). The haematopoietic system of vertebrates can be completely reconstituted with haematopoietic stem cells (HSCs) derived from, among others, the BM and umbilical cord blood (UCB). A cellular marker to identify those cells is the proteoglycan CD34, being the earliest identifiable HSC a $CD34^-$ fibroblast-like cell (Huss *et al.*, 2000), which express osteocalcin as a marker for mesenchymal progenitors and osteogenic precursors. In view of those recents data, it is conceivable that this could be a common precursor cell of hematopoiesis and mesengenic generating MSCs giving rise to cartilage, bone, and other mesenchymal organ systems.

At the end of the last decade new procedures to manipulate MSCs *ex vivo* have been develop, in order to create the opportunity to reimplant those capacitated cells into a host (Caplan and Bruder, 1997). Along that line, we have recently develop an *in vitro* procedure which includes a mesenchymal cell compatible collagen gel, impregnated with different GFs. We found that both, commercial hTGF-β1 and a genetically engineered factor, containing an auxiliary collagen binding domain (rhTGF-β1-F2), are able to select a cell population from the whole rat BM, capture it inside a collagen network, on the basis of it survival response to added exogenous GFs, and induce it toward the osteogenic lineage, expressing bone markers *in vitro* and when they were implanted *in vivo*. A possible therapeutic use in bone repair has been also recently discussed (Andrades *et al.*, 1999; 2001; 2002; Andrades and Becerra, 2002; Becerra *et al.*, 2001).

In this chapter, we review all these concepts from the point of view the usefulness of a new procedure to *in vitro* isolate, amplified and induce MSCs from different sources based on a collagen gel trap combined with modified GFs.

2. Materials and Methods

2.1. Genetic engineering of a recombinant hTGF-β1-F2 fusion protein

The rhTGF-β1-F2 fusion protein was engineered using a prokaryotic expression vector consisting of a 6xHis purification tag, an auxiliary vWF-derived collagen-binding domain (WREPSFMALS), and the carboxy-terminal 112-aminoacid sequence of human TGF-β1 active fragment (for more detail, see Andrades *et al.*, 1999).

2.2. Chemotaxis assay

To standardise the biological activities of the various GFs for our BM cell cultures, different concentrations of the GFs were evaluated using a chemotactic response assay. Rat BM cell chemotaxis was examined in modified Boyden chambers (Neuroprobe, Rockville, MD). The lower compartment of the chemotaxis chambers contained alpha-minimum essential medium (α-MEM) (negative control) or α-MEM to which GFs were added. To assay chemokinesis GFs were also added to the upper and lower compartments of each chamber. An 8 μm-pore size polycarbonate chemotaxis filter (Nucleopore, Pleasonton, CA), chosen after initial experiments to determine the pore size necessary for optimum response, separated the upper and lower well. Confluent cultures of primary rat BM cells were dissociated with 0.05% trypsin-0.02% EDTA, and cells were added to the upper wells at a concentration of 2×10^5 cells per ml of α-MEM medium. After 4 hours of incubation at 37°C in 5% CO_2, the filters were removed from the chambers, fixed in methanol, washed, stained with Wright's stain and mounted upside down on glass slides. Chemotaxis was quantitated by counting the number of cells that migrated in 20 light microscope high-power fields (×400) per filter. Chemotaxis was also determined with type I collagen gel in the lower wells mixed with the determined concentration of defined GF, in low-serum media or media containing 10% heat-inactivated fetal bovine serum (FBS). All samples were assayed in triplicate. Chemotactic activity was expressed as the mean ± SEM of the replicates.

2.3. Isolation and culture of BM stromal cells inside collagen gels

Isolation and cultures of BM stromal cells, and collagen gel preparations were performed as described previously (Andrades *et al.*, 1999). Briefly, fresh BM cells from 2-month-old male

Fischer 344 rats were used immediately after isolation. A single cell suspension was obtained by gently aspirating the disrupted marrow several times sequentially through 18, 20 and 22 gauge needles and filtered through a sterile 20-μm teflon sheet (Cell Strainer, Falcon, Lincoln Park, NJ) to exclude any tissue debris or cells clumps. For each experimental condition, 150 μl of collagen (0.35 mg/ml) were mixed with GF at concentrations previously selected by chemotaxis and assayed under the "trap-designed" conditions, as those concentrations that induced maximal chemotactic activity [Figs. 1(A) and (B)]. The concentrations of GFs used were:

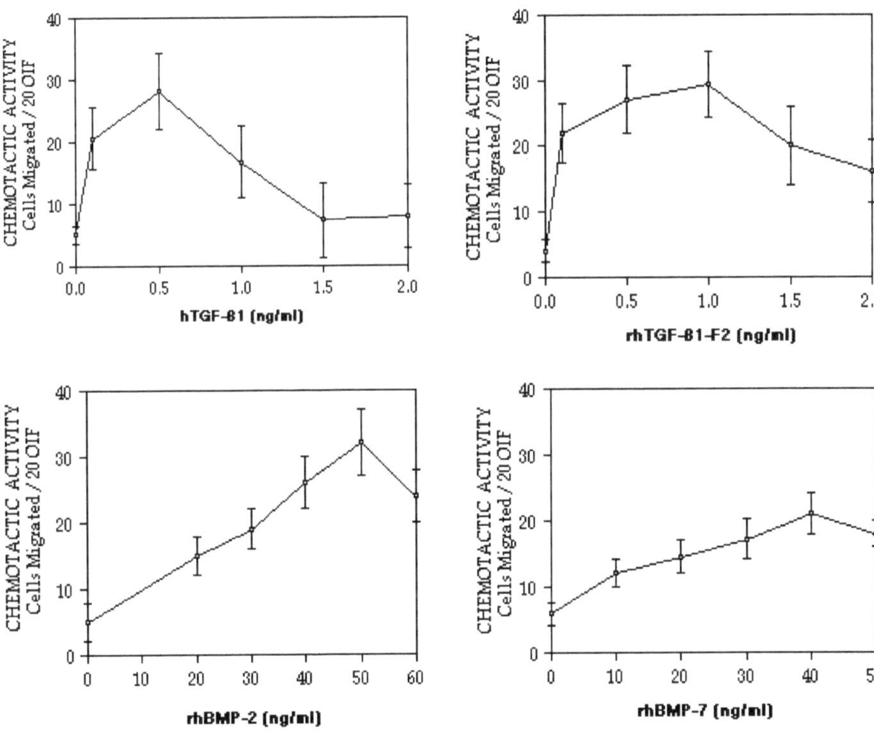

(A)

Fig. 1. (A) Chemotactic activity (cells migrated/20 OIF) of rat BM in the presence of different concentrations of GFs, and (B) under experimental conditions equivalents to S and A periods. OIF, light microscope high-power fields (×400). In B, GFs were used at the concentrations that gave rise maximum chemotactic response (see Materials and Methods for detail). Values are given as means ± SD for four filters.

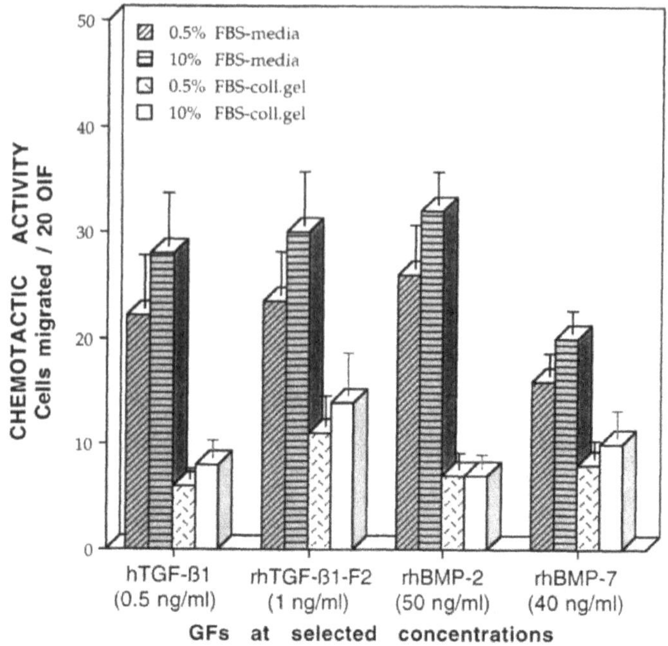

(B)

Fig. 1. (*Continued*)

hTGF-β1 0.5 ng/ml (R&D Systems, Minneapolis, MN), rhTGF-β1-F2 1 ng/ml, rhBMP-2 50 ng/ml (Genetics Institute, Cambridge, MA), and rhBMP-7 40 ng/ml (Creative Biomolecules, Boston, MA). Cells were mixed with the collagen-GF solution at a density of 2×10^6 cells/150 μl collagen/well in 48-well plates. The culture plates, 6 plates per experiment, were left 30 min at 37°C to allow the collagen to gel. Then, 150 μl/well of α-MEM containing the GF at the same concentration used in each respective experimental condition, were added on top of the collagen gel, and cultures maintained in a humidified atmosphere of 95% air and 5% CO_2 at 37°C and 100% relative humidity.

Figure 2 shows the schematic diagram of the experimental procedure. For the initial 10 days, cultures were carried out in 0.5% FBS in order to eliminate haematopoietic cells (selection period, S). The media was then changed to 10% FBS and cells

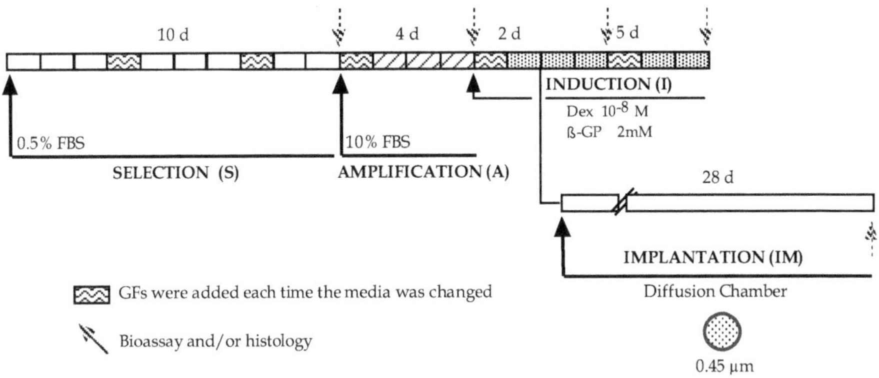

Fig. 2. Schematic diagram of the experimental procedure. See Materials and Methods for explanation.

cultured for an additional 4 days (amplification period, A). During the last 7 days of the *in vitro* period, 10^{-8} M dexamethasone (Dex, Sigma Chemical Co., St. Louis, MO) and 2 mM β-glycerophosphate (β-GP, Sigma) were added to the cultures to help with induction of osteogenesis (induction period, I). These inducers were added each time the media was changed. Eleven repetitions of the *in vitro* experiments were performed, each one with a different preparation of marrow-derived mesenchymal cells. The media were changed every 3 days and GFs added. Control cultures did not contain exogenous GFs. Cultures were monitored using an inverted phase-contrast Photomicroscope III (Zeiss, Oberkochen, Germany), and photographed using Kodak (Eastman Kodak Co., Rochester, NY) Technical PAN 400 ASA black and white film.

The culture medium consisted of α-MEM, containing 0.5% FBS for the initial 10 days of culture, and 10% up to the end of the *in vitro* experiment, 1% L-glutamine (GIBCO, Gaithersburg, MD), 1% penicillin/streptomycin (Sigma), and 0.5% fungizone (GIBCO).

2.4. Biochemical analysis

Cell number, expressed as DNA content, alkaline phosphatase (ALP) activity, osteocalcin (OC) expression, and calcium (Ca) content were measured as previously reported (Andrades *et al.*, 1999).

2.5. Cell culture studies

Cell morphology and colony appearance were documented at various times during culture. Mineralisation was examined by staining cultures *in situ* with von Kossa stain. Colony size was determined by measuring the diameter (mm) of colonies at days 10 (end S period), 14, 18 and 21 (A-I period), using a graded monocular eyepiece and a standard graticule. The number of colonies larger than 1 mm^2 were quantified by computer-assisted image analysis (Olympus CIA-102, Olympus Optical Co., Tokyo) by transiluminating plates on a light box and capturing the image on a television camera (Olympus Vanox-S).

Microscopic morphology was examined by fixing colonies in 10% phosphate buffered neutral formalin or Bouin's fixative solution, followed by dehydration in alcohol, embedding in paraffin, and sectioned at 5 μm. The sections were stained with haematoxylin-eosin (HE), picrosirius-haematoxylin (PSH) for detection of collagen under polarised light (Junqueira *et al.*, 1979), and 0.5% alcian blue (AB) (Fluka, Ronkonkoma, NY) in 0.1 N HCl, pH 1.0 (Montes and Junqueira, 1988), for detection of sulfated glycosaminoglycans (GAGs).

2.6. Measurement of chondrogenesis

The degree of chondrogenesis was evaluated by histochemical staining with AB. Colonies were released from the collagen gel either by shaking of the culture plates, or by collagenase digestion (0.05% collagenase type I, Worthington Biochemical Corp., Freehold, NJ) in Hank's balanced salts solution (Ca^{2+}- and Mg^{2+}-free) (GIBCO) for 5 min at 37°C. Colonies were washed with PBS, fixed for 10 min with 4% paraformaldehyde, stained with 0.05% AB in 5% acetic acid containing 400 mM $MgCl_2$ overnight, and rinsed with distilled water. The quantity of sulfated GAGs was estimated by measuring the amount of extractable dye. AB-stained cells were extracted with 4 mM guanidine-HCl for 2 hours at room temperature. Optical density of the extracted dye was measured at 580 nm in a multi-well plate reader (Molecular Devices).

2.7. Long-term haematopoietic fibroblast-like cell culture

Umbilical cord blood (UCB) samples were collected according to institutional guidelines from normal healthy donors, following procedures approved by the ethical committee of the medical center. Low-density mononuclear cells (MNC) were isolated from UCB by density centrifugation using Ficoll-Hypaque (Rafer S.L., Zaragoza, Spain). Cells were washed two times in medium 199 w/ Earles Salts (GIBCO) and then plated in each well of a 6-well plate at a density of 2×10^6 in the presence of McCoy's culture medium containing 12.5% FBS and 12.5% horse serum besides 1% L-glutamine (GIBCO), 1% penicillin/streptomycin (Sigma), and 0.5% Fungizone (GIBCO). The medium was changed every three days. During nearly 3 weeks in culture, cells were observed daily by an inverted phase-contrast Photomicroscope III and photographed as described.

2.8. *In vivo* studies

After 16 days of *in vitro* culture (Fig. 2), collagen gels were digested with collagenase (0.05% collagenase type I) as described by Andrades *et al.* (1999), and cells dissociated by subsequent digestion with 0.05% trypsin-0.02% EDTA at 37°C. Cells were centrifuged, washed, resuspended in serum-free α-MEM, counted and transferred to diffusion chambers. In this study, a total of 70 diffusion chambers, 14 mm diameter × 2 mm (Millipore Ltd., London), 0.45 μm pore-size filter, were implanted in 20 animals. In each animal, 7 diffusion chambers, filled with 5×10^5 cells processed *in vitro* under different experimental conditions, were implanted. Twenty eight-week-old male isogeneic (inbred strain) Fischer 344 rats were anaesthetised and surgery practised as described (Andrades *et al.*, 1999). Four weeks later the host rats were sacrificed by an overdose of anaesthesia, and the implants harvested and cut longitudinally in two halves, one for histology and one for biochemical analysis.

2.9. Histology

The plastic rings of the diffusion chambers were dissolved in toluene for 10 min. All procedures were carried out at room temperature. Routinely dehydration, paraffin embedding, cutting, and staining with AB, PSH, and HE, were performed. When necessary, digestions on tissue slides were practised with papain (Papain IV, Difco Laboratories, Detroit, MI) 0.5% dissolved in phosphate buffer, pH 4.7, containing 5 mM sodium bisulphite and 0.5 mM EDTA (Junqueira *et al.*, 1980), hyaluronidase (testicular hyaluronidase from Sigma) 0.05% dissolved in 0.1 M phosphate buffer pH 5.5 (Santamaría and Becerra, 1991), or chondroitinase AC (Miles Laboratories, Elkhart, IN) 1 U/ml in 0.05 M ethylenediamine-acetate buffer at pH 8.0 (Montes *et al.*, 1982).

2.10. Immunohistochemistry

Immunocytochemistry was carried out using antisera, anti-collagen I, II, III, IV and anti-laminin (Chemicon International Inc., Temecula, CA) as previously described (Santamaría *et al.*, 1992). To test the antisera, rat fetal ribs, ear cartilage, trachea, elastic artery, and kidney, were used as controls. Tissue sections were observed with a Nikon Microphot FXA (Nikon) and photographed using Kodak (Eastman Kodak Co.) Technical PAN 200 ASA colour film.

3. Results

3.1. Chemotactic response

All GFs assayed stimulated migration of BM cells to some extent in a dose-dependent manner [Fig. 1(A)]. Although BM cells showed higher chemotactic activity with a lower amount of hTGF-β1 than rhTGF-β1-F2, the modified recombinant factor was able to maintain elevated values of chemotactic activity over a wider concentration range. At 1 ng/ml rhTGF-β1-F2 stimulated chemotaxis to a greater extent than a similar concentration of hTGF-β1. Whereas migration was reduced to background levels at 1.5 ng/ml hTGF-β1, rhTGF-β1-F2 still induced chemotactic activity. However, maximum chemotaxis was the same in both cases. Cells showed greater chemotactic activity in the presence of rhBMP-2 than rhBMP-7. Recombinant hBMP-2 also stimulated chemotaxis with a marked peak at 50 ng/ml. Zigmond-Hirsch checkerboard analysis shows that all GFs tested stimulated chemotaxis and not chemokinesis (data not shown).

To simulate conditions used in the progenitor cell "trap-design", we measured the chemotactic activity of BM cells in low-serum (S period) and in 10% FBS media (A period), either with or without type I collagen matrices and added GFs [Fig. 1(B)]. Under these conditions, maximal chemotactic activity was exhibited in the presence of serum, and stimulation of chemotactic activity by GFs was markedly diminished when these were entrapped in collagen gels. In these circumstances, rhTGF-β1-F2 stimulated chemotaxis maximally, suggesting that the collagen binding domain imparts greater and/or prolonged biological activity to hTGF-β1.

3.2. *In vitro* cultures

At different points of the *in vitro* procedure (Fig. 2), cell number, expressed as DNA content, ALP activity, OC synthesis, and Ca content were measured. Chondrogenesis evaluation as well as morphological studies of the cultures were also carried out.

3.2.1. Biochemical analysis

As shown in Fig. 3(A), cultures exposed to hTGF-βs resulted in a slow but continuous cell death through the S period (*dot* and *encircled dot*). Cell proliferation was enhanced notably by 10% serum, which resulted in the amplification (A period) of the selected cell(s) population(s). Both hTGF-βs promoted cell survival, but maximal cell growth stimulation was obtained using rhTGF-β1-F2 at

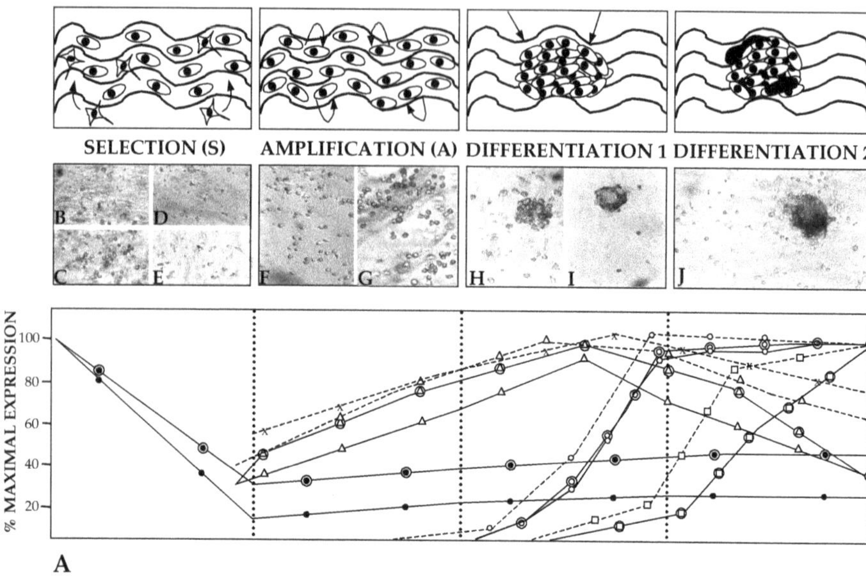

Fig. 3. Developmental sequence of the *in vitro* studies. (A) Temporal expression of cell growth- and differentiation-related markers during 21 days of culture. DNA: black dot; ALP: triangle; OC: white dot; Ca: square; Condrogenesis: cross. hTGF-β1 values: no encircled symbols; rhTGF-β1-F2: encircled symbols; rhBMP-2: dashed line. (B) hTGF-β1-, (C) rhTGF-β1-F2-, (D) rhBMP-2-, (E) rhBMP-7-cultured cells at day 10 observed under microscope. Round cells together with fibroblast-like cells are present at different levels of the collagen gel scaffold. (F) hTGF-β1-, (G) rhTGF-β1-F2-cultured cells at day 14. An homogeneous population of round blastoid cells can be observed. No fibroblast-like cells are present. (H) Only rhTGF-β1-F2-treated cells were able to form colonies concomitant with the first stages of an histogenesis process. (I) Histological paraffin section of a colony shows that it is formed by 10 to 20 cells integrated into a dense extracellular matrix. Hematoxylin-eosin. (J) Colonies start to mineralise at day 16 of culture.

The top panel represents a hypothetical model of the developmental sequence throught the *in vitro* experimental procedure. This model proposes that unrecognisable osteoprogenitor cells exist among the marrow nonadherent cell population. The selected progenitor cell population enriches it pool with fibroblast-like cells that, undifferentiated, modified their spread phenotype to round one, and it exerts mechanisms of proliferation in the presence of 10% serum (amplification period). Under morphogenic influences (collagen and bone-related regulators such as TGF-βs members), a histogenic process triggers the differentiation pathway which culminates with mineralisation events.

any time [Fig. 3(A), *encircled black dot*], according to the values of DNA synthesis obtained. ALP is a well-known early marker of bone cell differentiation. Because of the low cell number during the S period, values of ALP activity were low but detectable in the presence of both hTGF-βs [Fig. 3(A), △ *and* ⓐ]; however, cells cultured in the presence of rhTGF-β1-F2 (ⓐ) exhibited twofold higher levels of ALP activity. Once normal serum conditions were re-established (A period), values for rhTGF-β1-F2 were threefold greater than those for controls and onefold greater than those for hTGF-β1 (△). OC is an osteoblast-specific protein and a well-established marker of the mature osteoblasts phenotype. No detectable OC was found in controls. BM cells treated with hTGF-βs (∘ *and* ⊙) showed lower levels of OC during the early I period (dex and β-GP added) than rhBMP-2 (--∘--). The precipitated Ca analysed in cell cultures reflects the ordered accumulation of mineral in the ECM. No Ca was detected in controls or hTGF-β1-treated cells, being significant in BM cultures treated with rhTGF-β1-F2 (▣), although no Ca accumulation prior to the I period was detected. The highest values of Ca were for rhBMP-2 (--□--). Ca content was found to be dependent on the amount of inducers added and mineral deposition occurs during the late I stage (Fig. 3, late differentiation 1), considered the period of mineralisation.

3.2.2. Effect on chondrogenesis

AB staining for sulfated GAGs was used to evaluate chondrogenesis *in vitro*, since the production of this ECM component is a hallmark of chondrogenesis. While the intensity of AB staining increased over time in the presence of rhBMPs [Fig. 3(A), rhBMP-2 as --×--], no AB staining was detected in the presence of hTGF-βs. The number of AB-stained colonies reaches a maximum at day 18 (differentiation 2), after which the intensity of the stain decreases gradually. The AB-stained colonies appeared shortly after the A period, and quickly increased during the I period.

3.2.3. Morphological studies

Although TGF-β1 alone has been described to induce fibroblasts to form colonies in soft agar, in our study only rhTGF-β1-F2 was able to generate cell aggregates in the later S period [Fig. 3(C)]. The general appearance of cultures in the presence of hTGF-β1 were minor cell aggregates and abundant individual fibroblast-like cells visible by interference microscopy at different levels of the gel. Twice as many cell aggregates were formed at the end of this period in the hTGF-βs-group as in those treated with rhBMP-2 or rhBMP-7 [Figs. 3(D) and (E), respectively]. When cultures were returned to 10% FBS (A period), the number of round blastoid cells increased in cultures treated with hTGF-β1 [Fig. 3(F)], and abundant well-separated colonies of round blastoid cells were evident in cultures treated with rhTGF-β1-F2 [Figs. 3(G–I)]. This effect of rhTGF-β1-F2 on cell growth appears to be restricted to before the cells expressed the mature osteoblastic phenotype, when cultured in the presence of osteogenic inducers such as dex and β-GP (I period). These facts speak in favour of a better availability of the rhTGF-β1-F2 to the cells when combined with type I collagen, probably due to a longer half-life time of the ECM-targeted GF under these culture conditions. Only a slight increase over the initial number of colonies was observed in the presence of rhBMPs (data not shown). During the I period mineralisation took place, and the ECM of the colonies became mineralised [Fig. 3(J), differentiation 2], after the addition of the osteoinductive agentes.

3.2.3. Long-term haematopoietic fibroblast-like cell cultures

Cultures showed adherently growing fibroblast-like cells within 15–20 days (> 50% confluency of the tissue culture flask surface). Once a fibroblast-like layer is formed very few cells died, and a possible cell to cell interaction being crucial for viability and proliferation (Fig. 5). However, the efficacy of isolating these fibroblast-like cells from UCB by mean of plastic adherence is also dependent upon individual differences, probably due to the individual rate of these progenitor cells from the BM at the time of the blood collection. After a confluent adherent cell layer culture was established, for maintenance of long-term viable adherent cells, it was of pivotal importance to keep the confluency of the cells in culture between 70% and 90%. Of advantages was also the addition of long-term culture-conditioned medium to the established adherent layer to maintain high viability and to induce proliferation.

3.3.1. *In vivo* studies

After *in vitro* amplification and differentiation, the selected cells were implanted *in vivo*. Four weeks after implantation, observations were made on either two or three explants for each experimental conditions.

3.3.2. Histological studies

The study of the tissue formed inside the chambers was performed in each case following histochemical and immunocytochemical procedures, in order to understand the nature of such a tissue. A combination of structure and collagen types, proteoglycans (PGs), their interactions, alcianofilia and morphological features, have been used in this study.

Diffusion chamber controls were partially filled with loose fibrous connective tissue, tightly attached to each of the membrane filters, and containing collagen fibrils visualised by PSH stain (data not shown). Implants filled with cells treated with GFs, showed conspicuous differences among them. Cells treated with hTGF-β1 [Figs. 4(A–C)] and rhTGF-β1-F2 [Figs. 4(D–I)] formed several nodules of dense tissue (more extended in the rhTGF-β1-F2 case), located near the filters and surrounded by soft tissue, similar to the control. Nodules consisted of a tissue displaying mixed appearance: fibrous, cartilagenous and, even, osteoid. They showed a weak alcianofilia [Fig. 4(A)] and high collagen content [Fig. 4(D)]. Collagen displayed low interaction with PGs in the case of hTGF-β1 [Figs. 4(B) and (C)], and higher in rhTGF-β1-F2, as shown after papain digestion [Figs. 4(E) and (F)]. Tissue looks structured in

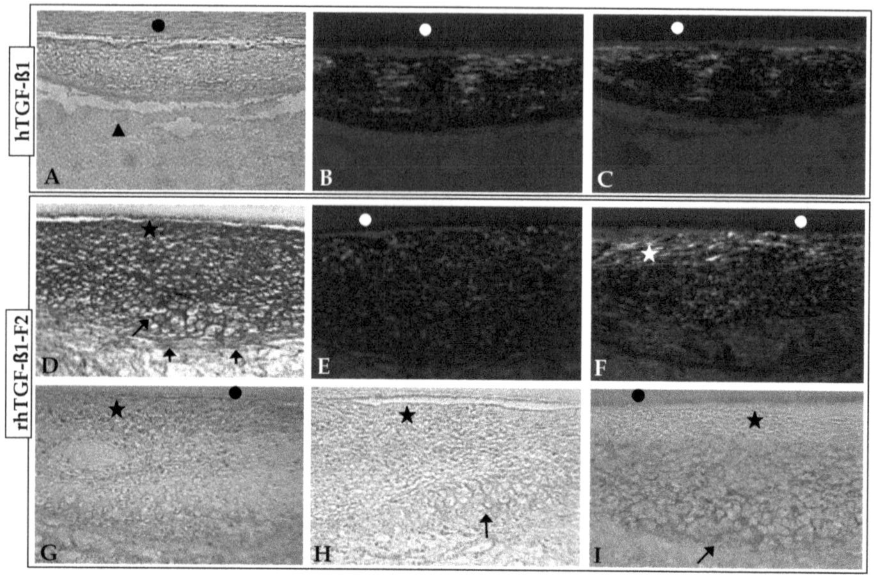

Fig. 4. Sections of tissue formed by cells treated with hTGF-β1 (A–C) and rhTGF-β1-F2 (D-I) implanted in diffusion chambers for 28 days. Contiguous sections of a nodule of tissue stained with AB (A), PSP (B) and PSP after papain digestion (C). Papain weakly modifies the birefringence of the collagen. Triangle indicates the fluid tissue surrounding the nodules. Contiguous sections of a nodule stained with PSH (D), PSP (E), PSP after papain digestion (F). Characteristics of a more mature tissue (stars) can be seen near the filter (dots) with an osteoid aspect rather than a immature area at the opposite side of the nodule (arrow). Small arrows show a fibrous perichondrium-like layer. Sections of nodules stained with anti-collagen I (G), anti-collagen II (H), and anti-collagen IV (I). In every case, differences between mature tissue (stars) close to the filter (dots) and immature areas (arrows) can be observed. Adapted from Andrades *et al.* (*Connect. Tissue Res.* in press) with permission.

this last case when observed under polarised light, and it even appears as osteoid near the wall of the filter [Fig. 4(F), ★]. As shown in Fig. 4(D), basofilia was more intense in those areas with cartilagenous morphology (↑). Near the wall, a more mature region was less cellular and more fibrous, looking like osteoid tissue. Moreover, the region around the chondroblast-like cells was weakly positive to anti-collagen II [Fig. 4(H), ↑], whereas anti-collagen I did not bind to any structure at the medial zone

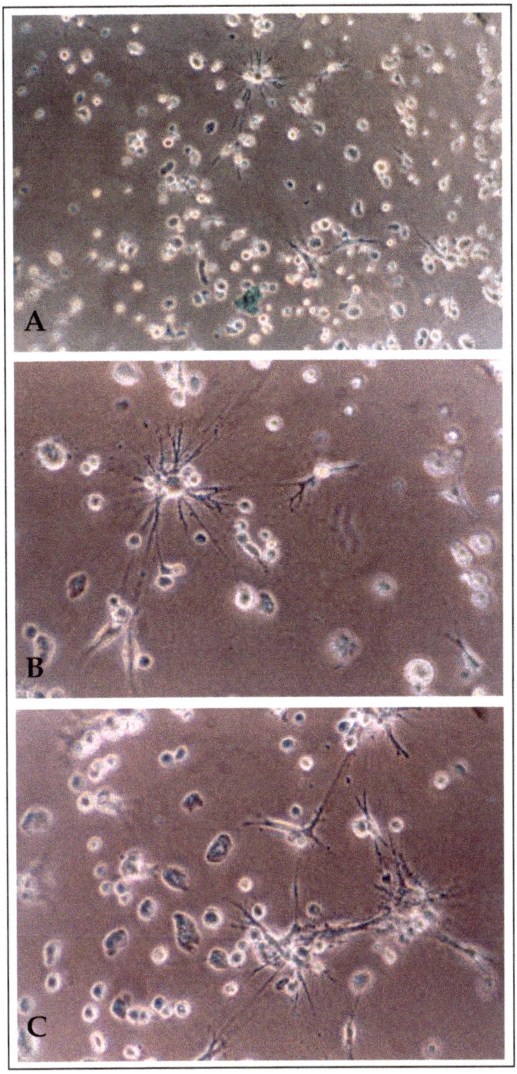

Fig. 5. Long-term cell culture from umbilical cord blood. (A) A 50% confluent culture of 18 days. (B) A detailed view of (A) shows adherently growing fibroblast-like cells that (C) interact among them. This cell population is a candidate to be investigated under the treatment of rhTGF-β1-F2 on the collagen scaffold system.

of the nodule, while in mature regions and at the periphery of the nodule, positive structures were present [Fig. 4(G), ★]. A very thin layer of fibers surrounding the cartilage-like area resembling a perichondrium, can also be detected [Fig. 4(D), *small arrows*]. Anti-laminin (data not shown) and anti-collagen IV [Fig. 4(I), ↑] antisera showed a similar pattern of distribution, revealing a network of fibers around the cells, being more conspicuous where the tissue appeared immature.

3.3.2. Biochemycal analysis

As shown in Table 1, maximum values for ALP were found in chambers filled with hTGF-βs-treated cells, particularly rhTGF-β1-F2: 6.7-fold greater than those for control chambers ($P < 0.001$) and double those of rhBMPs. Conversely, OC levels were highest in chambers implanted with cells treated with rhBMP-2, 14-fold greater than those for controls ($P < 0.01$), and were significantly lower in the hTGF-βs-treated cells. Values of Ca were very low and similar for both hTGF-βs. Maximum Ca content was detected in chambers with cells treated with rhBMP-2. No Ca was detected in controls.

Table 1. Biochemical analysis of the chambers content after the *in vivo* period of implantation, in the different experimental circumstances (Control, hTGF-β1, rhTGF-β1-F2, rhBMP-2, rhBMP-7). Levels of ALP (U/mg ww), OC expression (ng/mg dw), and Ca content (μg/mg dw).

	Diffusion chambers		
	ALP	OC	Ca
Control	4.1 ± 0.7	3.2 ± 0.9	0.0
hTGF-β1	20.5 ± 1.2	22.0 ± 1.9	1.4 ± 0.4
rhTGF-β1-F2	27.5 ± 1.8	25.0 ± 1.7	1.5 ± 0.4
rhBMP-2	11.5 ± 1.0	43.0 ± 2.5	3.6 ± 0.9
rhBMP-7	12.0 ± 1.0	32.0 ± 2.1	2.8 ± 0.7

Values are given as means ± SD for four samples.

4. Discussion

Mesenchymal stem cells are committed into the osteogenic lineage through a cascade of events that are not very well elucidated (Caplan 1991; Bruder *et al.*, 1998), where the structurally related family of osteoinductive factors, the TGF-βs, which include the BMPs, are implicated.

It has been described that TGF-βs are a family of GFs with an important role in different tissues. They regulate protein production, growth, adhesion and differentiation of cells. TGF-β has a very potent effect on cells associated with bone (Mundy, 1991). *In vitro* studies have shown that TGF-β stimulates proliferation, inhibits differentiation and increases matrix production in osteoblasts, and increases chondrogenesis. Sumner *et al.* (1995) with hydroxyapatite as a carrier, induced bone in dog humerus defects, and Moxham *et al.* (1996), using implants of DBM, have shown bone in segmental sheep tibial defects. Using an insoluble collagen matrix as a carrier in baboon, Ripamonti *et al.* (1996; 1997) have found bone formation in endochondral as well as in membranous sites. TGF-β is also capable of inducing bone when injected under the periosteum, and induce cartilage or bone ectopically (Ripamonti *et al.*, 2000). In every case, TGF-β induced bone using cells recruited within the host, from adjacent tissue, and probably already committed through the osteogenic lineage. However, in the present study, we have used uncommitted mesenchymal cells from BM, which were processed *in vitro* in the presence of a GF, and then introduced in the animal at ectopic sites, isolated from the influence of host cells. The presence of cartilage and bone in these implants demonstrates that the GFs used in our *in vitro* experimental conditions are able to select a mesenchymal cell population, amplify, and induce them along the chondro-osteogenic lineage. In controls, the starving situation together with the absence of exogenous GFs could not allow this cell population to survive and, therefore, be absent from the implanted population.

In our study, to test the effect of a collagen matrix combined with a collagen-binding domain rhTGF-β1-F2 on putative osteoprogenitor cells, freshly isolated BM cells were cultured into such a scaffold, in the presence of low serum. These conditions resulted in the selection (i.e. the survival) of a population of round blastoid cells. When normal serum culture conditions were re-established, this TGF-β-responsive cell population increased proliferating rates in the presence of rhTGF-β1-F2 and then differentiated into cells that expressed markers associated with the chondro/osteoblast lineage when stimulated by appropriate chemical modulators. Subsequently, large amounts of cartilage, and in some circumstances bone, appeared when those cells were implanted in chambers subcutaneously in rats.

In comparison, rhBMPs generated the highest initial values of ALP, OC, matrix mineralisation, and chondrogenesis. Comparing rhTGF-β1-F2-collagen with rhBMPs-collagen, the number of cells recruited in the presence of rhTGF-β1-F2, as well as the number and size of colonies, were considerably greater. The prompt mineralisation of colonies formed under rhBMPs treatment could limit their growth and may also explain the decrease of AB staining of GAGs found during the differentiation period. Alternatively, the rhBMP-selected cells may represent a more committed stem cell. They express skeletal markers and form hard tissues when implanted (Andrades *et al.*, 1999; Becerra *et al.*, 2001). We observed that cells cultured in the presence of rhBMP-2 or rhBMP-7 were able to form large amounts of cartilage and bone, in diffusion chambers (Andrades *et al.*, 2001). Values of OC synthesis and Ca content were in concordance with the histology. However, ALP activity was much less than that obtained with rhTGF-β1-F2, probably due to the earlier expression of this enzyme at the initiation of the skeletogenic pathway.

Taken together, these facts suggest that rhTGF-β1-F2 *in vitro* applied to a bovine collagen matrix as vehicle and delivery system could be of advantage in promoting the survival, proliferation, differentiation, and colony mineralisation of the osteogenic precursor cell population. Differences between rhTGF-β1-F2 and the commercial hTGF-β1 may be due to the slow release of the collagen binding factor from the collagen fibrillar network to which it is bound, resulting in a longer half-life. The rhBMPs appear to promote the survival and differentiation of committed osteogenic precursors into more mature osteoblasts. In fact, the results reported by our group suggest that rhBMP-2 and -7 select as well as stimulate the differentiation and maturation of committed chondro/osteoprogenitors arising from BM, while rhTGF-β1-F2 is selective for a more proliferative, less differentiated MSC, as it has been postulated recently (Andrades and Becerra, 2002).

On the basis of there is a common progenitor cell of CD34$^-$ HSC and MSC (Pittinger *et al.*, 1999), and CD34$^-$ fibroblast-like cells can be the source for a variety of specific tissues (Huss, 2000; Huss *et al.*, 2000), our laboratory is currently working under the hypothesis of those cells will provide a new tool in the treatment of diseases which affect the mesenchymal organ systems. Since UCB seems to be an excellent source of such cells (Erices *et al.*, 2000; Gutiérrez-Rodríguez *et al.*, 2000), we have initiated studies aimed at the biological characterisation of UCB-MSC derived cell populations. With this in mind, the goal of such as studies is to investigate the effect of rhTGF-β1-F2 on the collagen-trap system to enable CD34$^-$ fibroblast-like cells to differentiate into MSC and from this to the chondro/osteogenic cascade (Fig. 6). As it has been described by others (Huss *et al.*, 1997; Zvaifler *et al.*, 2000), our cultures of circulating MNC gave rise to a population of cells which change their morphological features (from round to fibroblast-like cells) and growth characteristics (from suspension to adherent growth), and were able to be maintained in long-term culture conditions without growth ceasing. This is an indication of the presence of appropriate self-GFs in the conditioned medium as well as the long-term supporting stroma in culture. A group of experiments need to be done in order to catch the "precursor" stem cell from the CD34$^-$ adherently growing fibroblast-like haematopoietic progenitor cells, upon the appropriate culture combination between rhTGF-β1-F2 and collagen matrices. The putative stem cell might differentiate into a "repopulating" stem cell that can differentiate into multiple lineages, such as muscle, tendon, ligament, endothelium, cartilage, bone or even nervous ones, according to the tissular enviroment and molecular induction.

In summary, our results suggest that TGF-β1 plays a crucial role in early stages of osteogenic commitment and differentiation, while

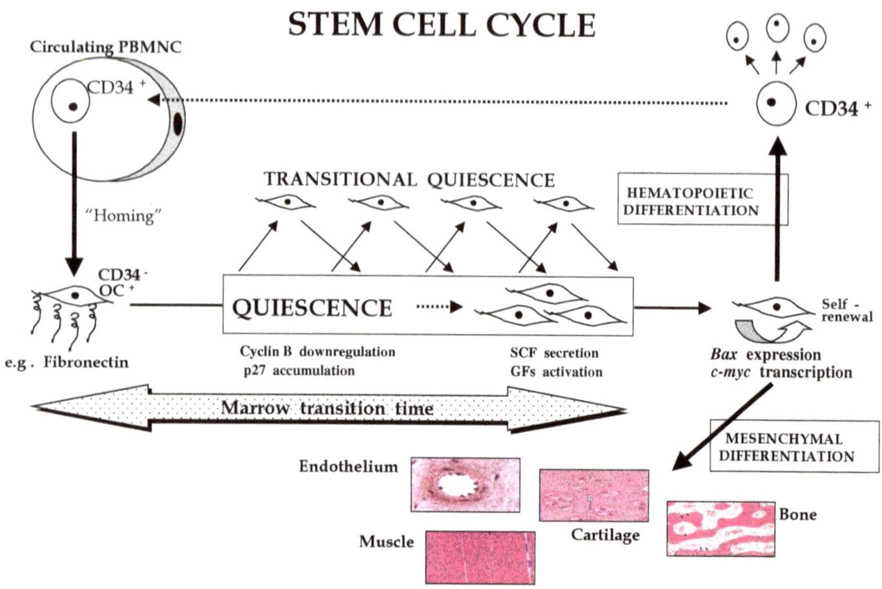

Fig. 6. Taking into account the information published by Huss *et al.*, expressed in part in this figure, as well as our recent preliminary results, the forthcoming objetive of our group is to investigate the possible relationship between CD34$^+$ and MSCs and to catch the adherently growing fibroblast-like haematopoietic progenitor cell following our *in vitro* culture technology. Under the appropiate combination between rhTGF-β1-F2 and type I collagen, this putative stem cell might initiate a mesenchymal differentiation into multiple lineages such as cartilage and bone, among others.

rhBMPs promote the survival and maturation of a less proliferative and more differentiated population of osteoprogenitor cells, as recently proposed by others (Van den Bergh *et al.*, 2000). The possible mechanism of different efficacy between commercial and modified TGF-β1 can be explained as follow: the collagen binding domain incorporates to the factor the ability to bind collagen type I specifically (an ECM-targeted GF, which means the possibility to deliver the GF to specific targets). This allows a slow release of the factor from the collagen matrix to which it is bound and therefore providing a longer half-life and better availability to the target cells. The collagen binding domain may impart greater and/ or prolonged biological activity to the fusion protein (Gordon *et al.*, 1997; Andrades *et al.*, 1999; Becerra *et al.*, 2001). We support the

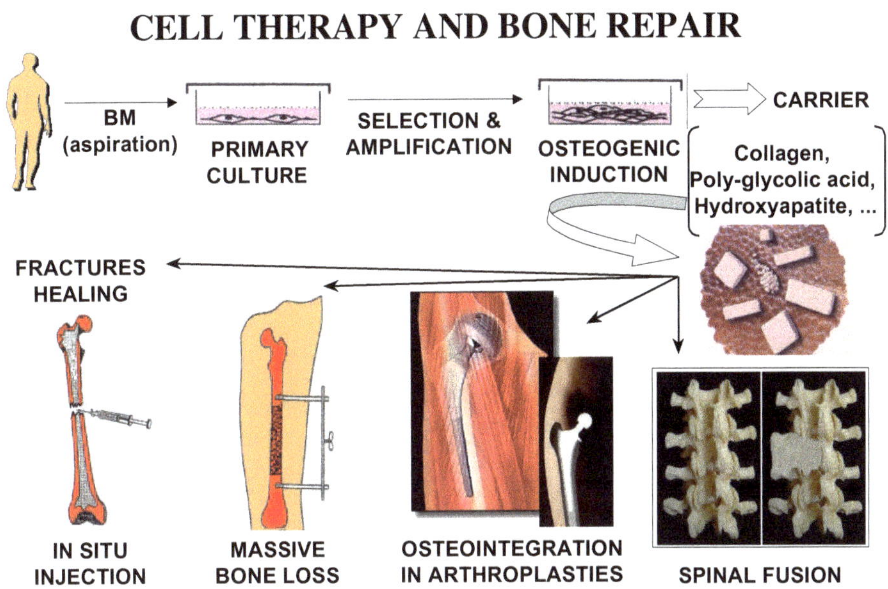

Fig. 7. With the aim to translate our experimental results to the clinical of bone repair, we are using our *in vitro* culture procedure containing human bone marrow cells. The committed cells, once induced to the chondro/osteogenic lineage, are adsorbed to hydroxyapatite implants as one of the most used carriers. These composites can be used to afford fracture healing programmes either by *in situ* injections into nonhealing small fractures, by reconstruction of traumatic bone loss, by recovering of prothesis to be used in osteointregration arthroplasties, or even to afford spinal fusion initiatives.

concept that osteogenic precursor cells can be selected from a mixed population of BM-MSCs by virtue of their distinctive survival responses in the presence of defined GFs and type I collagen, as a vehicle and delivery system, which may have potentially important implications in terms of developing strategic biomatrices for wound applications. Those selected cells exhibit unique properties in the chondro/osteogenic lineage that can ultimately be utilised to therapeutic advantages (Fig. 7).

5. Acknowledgements

The authors are grateful for the financial support of the Ministerio de Educación y Cultura (SAF99-0133) and Junta de Andalucía (CVI

0217, Consejería de Salud), Spain. The authors also wish to thank Dr. Marcel E. Nimni for his constructive criticisms of their study and for providing them with the rhTGF-β1-F_2. Special thanks go to Dr. Isidro Prat and Dr. María del Carmen Hernández (Blood Transfusion Center and Tissue Bank of Málaga, Spain) for making umbilical cord blood samples available for our laboratory; and Dr. Enrigue Guerado (Costa del Sol Hospital, Marbella) for helping the authors with clinical – orthopaedic concepts.

6. References

ANDRADES, J.A. and BECERRA, J. (2002). Bone morphogenetic proteins, collagen matrices and mesenchymal stem cells. In: *Advances in Skeletal Reconstruction Using Bone Morphogenetic Proteins*, T.S. Lindholm, ed., World Scientific Publishing Co. Pte. Ltd., Singapore, Vol. 2, pp. 54–62.

ANDRADES, J.A., HAN, B., BECERRA, J., SORGENTE, N., HALL, F.L. and NIMNI, M.E. (1999). Recombinant human TGF-β1 fusion protein with collagen-binding domain promotes migration, growth, and differentiation of bone marrow mesenchymal cells, *Exp. Cell. Res.* **250**, 485–498.

ANDRADES, J.A., HAN, B., NIMNI, M.E., HALL, F.L., ERTL, D.C., SIMPKINS, R.T. and BECERRA, J. Bone marrow cells induced *in vitro* by a modified human TGF-b1 form cartilage and bone when they are implanted *in vivo, Connect. Tissue Res.* (in press).

ANDRADES, J.A., SANTAMARÍA, J.A., NIMNI, M.E. and BECERRA, J. (2001). Selection and amplification of a bone marrow cell population and its induction to the chondro-osteogenic lineage by rhOP-1: An *in vitro* and *in vivo* study, *Int. J. Dev. Biol.* **45**, 689–693.

BAX, B.E., WOZNEY, J.M. and ASHHURST, D.E. (1999). Bone morphogenetic protein-2 increases the rate of callus formation after fracture of the rabbit tibia, *Calcif. Tissue Int.* **65**, 83–89.

BECERRA, J., ANDRADES, J.A., SANTAMARÍA, J.A., CIFUENTES, M. and GUERADO, E. (2001). Regeneración ósea, terapia celular e ingeniería tisular, *Med. Clin. (Barc.)* **116**, 23–34.

BOSTROM, M.P. and ASNIS, P. (1998). Transforming growth factor beta in fracture repair, *Clin. Orthop.* **355**(Suppl.), 124–131.

BRUDER, S.P., JAISWAL, N., RICALTON, N.S., MOSCA, J.D., KRAUS, K.H. and KADIYALA, S. (1998). Mesenchymal stem cells in osteobiology and applied bone regeneration, *Clin. Orthop. Rel. Res.* **355**(Suppl.), 247–256.

CAPLAN, A.I. (1991). Mesenchymal stem cells, *J. Orthop. Res.* **9**, 641–650.

CAPLAN, A.I. and BRUDER, S.P. (1997). Cell and molecular engineering of bone regeneration. In: *Textbook of Tissue Engineering*, R. Lanza, R. Langer and W. Chick, eds., RG Landes Co., Georgetown, pp. 603–618.

CENTRELLA, M., McCARTHY, T.L. and CANALIS, E. (1988). Skeletal tissue and transforming growth factor beta, *FASEB J.* **2**, 3066–3073.

ERICES, A., CONGET, P. and MINGUELL, J.L. (2000). Mesenchymal progenitor cells in human umbilical cord blood, *Brit. J. Haematol.* **109**, 235–242.

GAO, T.J., LINDHOLM, T.S., KOMMONEN, B., RAGNI, P., PARONZINI, A., LINDHOLM, T.C., JALOVAARA, P. and URIST, M.R. (1997). The use of a coral composite implant containing bone morphogenetic protein to repair a segmental tibial defect in sheep, *Int. Orthop.* **21**, 194–200.

GORDON, E.M., SKOTZKO, M., KUNDU, R.K., HAN, B., ANDRADES, J.A., NIMNI, M.E., ANDERSON, W.F. and HALL, F.L. (1997). Capture and expansion of bone marrow-derived mesenchymal progenitor cells with a transforming growth factor-β1-von Willebrand's factor fusion protein for retrovirus-mediated delivery of coagulation factor IX, *Hum. Gene Ther.* **8**, 1385–1394.

GROENEVELD, H.H. and BURGER, E.H. (2000). Bone morphogenetic proteins in human bone regeneration, *Eur. J. Endocrinol.* **142**, 9–21.

GROENEVELD, H.H., VAN DEN BERGH, J.P.A., HOLZMANN, P., TEN BRUGGENKATE, C.M., TUINZING, D.B. and BURGER, E.H. (1999). Mineralisation processes in demineralised bone matrix grafts in sinus floor elevations, *J. Biomed. Mater. Res.* **48**, 393–402.

GUTIÉRREZ-RODRÍGUEZ, M., REYES MALDONADO, E. and MAYANI, H. (2000). Characterisation of the adherent cells develop in Dexter-type long-term cultures from human umbilical cord blood, *Stem Cells* **18**, 46–52.

HECHT, B.P., FISCHGRUND, J.S., HERKOWITZ, H.N., PENMAN, L., TOTH, J.M. and SHIRKHODA, A. (1999). The use of recombinant human bone morphogenetic protein-2 (rhBMP-2) to promote spinal fusion in a nonhuman primate anterior interbody fusion model, *Spine* **24**, 629–636.

HORWITZ, E.M., PROCKOP, D.J., FITZPATRICK, L.A., KOO, W.W.K., GORDON, P.L., NEEL, M., SUSSMAN, M., ORCHARD, P., MARX, J.C., PYERITZ, R.E. and BRENNER, M.K. (1999). Transplantability and therapeutic effects of bone marrow derived mesenchymal cells in children with osteogenesis imperfecta, *Nat. Med.* **5**, 309–313.

HUSS, R. (1997). New definition and methods for isolation of the earliest peripheral blood-derived haematopoietic stem cells. In: *Transfusionsmedizin 1996/97. Beitr Infusionsther Transfusionsmed*, W. Sibrowski, W. Tangel and N. Müller, eds., Basel, Karger, Vol. 34, pp. 128–132.

HUSS, R. (2000). Isolation of primary and immortalized $CD34^-$ haematopoietic and mesenchymal stem cells from various sources, *Stem Cells* **18**, 1–9.

HUSS, R., GATSIOS, P., GRAEVE, L., LANGE, C., EISSNER, G., KOLB, H.J., THALMEIER, K. and HEINRICH, P.C. (2000).

Quiescence of CD34-negative haematopoietic stem cells is mediated by downregulation of cyclin B and no STAT activation, *Cytokine* **12**, 1195–1204.

JUNQUEIRA, L.C.U., BIGNOLAS, G. and BRENTANI, R.R. (1979). Picrosirius staining plus polarisation microscopy, a specific method for collagen detection in tissue sections, *Histochem. J.* **11**, 447–455.

JUNQUEIRA, L.C.U., BIGNOLAS, G., MOURAO, P.A. and BONETTI, S.S. (1980). Quantitation of collagen-proteoglycan interaction in tissue sections, *Connect Tissue Res.* **7**, 91–96.

KHOR, E. (1997). Methods for the treatment of collagenous tissues for bioprostheses, *Biomaterials* **18**, 95–105.

MONTES, G.S., BECERRA, J., TOLEDO, O.M., GORDILHO, M.A. and JUNQUEIRA, L.C. (1982). Fine structure and histochemistry of the tail fin ray in teleosts, *Histochem.* **75**, 363–376.

MONTES, G.S. and JUNQUEIRA, L.C.U. (1988). Histochemical localization of collagen and of proteoglycans in tissues. In: *Collagen*, M.E. Nimni, eds., CRC Press, Boca Raton, Vol. 2, pp. 41–72.

MOXHAM, J.P., KIBBLEWHITE, D.J., DVORAK, M., PEREY, B., TENCER, A.F., BRUCE, A.G. and STRONG, D.M. (1996). TGF-β1 forms functionally normal bone in a segmental sheep tibial diaphyseal defect, *J. Otolaryngol.* **25**, 388–392.

MUNDY, G.R. (1991). The effects of TGF-β on bone. Clinical applications of TGF-β, *Ciba Foundation Symposium* **157**, 137.

MUTHUKUMARAN, N., MA, S. and REDDI, A.H. (1988). Dose-dependence of and threshold for optimal bone induction by collagenous bone matrix and osteogenin-enriched fraction, *Coll. Relat. Res.* **8**, 433–441.

NIMNI, M.E. (1997). Polypeptide growth factors: Targeted delivery systems, *Biomaterials* **18**, 1201–1225.

PITTINGER, M.F., MACKAY, A.M., BECK, S.C., JAISWAL, R.K., DOUGLAS, R., MOSCA, J.D., MOORMAN, M.A., SIMONETTI, D.W., CRAIG, S. and MARSHAK, D.R. (1999). Multilineage potential of adult human mesenchymal stem cells, *Science* **284**, 318–321.

RAO, K.P. (1995). Recent developments of collagen-based materials for medical applications and drug delivery systems, *J. Biomater. Sci.* **7**, 623–645.

REDDI, A.H. (1992). Regulation of cartilage and bone differentiation by bone morphogenetic proteins, *Curr. Opin. Cell. Biol.* **4**, 850–855.

RIPAMONTI, U., VAN DEN HEEVER, B., SAMPATH, T.K., TUCKER, M.M., RUEGER, D.C. and REDDI, A.H. (1996). Complete regeneration of bone in the baboon by recombinant human osteogenic protein-1 (hOP-1, bone morphogenetic protein-7), *Growth Factors* **13**, 273–289.

RIPAMONTI, U., DUNEAS, N., VAN DEN HEEVER, B., BOSCH, C. and CROOKS, J. (1997). Recombinant transforming growth factor-beta 1 induces endochondral bone in the baboon and synergizes with recombinant osteogenic protein-1 (bone morphogenetic protein-7) to initiate rapid bone formation, *J. Bone Miner. Res.* **12**, 1584–1595.

RIPAMONTI, U., CROOKS, J., MATSABA, T. and TASKER, J. (2000). Induction of endochondral bone formation by recombinant human transforming growth factor-beta2 in the baboon (*Papio ursinus*), *Growth Factors* **17**, 269–285.

ROBERTS, A.B. and SPORN, M.B. (1989). Regulation of endothelial cell growth, architecture, and matrix synthesis by TGF-beta, *Am. Rev. Respir. Dis.* **140**, 1126–1128.

ROBERTS, A.B. and SPORN, M.B. (1992). Differential expression of the TGF-beta isoforms in embryogenesis suggests specific roles in developing and adult tissues, *Mol. Reprod. Dev.* **32**, 91–98.

SANTAMARÍA, J.A. and BECERRA, J. (1991). Tail fin regeneration in teleosts: Cell-extracellular matrix interaction in blastemal differentiation, *J. Anat.* **176**, 9–21.

SANTAMARÍA, J.A., MARÍ-BEFFA, M. and BECERRA, J. (1992). Interactions of the lepidotrichial matrix components during tail fin regeneration in teleosts, *Differentiation* **49**, 143–150.

SHI, S., KIRK, M. and KAHN, A.J. (1996). The role of type I collagen in the regulation of the osteoblast phenotype, *J. Bone Miner. Res.* **11**, 1139–1145.

SUMNER, D.R., TURNER, T.M., PURCHIO, A.F., GOMBOTZ, W.R., URBAN, R.M. and GALANTE, J.O. (1995). Enhancement of bone ingrowth by transforming growth factor-beta, *J. Bone Joint Surg. Am.* **77**, 1135–1147.

URIST, M.R. (1965). Bone: Formation by autoinduction, *Science* **150**, 893–899.

VAN DEN BERGH, J.P., TEN BRUGGENKATE, C.M., GROENEVELD, H.H., BURGER, E.H. and TUINZING, D.B. (2000). Recombinant human bone morphogenetic protein-7 in maxillary sinus floor elevation surgery in 3 patients compared to autogenous bone grafts. A clinical pilot study, *J. Clin. Periodontol.* **27**, 627–636.

WELCH, R.D., JONES, A.L., BUCHOLZ, R.W., REINERT, C.M., TJIA, J.S., PIERCE, W.A., WOZNEY, J.M. and LI, X.J. (1998). Effect of recombinant human bone morphogenetic protein-2 on fracture healing in a goat tibial fracture model, *J. Bone Miner. Res.* **13**, 1483–1490.

ZVAIFLER, N.J., MARINOVA-MUTAFCHIEVA, L., ADAMS, G., EDWARDS, C.J., MOSS, J., BURGER, J.A. and MAINI, R.N. (2000). Mesenchymal precursor cells in the blood of normal individuals, *Arthritis Res.* **2**, 477–488.

SECTION IV:
VASCULAR GRAFTS

Advances in Tissue Banking Vol. 6

18 EXPERIMENTAL STUDIES ON ARTERIAL HOMOGRAFTS

J. BUJÁN and J. BELLÓN

Departament of Morphological Sciences and Surgery
Faculty of Medicine, University of Alcalá
Ctra N-II Km 33,6, 28871-Alcalá de Henares
Madrid, Spain

1. Introduction

In arterial revascularisation or reconstruction procedures, best results have been achieved using autologous material. More specifically, in the case of small and medium-calibre vessels, the ideal substitute is the internal saphenous vein. The problem arises when this vessel may not be employed because it has been used in previous heart or lower limb revascularisation surgery, or when the vein is varicosed, post-phlebitic, or of poor quality or insufficient size. This may occur in over 20% of patients (Kent *et al.*, 1989). The second option selected by many vascular surgeons is the use of the contralateral saphenous vein or other veins such as the external saphenous, cephalic or basilic. Nevertheless, in 10% of cases the use of autologous material is not possible (Gentile *et al.*, 1996).

Arterial replacement is, therefore, a serious problem in patients with no vessels available for grafting or bypass procedures and for several years now, possible alternatives have been the subject of ongoing research.

Historically, the first attempt to overcome the problem took the form of the arterial allograft (Carrell *et al.*, 1912; Oudot *et al.*, 1951). However, the preservation methods used and immunological rejection provoked substantial deterioration of the graft, leading to severe inflammatory reactions with secondary fibrosis, calcifications and the formation of aneurysms (Szilagyi *et al.*, 1970). These limitations resulted in a gradual decline of their clinical use.

The discovery of prosthetic materials that were biologically well tolerated was a significant development in arterial replacement surgery. However, in small and medium-calibre vessels, the rate of failure of these prostheses is high (Taylor *et al.*, 1992; Abbott *et al.*, 1997) and mainly attributable to thrombosis and stenosis at the sites of anastomosis, such that the goal of maintaining vascular patency in the graft zone is not achieved (Sottiurai *et al.*, 1988; Da Gama *et al.*, 1994). Moreover, the prosthetic material is susceptible to a further severe and difficult to resolve complication — graft infection — which eventually necessitates its retrieval and replacement (Moore and Cole, 1991; Koskas *et al.*, 1996; Verhelst *et al.*, 2000).

The emergence of tissue and organ banks has contributed towards the development of new alternative vessel substitutes, as are cryopreserved vessels or **HOMOGRAFTS**.

Cryopreserved veins are presently the subject of clinical studies and have mainly been applied in limb salvage surgery (Posner *et al.*, 1996; Buckley *et al.*, 2000; Harris *et al.*, 2001) and in difficult vein access for haemodialysis (Matsuura *et al.*, 2000). However, the number of publications on arterial homografts steadily increases, with satisfactory results obtained and applications in human clinical practice (Eugene *et al.*, 1998; Wengerter *et al.*, 1999).

Cryopreserved arterial homografts are especially useful in cases of an infected prosthetic graft. Indeed, according to several authors, this may be the most effective treatment for this complication and mycotic aneurysms (Vogt *et al.*, 1998, 1999; Desgranges *et al.*, 1998; Coselli *et al.*, 1999; Verhelst *et al.*, 2000). They may also be used in revascularisation procedures (Gournier *et al.*, 1993; 1955; Callow, 1996; Alonso *et al.*, 1999; Castier *et al.*, 1999) and in cases of aortoenteric fistulae (Vogt *et al.*, 1997). Another important application is their use in patients with multiple vascular access failure in haemodialysis due to thrombosis or infection (Takamoto *et al.*, 1998). Moreover, cryopreserved grafts may be reoperated on for their renewal or in the case of a thrombectomy (Matsuura *et al.*, 2000).

Vascular cryopreservation is thus an interesting option in the search for arterial substitutes. However, the efficiency of cryopreserved arterial grafts is not yet that desired, mainly because of spontaneous fracturing at the time of thawing or of implant (Lehalle *et al.*, 1997; Pegg *et al.*, 1997). Post-implant, degenerative or immunological complications (Couvelard *et al.*, 1995; Vischjager *et al.*, 1996; Leseche *et al.*, 1997; Castier *et al.*, 1999; Turley, 1999) have also been described.

The use of the experimental model is essential to improve current knowledge and understanding of the different methods and techniques designed to improve the viability and behaviour of these vessels in clinical practice. These models range from those described by authors such as Ellis *et al.* (1991), Muller-Schweinitzer *et al.* (1994) and Rossett *et al.* (1996), who evaluated the functional activity of cryopreserved arteries using human vessels, to those of workers such as Takeishi *et al.* (1994), who morphologically evaluated femoral arteries in allogenic implants performed in the rat. Experiments have also been undertaken in the rabbit (Pegg *et al.*, 1989; Fini e*t al.*, 1993; Song *et al.*, 1994) and dog (Ku *et al.*, 1990; 1994). In this last model, coronary arteries were frozen at −75°C to perform pharmacological studies of vessel constriction/dilation responses to cryopreservation. In 1994, Müller-Schweinitzer *et al.* performed an *in vitro* investigation on the contractile capacity of the cryopreserved dog femoral artery. More recent investigations have opted for the use of pig arteries to test different methods of cryopreservation (Arnaud, 2000) and thawing (Pascual *et al.*, 2000).

These and many other research efforts have provided the foundations on which to base clinical strategies, although the literature is full of discrepancies which we will discuss in this chapter.

1.1. Source of vessels for cryopreservation

Theoretically, vessels from cadavers or organ donors are the most easily accessible (Müller-Schweinitzer *et al.*, 1988). However, the former are not recommended for use because of irreversible alterations. Further, the number of autopsies in hospitals has decreased resulting in a lack of suitable donors (Moriyama *et al.*, 2000). This leaves functional vessels at the time of extraction as the best current option (Dinis da Gama *et al.*, 1994; Rosset *et al.*, 1996).

1.2. Transport and cryoprotecting media

The second point of possible controversy is how to transport these vessels and which protecting agents to use. Direct freezing of vessels in saline solution plus antibiotics yields poor results (Tice *et al.*, 1972; Ochsner *et al.*, 1984), with the formation of intracellular ice crystals. Thus, one has to consider not only the medium in which cryopreservation is conducted but also that in which the harvested vessels are immersed and transported to the cryopreservation unit. The findings of studies comparing University of Wisconsin solution (UW) and minimal essential medium (MEM) (Buján *et al.*, 2000), have served to minimise the changes produced in response to harvesting and subsequent cryopreservation. Also, the progression of intimal hyperplasia after implant may be diminished by the use of solutions such as those described by Rigol *et al.* (2000) or the addition of substances including polyethylene glycol and glutathione to the cryopreservation medium (Davies *et al.*, 1999).

Reducing the temperature below the point at which water freezes provoked the development and use of different cryprotecting solutions such as dimethylsulphoxide, ethylene glycol, glycerol, propane-1,2 diol and sucrose (Pollock *et al.*, 1991; Wusteman *et al.*, 1996; Shirakashi *et al.*, 1998). Of these, best results have been achieved with dimethylsulphoxide (DMSO) in different vehicle solutions at concentrations in the range 10 to 15% (Bateson *et al.*, 1994; Song *et al.*, 1995). Recent studies on its toxicity (Buján *et al.*,

2000; Ruddle *et al.*, 2000), indicate that the approximate 10% cell damage attributable to the DMSO, is an effect compensated for by its benefits.

An important consideration related to the use of a cryoprotectant is the time needed for the vessel to adapt to this type of solution at room temperature before the start of the freezing process. Most authors who mention this time period (Takeishi *et al.*, 1994; Ku *et al.*, 1994; Müller-Schweinitzer, 1998) apply a mean equilibrating time of approximately one hour.

1.3. Freezing

It is universally accepted in cryobiology that the freezing process must be slow in that the temperature needs to be reduced in a gradual manner (Bank *et al.*, 1987). Accordingly, mean accepted cooling rates are approximately 1°C/min, or even slower, until a temperature of –70°C to –120°C. Today, there are biological freezers in which this process is fully automated (Pomar *et al.*, 1996; Rigol *et al.*, 1997). Finally, a new preservation technique — vitrification — is being evaluated and involves the formation of a noncrystalline solid structure instead of ice (Brockbank *et al.*, 2000; Arnaud, 2000).

1.4. Maintenance and storage of cryopreserved vessels

Some authors indicate that maintenance of vessels at a temperature of –80°C gives rise to better results when specimens continue to cool down in liquid nitrogen vapour until a temperature of –140°C to –160°C (Ku *et al.*, 1994; Müller-Schweinitzer *et al.*, 1994; 1998). However, most researchers store vessels at –196°C by immersion in liquid or gaseous nitrogen.

Storage time is another factor yet to be standardised. These times, often undefined, range from a few days to several months (Fujitani *et al.*, 1992; Rossett *et al.*, 1996; Müller-Schweinitzer *et al.*, 1998). Some authors report that the cryopreservation time can affect the preserving of vessel structure (Dinis da Gama *et al.*, 1994; Rigol, 2000).

1.5. Thawing

Damage to vessels is been mainly attributed to the three factors mentioned above. Nevertheless, in the last few years a further factor has been identified as possibly responsible for this damage. This factor is the thawing process. Thawing is generally rapidly undertaken in a few minutes at 37°C (Song *et al.*, 1994; Knosalla *et al.*, 1998; Bellón *et al.*, 1999; Buján *et al.*, 2000; Rigol *et al.*, 2000; Pukaci *et al.*, 2000), although some workers have observed improved preservation of vessel structure when the thawing process is conducted in a gradual manner (Pegg *et al.*, 1997; Müller-Schweinitzer *et al.*, 1998).

In some investigations it was noted that rapid thawing induces damage particularly to the innermost third of the arterial wall. This damage would appear to be associated with a greater incidence of spontaneous fractures and especially with the appearance of microfractures. Several factors are related to the formation of these fractures and a critical range for their appearance (–100 to –150°C) has been established (Hunt *et al.*, 1994; Song *et al.*, 1994; Pegg *et al.*, 1997; Buján *et al.*, 2000).

2. Vascular Homografts

As may be inferred from the literature, there is no widely accepted protocol for cryopreserving and thawing arterial specimens. One should not forget that many of the different results obtained may be attributable to the different vessel types employed (Muller-Schweinitzer, 1994; Rosset, 1996) or even within the same vessel there are site differences (Arnaud, 2000). Needless to mention inter-species differences. Thus, for instance, if we examine the common iliac artery in the pig and rat, based on histological differences, we can classify the first as a muscular artery and the second as an elastic type artery [Figs. 1(a) and (b)]; this could obviously lead to conflicting results.

The choice of vascular setting for our research, the common iliac artery, was based on the past experience of our team (Bujan *et al.*, 1992; Bellón *et al.*, 1994; Bellon *et al.*, 1996; Bujan *et al.*, 1997).

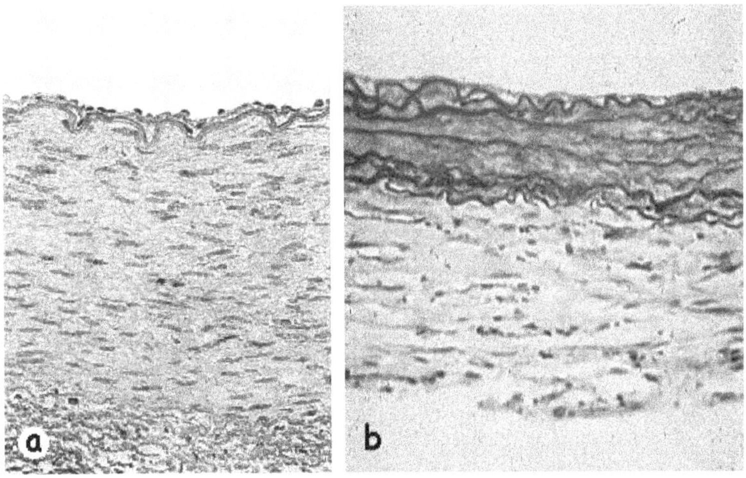

Fig. 1. (a) Iliac artery of the pig. Distribution of the elastic laminar component showing a lamina at the inside border and an abundance of laminae in the tunica adventitia (LM x200). (b) Iliac artery of the rat. Elastic laminae are distributed throughout the medial layer (LM x400).

This artery has a suitable lumen diameter and is not prone to vascular spasm, which at times causes delays and drawbacks when undertaking surgical techniques. Female recipients were selected due to the ease of grafting in the iliac region.

In *in vitro* studies on cryopreserved rat iliac arteries, we observed the appearance of changes, particularly endothelial, which we tried to reverse by immersing the iliac specimens in an organ bath. Contrary to the findings of other authors in fresh arteries that developed intimal hyperplasia, we observed that the changes induced by cryopreservation were irreversible, with signs of accelerated degradation (Bellón *et al.*, 1998).

We also conducted viability trials using different freezing media (MEM and UW) and temperatures (–80° and –196°C). Finally, we evaluated the slow thawing method by comparing it with the conventional rapid procedure. Using pig iliac arteries as the experimental model, results showed that the rate of total fractures and microfractures in the artery wall when frozen at –196°C exceeded 90% of the specimens processed, regardless of the freezing solution

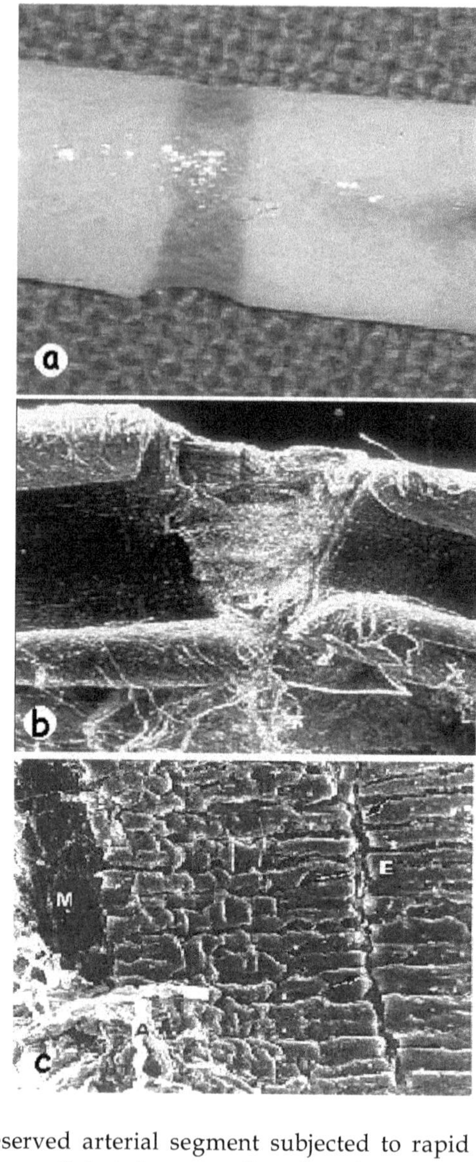

Fig. 2. (a) Cryopreserved arterial segment subjected to rapid thawing. Note the fracture in the artery wall that does not affect the adventitial layer. (b) View of the previous specimen after cutting open longitudinally. Detailed view of the sectioned artery wall. The base is composed of elastic fibres of the adventitial layer (SEM x20). (c) Same detail as above at greater magnification showing the edge of the fracture in the tunica media and several microfractures at the luminal surface close to the main fracture line. These microfractures typically affect the endothelium and internal elastic lamina (SEM x500).

used [Figs. 2(a)–(c)]. However, if thawing was conducted in a slow, gradual manner, the number of fractures significantly fell to around 40%.

There were no significant differences in terms of structural or ultrastructural changes in the vessel wall when the two (MEM, UW) cryopreservation solutions were compared. We only observed the accumulation of liquid in perielastic regions, possible corresponding to sites susceptible to fracturing during rapid warming. When the degree of cell damage induced by the cryopreservation process was analysed, we noted apoptosis values of some 36% for arterial segments preserved in Wisconsin solution and a smaller proportion, about 12%, when MEM was used. The enzyme activity of the extracellular matrix, determined by immunohistochemical techniques using antibodies specific for different metalloproteinases (MMPs), indicated enhanced MMP-2 expression after the cryopreservation process. This enzyme appears to be a good indicator of the degradative state of the matrix (Table 1).

Given all these previous studies, we were able to base the design of our *in vivo* grafting experiments on some of the already established factors. Thus, we applied the method of freezing in nitrogen vapour at a temperature of around –145°C. In each case,

Table 1. Metalloproteinases (MMP) expression during both control and thawing processes in porcines arteries. Stain criterium (–) < 10%, (+) 10–25%; (++) 25–50%, and (+++) > 50%.

		MMP-1	MMP-2	MMP-3	MMP-9
Control	Endothelium	+	–+	–	–
	Media	–	–+	–	–
	Adventitia	–	–	–	–
Slow thawing	Endothelium	++	++	–	–
	Media	+	++	–	–
	Adventitia	+	+	–	–
Rapid thawing	Endothelium	–	+	–	–
	Media	–	+	–	–
	Adventitia	–	+	–	–

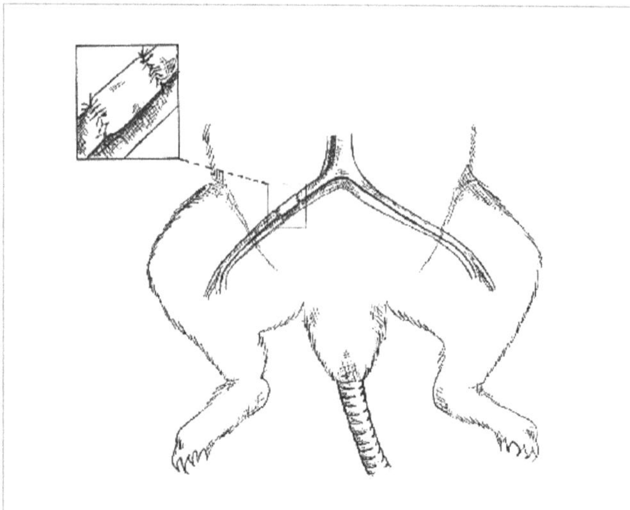

Fig. 3. Graft model in rat common iliac artery.

a slow, automated thawing procedure was undertaken such that the temperature was gradually lowered.

The next step was choosing the experimental animal model. We had investigated pig and rat iliac arteries and now was the time to commence implant studies in an experimental animal (Fig. 3). Today, the experimental animal needs to fulfil a series of requirements including ease of handling and low cost. Although there are many interspecies differences due to a range of variables from the structure of the arterial wall to the organism's coagulation system, which notably differs among species, the rat is the most widely used model in vascular research. Perhaps, the only drawback associated with this model is that the vessels are of a small luminal diameter and to perform a surgical procedure, one has to resort techniques of microsurgery.

At these initial stages, vascular transplants were performed in syngenic rats, since our interest lay in evaluating behavioural and morphological changes arising after implant related to the use of fresh or cryopreserved grafts, with no interference due to rejection.

Vascular viability was 73% in the group of cryopreserved grafts, compared to the 100% recorded for the fresh autografts. Similar results were reported by Takeishi *et al.* (1994) and somewhat better ones by Vischjager *et al.* (1996) in a dog model. This difference between fresh and cryopreserved grafts is probably related to the luminal surface, and particularly the behaviour of the endothelial monolayer. In the fresh autografts,

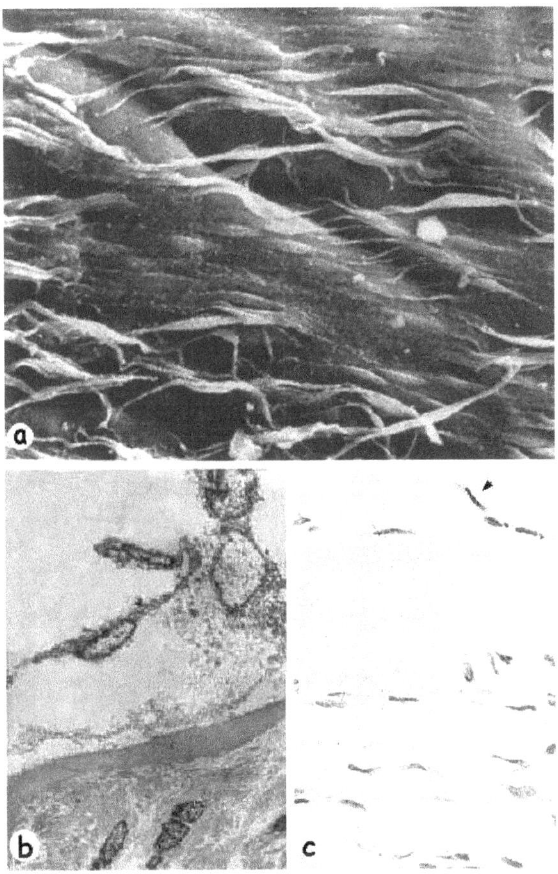

Fig. 4. (a) After grafting, damaged endothelial cells come away from the matrix, take on a fusiform shape and eventually become fully detached and free in the lumen (SEM x1000). (b) In the homograft, endothelial cells become detached. The ultrastructure indicates cell damage (TEM x3000). (c) Apoptotic cells showing dark nuclei (↑) indicating TUNEL-positivity (LM x400).

apparently undamaged endothelial cells remain on the luminal surface, although due to the hypoxia and ischaemia of the surgical procedure itself, changes are provoked that probably affect their capacity to adhere to the substrate. This leads to their detachment once arterial blood flow to the area is restored. From a morphological perspective, this detachment is accompanied by a change in cell shape to a fusiform cell that becomes fully detached and passes into the lumen [Fig. 4(a)]. The extracellular matrix, which becomes exposed to the blood interface, allows the deposition of platelets and red blood cells as a monolayer but there is no platelet aggregation. Consequently, the platelets themselves by secreting growth factors induce, among other effects, the migration of neighbouring healthy endothelial cells, contributing towards re-establishing the endothelial monolayer. In contrast, the luminal surface of the cryopreserved graft shows substantial endothelial cell damage compared to that shown by fresh grafts [Fig. 4(b)]. This cell damage is ultrastructurally characterised by peripheral condensation of chromatin, dilation of the rough endoplasmic reticulum, cytoplasmic microvesiculation and alterations to mitochondria, which appear swollen with broken crests. This description is the typical morphological picture of cell apoptosis and is also confirmed by the detection of broken ends of DNA using the TUNEL method [Fig. 4(c) and Table 2]. This type of cell damage

Table 2. Percentage of TUNEL-positive cells (means ± SD).

Groups	Time (days)	Intima	Media	Adventitia
Control	–	2.06 ± 0.20	0.25 ± 0.02	1.25 ± 0.9
Fresh autografts	14		–	–
	30		1.87 ± 0.06	–
	90		4.96 ± 0.99	–
Cryopreserved homografts	14	8.94 ± 1.01	4.37 ± 0.12	2.48 ± 0.04
	30	8.47 ± 0.67	7.03 ± 0.3	7.07 ± 0.53
	90	6.60 ± 0.32	11.22 ± 0.67	5.54 ± 1.08

may be partly responsible for an increase in the amount of blood deposited on the luminal surface during the initial stages after grafting, thus increasing the rate of thrombosis associated with cryopreserved vessels. On other occasions, the complication noted was haemorrhage (16.6%), possible due to the change in compliancy provoked by cryopreservation.

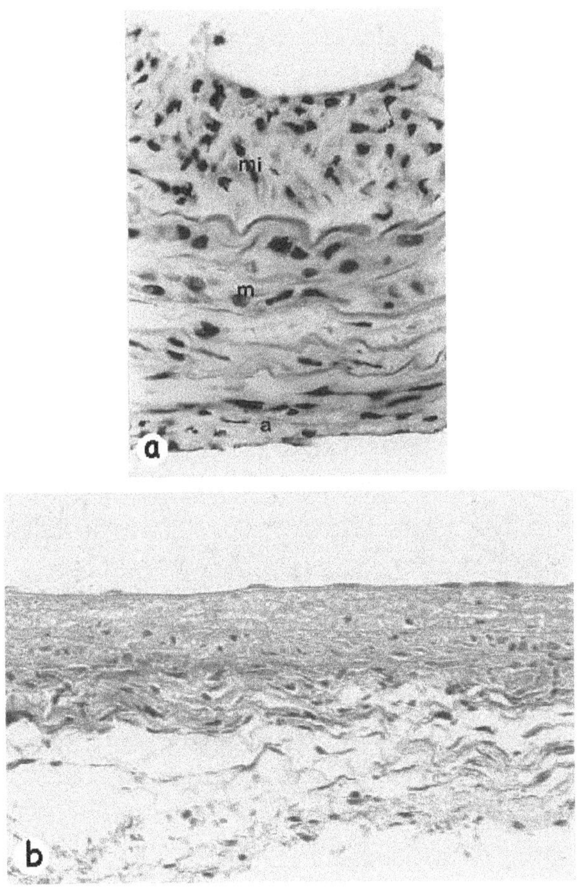

Fig. 5. (a) The reparatory response to the autograft at two weeks may be observed as the appearance of a new layer denoted the myointima which overlies the micrograft's internal elastic lamina. Note the medial layer of the autograft filled with smooth muscle cells (LM x630). (b) Absence of intimal hyperplasia in a cryopreserved micrograft two weeks after grafting. Endothelialisation is complete and there is marked cell loss in the medial layer (LM x400).

The region affected by the graft shows different behaviour. Thus, at the adjacent arterial ends, or recipient ends, the reparatory response is similar for fresh and cryopreserved grafts and involves centripetal expansion of intimal hyperplasia towards the graft. However, notable differences arise within the micrograft. While in the fresh vessel, this hyperplasia invades the entire surface forming a concentric layer over the luminal surface [Fig. 5(a)], when the graft is cryopreserved, intimal hyperplasia from the recipient ends timidly advances at the sites of anastomosis and the graft becomes recoated with regenerating endothelial cells on the luminal surface (the reader is reminded of the characteristics of these cells; they are larger in size and, on SEM examination, show abundant microvilli and well-defined cell borders) [Fig. 5(b)]. This delay in the appearance of intimal hyperplasia over the graft also occurs in the autograft model when this response is blocked with pharmacological agents and vascular prostheses of polytetrafluoroethylene (PTFE) (Bellón *et al.*, 1996; 1998).

The second event is cell loss in the tunica media first described in a paper by Pacholevic *et al.* (1996). The smooth muscle cells, which in previous post-cryopreservation investigations showed a certain degree of damage, are incapable of recovery after placement in the arterial setting. This phenomenon in fresh grafts occurs over the long term, yet becomes evident at two weeks in cryopreserved grafts and is also favoured by the part to be played by white blood cells. Both neutrophils and macrophages infiltrate the arterial wall and contribute towards the degradation of the extracellular matrix and mopping up of the remaining cell debris [Figs. 6(a)–(d)].

The physical changes suffered by the cryopreserved vessel are not immediately detectable after grafting. Determinations made of the elasticity or dilation of specimens following cryopreservation have failed to show any significant changes with respect to fresh vessels. However, after implant, we noted a marked lengthening of the cryopreserved vessel of up to 15%, compared to fresh grafts. This finding may be somehow related to the loss of cells; cells are not renewed and the elastic properties of the wall are gradually lost. Most evidence is shown by the thickness of the medial layer, where the changes incurred amount to 20% attributable to the

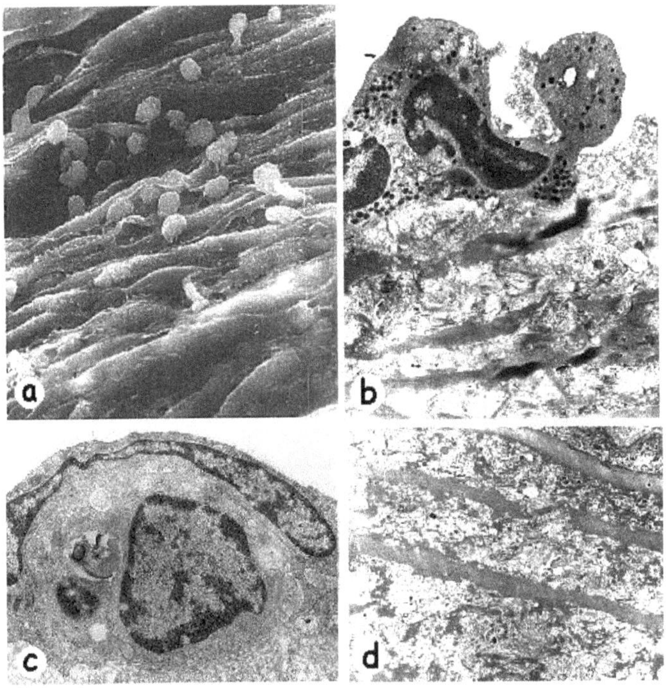

Fig. 6. (a) Image of the luminal surface of a homograft (SEM x1000). Note the adhesion and infiltration of white blood cells. (b) Neutrophils penetrate the homograft two weeks after implant (TEM x7000). (c) Presence of a monocyte-macrophage beneath the endothelium (TEM x12 000). (d) Ultrastructural appearance of the medial layer showing the absence of cells, matrix degradation and disrupted elastic laminae (TEM x3000).

cryopreservation process itself and to 50% related to grafting (Fig. 7). The mid- and long-term response is similar. Thus, the flap of initimal hyperplasia at the healthy ends colonises the entire graft and a neolayer is gradually formed, covering the cryopreserved segment and taking on the functions of the medial muscular layer (Bellón *et al.*, 1994; 1998) [Figs. 8(a) and (b)].

It may be summarised that these experimental studies on homografts have served to establish the reparatory response to a grafted cryopreserved vessel. Most significant differences with respect to fresh vessel grafts are changes in length, thinning of the wall and

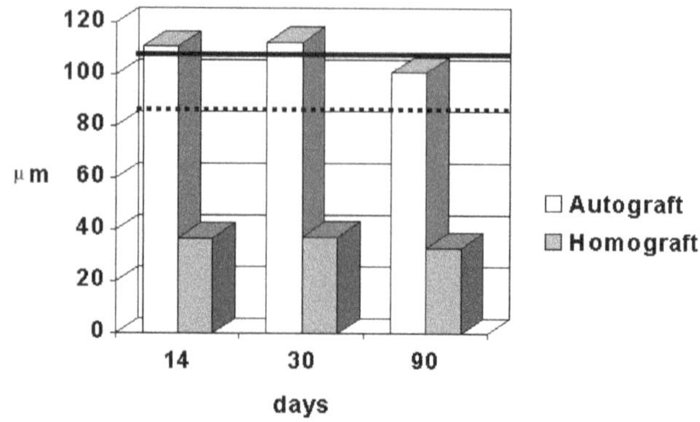

Fig. 7. Thickness of the medial layer in the grafted arterial segment. ....: mean medial thickness of normal rat iliac arteries. ——— : mean medial thickness of cryopreserved arteries prior to implant. Significant differences between groups were shown by the Mann-Whitney U-test at the level $^{*}p < 0.005$.

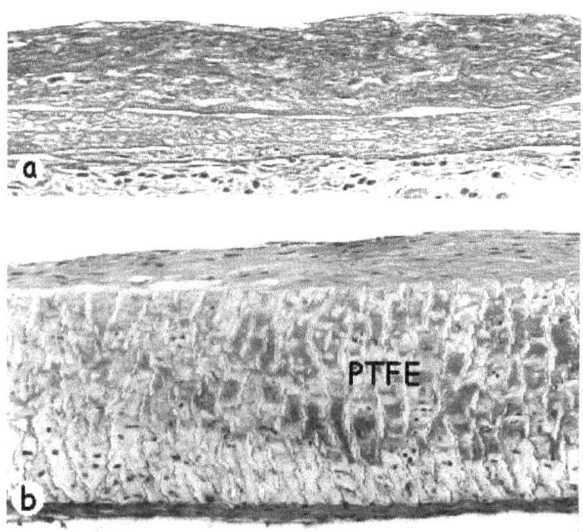

Fig. 8. (a) Three months after grafting, the cryopreserved graft shows medial atrophy. Only the original elastic laminae remain. This layer is now covered by myointimal and adventitial neolayers (LM x400). (b) Longitudinal section of a polytetrafluoroethylene (PTFE) graft showing a similar appearance to a cryopreserved homograft. The biomaterial appears "sandwiched" between the myointimal and adventitial layers (LM x400).

transformation of the cryopreserved vessel into a biotube which is as well tolerated by the recipient organism as a fresh autograft.

The problems associated with the clinical use of cryopreserved vessels might be resolved when greater insight has been gained into the main changes identified that take place after implant. It is proposed that further research efforts should centre on avoiding the appearance of microfractures during the preservation process and improving the graft/blood interface during the initial stages of implant. Attempts should also be made to avoid rapid degradation of the graft and to stimulate a fibrous response to ensure graft patency in the long-term.

3. Acknowledgements

This work was supported by CICYT-SAF 96-0196 and by FIS research grant 01/0431.

4. References

ABBOTT, W.M. (1997). Prosthetic above-knee femoral-popliteal bypass: Indications and choice of graft, *Semin. Vasc. Surg.* **10**, 3–7.

ALONSO, M., SEGURA, R.J., PRADA, C., CAEIRO, S., CACHALDORA, J.A., DÍAZ, E., LUJÁN, S., CAL, L. and VIDAL, J. (1999). Cryopreserved arterial homografts:preliminary results in infrageniculate arterial reconstructions, *Ann. Vasc. Surg.* **13**, 261–267.

ARNAUD, F. (2000). Endothelial and smooth muscle changes of the thoracic and abdominal aorta with various types of cryopreservation, *J. Surg. Res.* **89**, 147–154.

BANK, H.L. and BROCKBANK, K.G. (1987). Basic principles of cryobiology, *J. Cardiac. Surg.* **1**(Suppl. 3), 137.

BATESON, E.A.J., BUSZLA, A.L., PEGG, D.E. and TAYLOR, M.J. (1994). Permeation of rabbit common carotid arteries with dimethyl sulfoxide, *Criobiology* **31**, 393–397.

BELLÓN, J.M., BUJÁN, J., HERNANDO, A., HONDUVILLA, N.G. and JURADO, F. (1994). Arterial autografts and PTFE vascular microprostheses: Similarities in the healing process, *Eur. J. Vasc. Surg.* **8**, 694–702.

BELLÓN, J.M., JURADO, F., DE MIGUEL, M.P., FRAILE, B. and BUJÁN. J. (1996). Long-term behavior of an arterial autograft: A new role for intimal hyperplasia? *Int. J. Microcirc. Clin. Exp.* **16**, 240–249.

BELLON, J.M., BUJAN, J., JURADO, F., DOMINGUEZ, B., GIMENO M.J., HONDUVILLA, N. and HERNANDO, A. (1996). Inhibitor of angiotensin-converting enzyme modifies myointimal origin in an arterial autograft model, *J. Cardiovasc. Pharmacol.* **28**, 285–293.

BELLÓN, J.M., BUJÁN, J., JURADO, F., CONTRERAS, L.A. and HERNANDO, A. (1998). Long-term evaluation of the behavior of a polytetrafluoroethylene microprosthesis in the rat iliac artery: Myointimal regression, *J. Reconstr. Microsurg.* **14**, 251–258.

BELLON, J.M., GIMENO, M.J., PASCUAL, G., GARCIA-HONDUVILLA, N., DOMINGUEZ, B. and BUJAN, J. (1999). Arterial damage induced by cryopreservation is irreversible following organ culture, *Eur. J. Endovasc. Surg.* **17**, 136–143.

BROCKBANK, K.G., LIGHTFOOT, F.G., SONG, Y.C. and TAYLOR, M.J. (2000). Intersticial ice formation in cryopreserved homografts: A possible cause of tissue deterioration and calcification *in vivo*, *J. Heart Valve Dis.* **9**, 200–206.

BUCKLEY, C.J., ABERNATHY, S., LEE, S.D., ARKO, F.R., PATTTERSON, D.E. and MANNING, L.G. (2000). Suggested treatment protocol for improving patency of femoral-infrapopliteal cryopreserved saphenous vein allografts, *J. Vasc. Surg.* **32**, 731–738.

BUJAN, J., BELLON, J.M., GOLITTSIN, A., GIANONATTI, C. and TUREGANO. F. (1992). Lack of intimal hyperplasia response in an experimental model of non-endothelial vascular wall damage, *Histol. Histopathol.* **7**, 693–702.

BUJÁN, J., BELLÓN, J.M., JURADO, F., HONDUVILLA, N., GIMENO, M.J., ESCUDERO, C., GONZALEZ, J. and ALVAREZ, M.V. (1997). A monoclonal anti-platelet fibrinogen receptor antibody (P 37) prevents intimal hyperplasia in a model of arterial autograft in rats, *Platelets* **8**, 337–347.

BUJAN, J., PASCUAL, G., HONDUVILLA, N., GIMENO, M.J., JURADO, F., CARRERA-SAN MARTÍN, A. and BELLÓN, J.M. (2000). Rapid thawing increases the fragility of the cryopreserved arterial wall, *Eur. J. Vasc. Endovasc. Surg.* **20**, 13–20.

CALLOW, A.D. (1996) Arterial homografts, *Eur. J. Endovasc. Surg.* **12**, 278–281

CARRELL, A. (1912). Ultimate result of aortic transplantation, *J. Exp. Med.* **15**, 389–398.

CASTIER, Y., LESECHE, G., PALOMBI, T., PETIT, M.-D. and CERCEREAU, O. (1999). Early experience with cryopreserved arterial allografts in below-knee revascularization for limb salvage, *Am. J. Surg.* **177**, 197–202.

COSELLI, J.S., KÖKSOY, C. and LEMAIRE, S.A. (1999). Management of thoracic aortic graft infections, *Ann. Thorac. Surg.* **67**, 1990–1993.

COUVELARD, A., LESECHE, G., SCOAZEK, J. and GROUSSARD, O. (1995). Human allograft vein failure: Immunohistochemical argumemts supporting the involvement of an immune-mediated mechanism, *Hum. Pathol.* **26**, 1313–1320.

DAVIES, M.G., HUYHN, T.T., FULTON, G.J., SVENDSEN, E., BROCKBAND, F.G. and HAGEN, P.O. (1999). Controlling transplant vasculopathy in cryopreserved vein grafts with polyethylene glycol and glutathione during transport, *Eur. J. Vasc. Endovasc. Surg.* **17**, 493–500.

DESGRANGES, P., BEAJUAN, F., BRUNET, S., CAVILLON, A., QVARFORDT, P., MELLIERE, D. and BECQUEMIN, J.P. (1998).

Cryopreserved arterial allografts used for the treatment of infected vascular grafts, *Ann. Vasc. Surg.* **12**, 583–588.

DINIS DA GAMA, A., SARMENTO, C., VIEIRA, T. and DO CARMO, G.X. (1994). The use of arterial allografts for vascular reconstruction in patients receiving immunosuppression for organ transplantation, *J. Vasc. Surg.* **20**, 271–278.

ELLIS, P. and MÜLLER-SCHWEINITZER, E. (1991). Maintenance of functional activity of human pulmonary arteries after cryopreservation, *Br. J. Pharmacol.* **103**, 1377–1380.

EUGENE, M. and GEROTA, J. (1998). Cryopreserved aortic allografts replacement of infected prosthetic grafts in man: Processing and clinical results, *Transpl. Int.* **11**, S452–S454.

FINI, M., FAGGIOLI, G.L., GARGIULO, M., STELLA, A., D'ADDATO, M., PASQUINELLI, G., PREDA, P., CORBASCIO, M. and GIARDINO, R. (1993). Experimental evaluation of long-term cryopreserved veins for vascular prosthesis, *EUSR* (Abstracts), 102–103.

FUJITANI, R.M., BASSIOUNY, H.S., GEWERTZ, B.L., GLAGOV, S. and ZARINS, C.K. (1992). Cryopreserved saphenous vein allogenic homografts: An alternative conduit in lower extremity arterial reconstruction in infected fields, *J. Vasc. Surg.* **15**, 519–526.

GENTILE, A.T., LEE, R.W., MONETA, G.L., TAYLOR, L.M., EDWARDS, J.M. and PORTER, J.M. (1996). Results of bypass to the popliteal and tibial arteries with alternative sources of autogeneus vein, *J. Vasc. Surg.* **23**, 272–280.

GOURNIER, J.P., FAVRE, J.P. and GAY, J.L. (1995). Cryopreserved arterial allografts for limb salvage in the absence of suitable saphenous vein:two-year results in 20 cases, *Ann. Vasc. Surg.* **9**(Supp.), S7–14.

HARRIS, L., O´BRIE-IRR, M. and RICOTTA, J. (2001). Long-term assessment of cryopreserved vein bypass grafting sucess, *J. Vasc. Surg.* **33**, 528–532.

HUNT, C.J., SONG, Y.C., BATESON, E.A.J. and PEGG, D.E. (1994). Fractures in cryopreserved arteries, *Cryobiology* **31**, 506–515.

KENT, K.C., WHITTEMORE, A.D. and MANNICK, J.A. (1989). Short-term and midterm results of an autogenous tissue policy for infrainguinal reconstruction, *J. Vasc. Surg.* **9**, 107–114.

KNOSALLA, C., GOÉAU-BRISSONNIÉRE, O., LEFLON, V., BRUNEVAL, P., EUGENE, M., PECHERE, J.C., KOSKAS, F., NICOLAS, M.H., LESCHI, J.P., GEROTA, J. and KIEFFER, E. (1998). Treatment of vascular graft infection by *in situ* replacement with cryopreserved aortic allografts: an experimental study, *J. Vasc. Surg.* **27**, 689–698.

KOSKAS, F., PLISSONIER, D., BAHNINI, A., RUOTOLO, C. and KIEFFER, E. (1996). *In situ* arterial allografting for aortoiliac graft infection: A 6-year experience, *Cardiovasc. Surg.* **4**, 495–499.

KU, D.D., WILLIS, W.L, and CAULFIELD, J.B. (1990). Retention of endothelium-dependent vasodilatory responses in canine coronary arteries following cryopreservation, *Cryobiology* **27**, 511–520.

KU, D.D., LIU, Q., NORTON, P. and CAULFIELD, J.B. (1994). Cryopreservation of coronary endothelium and endothelial-mediated responses, *Cryobiology* **31**, 82–89.

LEHALLE, B., GESCHIER, C., FIEVE, G. and STOLTZ, J.F. (1997). Early rupture and degeneration of cryopreserved arterial allografts, *J. Vasc. Surg.* **25**, 751–752.

LESECHE, G., PENNA, C., BOUTTIER, S., JOUBERT, S. and ANDREASSIAN, B. (1997). Femorodistal bypass using cryopreserved venous allografts for limb salvage, *Ann. Vasc. Surg.* **11**, 230–236.

MATSUURA, J.H., JOHANSEN, K.H., ROSENTHAL, D., CLARK, M.D., CLARKE, K.A. and KIRBY, L.B.K. (2000). Cryopreserved femoral vein grafts for difficult hemodialysis access, *Ann. Vasc. Surg.* **14**, 50–55.

MOORE, W.S. and COLE, C.W. (1991). Infection in prosthetic vascular grafts. In: *Vascular Surgery: A Comprehensive Review*, Moore W.S., ed., 3rd edn., W.B. Saunders, Philadelphia, pp. 598–609.

MORIYAMA, S., UTOH, J., SUN, L.B., OKAMOTO, K., KUNIMOTO, R., HIROSE, K. and KITAMURA, N. (2000). Clinical controversy concerning the use of cryopreserved allografts in cardiovascular surgery, *Transplantation Proceedings* **32**, 336–338.

MÜLLER-SCHWEINITZER, E. (1988). Cryopreservation: A useful technique for storing tissues for phamacological investigations, *TIPS* **9**, 221–223.

MÜLLER-SCHWEINITZER, E. (1994). Applications for cryopreserved blood vessels in pharmacological research, *Cryobiology* **31**, 57–62.

MÜLLER-SCHWEINITZER, E., STULTZ, P., STRIFFELER, H. and HAEFELI, W.E. (1998). Functional activity and transmembrane signaling mechanisms after cryopreservation of human internal mammary arteries, *J. Vasc. Surg.* **27**, 528–537.

OCHSNER, J.L., LAWSON, J.D., ESKIND, S.J., MILLS, N.L. and DE CAMP, P.T. (1984). Homologous veins as an arterial substitute: Long-term results, *J. Vasc. Surg.* **1**, 306–313.

OUDOT, J. (1951). La greffe vasculaire dans les thromboses du carrefour aortique, *Presse. Med.* **59**, 234–236.

PASCUAL, G., GARCIA-HONDUVILLA, N., GIMENO, M.J., JURADO, F., TUREGANO, F., BELLON, J.M. and BUJAN, J. (2000). Slow thawing maintain cryopreserved arterial wall viability, *Angiología* **1**, 25–32.

PEGG, D.E. (1989). Viability assays for preserved cells, tissues and organs, *Cryobiology* **26**, 212–231.

PEGG, D.E., WUSTEMAN, M.C. and BOYLAN, S. (1997). Fractures in cryopreserved elastic arteries, *Cryobioloy* **34**, 183–192.

POLLOCK, G.A., HAMLYN, L., MAGUIRE, S.H., STEWART-RICHARDSON, P.A. and HARDIE, I.R. (1991). Effects of four cryoprotectants in combination with two vehicle solutions on cultured vascular endothelial cells, *Cryobiology* **28**, 413–421.

POMAR, J.L., RIGOL, M., AGUSTI, E., HERAS, M., ROIG, E. and SANZ, G. (1996). Early functional findings of small and medium size cryopreserved arteries in an experimental model, *Cardiovasc. Surg.* **4**(Suppl. 1), 51.

POSNER, M.P., MAKHOUL, R.G., ALTMAN, M., KIMBALL, P., COHEN, N. and SOBEL, M. (1996). Early results of infrageniculate arterial reconstruction using cryopreserved homograft saphenous conduit (cadvein) and combination low-dose systemic inmunosuppresion, *J. Am. Coll. Surg.* **183**, 208–216.

PUKACKI, F., JANKOWSKI, T., GABRIEL, M., OSZKINIS, G., KRASINSKY, Z. and ZAPALSKI, S. (2000). The mechanical properties of fresh and cryopreserved arterial homografts, *Eur. J. Vasc. Endovasc. Surg.* **20**, 21–24.

RIGOL, M., HERAS, M., ROIG, E., MARTÍNEZ, A., AGUSTI, E., POMAR, J.L. and SANZ, G. (1997). Effects of cryopreservation on vascular performance: A new method of preserving arterial grafts in a porcine model, *Eur. Heart J.* **18**, 306.

RIGOL, M., HERAS, M., MARTÍNEZ, A., ZURBANO, M.J., AGUSTI, E., ROIG, E., POMAR, J.L. and SANZ, G. (2000). Changes in the cooling rate and medium improve the vascular function in cryopreserved porcine femoral arteries, *J. Vasc. Surg.* **31**, 1018–1025.

ROSSET, E., FRIGGI, A., RIEU, R., ROLLAND, P.H., NOVAKOVITCH, G. and CHOUX, R. (1996). Mechanical properties of the arteries. Effects of cryopreservation, *Chirurgie* **121**, 285–297.

RUDDLE, A.C., GEORGE, S.J., ARMITAGE, W.J., ALEXANDER, E.L. and MICHELL, D.C. (2000). A simplified technique for the cryopreservation of vein allografts, *Eur. J. Vasc. Endovasc. Surg.* **19**, 233–237

SHIRAKASHI, R. and TANASAWA, I. (1998). Method of designing pre-freezing protocol in cryopreservation of biological materials, *Ann. NY Acad. Sci.* **11**, 175–182.

SONG, Y.C., HUNT, C.J. and PEGG, D.E. (1994). Cryopreservation of the common cartotid artery of the rabbit, *Cryobiology* **31**, 317–329.

SONG, Y.C., PEGG, D.E. and HUNT, C.J. (1995). Cryopreservation of the common carotid artery of the rabbit optimization of dimethyl sulfoxide concentration and cooling rate, *Cryobiology* **32**, 405–421.

SOTTIURAI, V.S., SUE, S.L., FEINBERG, E.L., BRINZAGE, W.L., TRAN, A.T. and BATSON, R.C. (1988). Distal anastomotic intimal hyperplasia: Biogenesis and etiology, *Eur. J. Vasc. Surg.* **2**, 245–256.

SZILAGY, D.E., RODRIGUEZ, F.T., SMITH, R.F. and ELLIOT, J.P. (1970). Late fate of arterial allografts. Observations 6 to 15 years after implantation, *Arch. Surg.* **101**, 721–733.

TAKEISHI, M., HIRASE, Y. and KOJIMA, T. (1994). Experimental study of cryopreserved allogeneic transfer of vessel: Preliminary report, *Microsurgery* **15**, 55–62.

TAKAMOTO, S., NAKAJIMA, S., OKITA, Y., MOTOMURA, N., YUASA, M., SADA, M., MAKITA, M. and TSUJI, T. (1998). Cryopreserved femoral arterial allografts for vascular acces in hemodiálisis, *Transplantation Proceedings* **30**, 3917–3919.

TURLEY, K. (1999). Aortic allografts: Reconstruction of right ventricle-pulmonary artery continuity? (Updated), *Ann. Thorac. Surg.* **68**, 289–290.

TAYLOR, R.S., LOH, A., MCFARLAND, R.J., COX, M. and CHESTER, J.F. (1992). Improved technique for polytetrafluoroathylene bypass grafting: Long-term results using anastomotic vein patches, *Br. J. Surg.* **79**, 348–354.

TICE, D.A. and ZERBINO, V.R. (1972). Clinical experience with preserved human allografts for vascular reconstruction, *Surgery* **72**, 260–267.

VERHELST, R., LACROIX, V., VRAUX, H., LAVIGNE, J.P., VANDAMME, H., LIMET, R., NEVELSTEEN, A., BELLENS, B., VASSEUR, M.A., WOZNIAK, B. and GOFFIN, Y. (2000). Use of cryopreserved arterial homografts for management of infected prosthetic grafts: A multicentric study, *Ann. Vasc. Surg.* **14**, 602–607.

VISCHJAGER, M., Van GULIC, T., MARLE, J., PFAFFENDORF, M. and JACOBS, M.J. (1996). Function of cryopreseved arterial allografts under immuosupresssive protection with cyclosporine A, *J. Vasc. Surg.* **24**, 876–872.

VOGT, P.R., PFAMMATTER, T., SCHLUMPF, R., GENONI, M., KÜNZLI, A., CANDINAS, D., ZÜND, G. and TURINA, M. (1997). *In situ* repair of aortobronchial, aortoesophageal, and aortoenteric fistulae with cryopreserved aortic homorafts, *J. Vasc. Surg.* **26**, 11–17.

VOGT, P.R., BRUNNER-LA ROCCA, H.P., CARRELL, T., VON SEGESSER, L.K., RUEF, C., DEBATIN, J., SEIFERT, B., KIOWSKI, W. and TURINA, M.L. (1998). Cryopreserved arterial allografts in the treatment of major vascular infection: A comparison with conventional surgical techniques, *J. Thorac. Cardiovasc. Surg.* **116**, 965–972.

VOGT, P.R. and TURINA, M.L. (1999). Management of infected aortic grafts: Development of less invasive surgery using cryopreserved homografts, *Ann. Thorac. Surg.* **67**, 1986–1989.

ENGERTER, K. and DARDIK, H. (1999). Biological vascular grafts, *Semin. Vasc. Surg.* **12**, 46–51.

WUSTERMAN, M., BOYLAN, S. and PEGG, D.E. (1996). Toxicity of ethyleneglycol when used as a cryoprotectant for rabbit common carotid arteries, *Cryo Letters* **17**, 85–92.

Advances in Tissue Banking Vol. 6

19 POSSIBLE FACTORS THAT INFLUENCE THE STRUCTURE OF VASCULAR SEGMENT: CRYOPRESERVATION AND THAWING. IMPORTANCE OF THE QUALITY OF THE CONTROL USED

Mª ESTHER RENDAL VÁZQUEZ,[1] R.O. FERNÁNDEZ MALLO,[1] M. RODRÍGUEZ CABARCOS,[1] J. SÁNCHEZ IBÁÑEZ,[2] R. SEGURA IGLESIAS,[3] G. MATHEU CAPÓ,[4] P. FILGUEIRA FERNÁNDEZ,[4] S. PÉRTEGA DÍAZ,[5] C. RODRÍGUEZ FERNÁNDEZ[1] and C. ANDIÓN NÚÑEZ[1]

[1]Cryobiology Unit, [2]Transplant Coordination Office,
[3]Department of Vascular Surgery,
[4]Pathology Department, [5]Statistics Department,
Complejo Hospitalario Juan Canalejo, La Coruña, Spain

1. Introduction

Cryopreservation of tissues to –196°C offers the prospect of virtually indefinite storage. Improvements in the preservation techniques, such as the introduction of cryoprotectant agents and a controlled rate of freezing, have enabled the development of human blood vessel banks. The recent interest in cryopreservation techniques has allowed, in the last few years, their development and the use of cryopreserved tissues in different fields of surgery with promising results (Faggioli *et al.*, 1994). There are several reasons for this resurgence. One is the usefulness of these grafts in certain clinical situations in patients who are not eligible for autologous grafting because of the unavailability of a saphenous vein. Vascular surgeons have shown renewed interest in the utilisation of cryopreserved allografts (Goffin *et al.*, 1998; Zapalski *et al.*, 1998).

Although during the last few years the technique of cryopreservation of isolated tissues in nitrogen has been improved as has been shown by the effective preservation of various biochemical and smooth muscle cell functions. However, significant preservation of the endothelial cell lining during the freezing/thawing procedure is still unsatisfactory (Bellón *et al.*, 1995; Ku *et al.*, 1994; Nataf *et al.*, 1994). Changes in the structure may also lead to implant failure. Vessels affected in this manner are unable to respond to the physical requirements of the vascular receptor system. The occurrence of fractures in the frozen and thawed grafts is another great problem. The fractures run perpendicular to the vessel wall and are invariably at right angles to the long axis of the vessel. They usually extend around most of the vessel circumference and have clean-cut edges, penetrating into the intima and the media (Hunt *et al.*, 1994; Pegg *et al.*, 1997).

Numerous groups are now obtaining and freezing arteries prior to storage, either in the vapour phase of liquid nitrogen (–150°C) or directly in liquid nitrogen (–196°C). There is no clear consensus as to which of the two methods of storage is preferable (Adam *et al.*, 1990; Buján *et al.*, 2000; Müller-Schweinitzer *et al.*, 1998; Stanke *et al.*, 1998). Likewise, there is no clear consensus as to which of the two methods of thawing is better. A series of experiments are performed to investigate the influence of rapid thawing at 100°C/min or slow thawing at 15°C/min on the fractures and the wall of the grafts that have been frozen (Bellón *et al.*, 1999; Hunt *et al.*, 1994; Lehalle *et al.*, 1997; Müller-Schweinitzer *et al.*, 2000; Takeishi *et al.*, 1994).

The experiments described in this paper explore the influence of accumulation in the gas phase, the liquid phase or accidental immersion in liquid nitrogen on the presence of fractures or on the presence of changes in the structure of the wall of the vessel together with the influence of the thawing on the presence of fractures or on the presence of changes in the structure of the wall of the vessel.

2. Material and Methods

2.1. Tissue preparation

Samples were carefully harvested from 15 pigs. The tissues were immediately placed in RPMI solution (RPMI 1640 medium, with 25 mm hepes buffer and with L-glutamine), pH = 7.4, at room temperature and transported to the laboratory (a period of 5 min from the harvesting of tissues in the unit of experimental surgery to the place tissues process in the unit of cryobiology), where incubation in antibiotic solution was carried out (vancomycin 50 μgr/ml, metronidazol 50 μgr/ml, amikacin 50 μgr/ml, amphotericin 5 μgr/ml), for 20 hours at 4°C.

2.2. Experimental design

Each aorta was then cut into equal segments of approximately 2 cm length and were randomly assigned to each study group. The control group consisted of a segment of fresh unfrozen aorta after transport in RPMI at room temperature from the unit of experimental surgery to the unit of cryobiology. Three cryopreservation groups were established: the gas phase (–150°C), the liquid phase (–196°C) and the gas/liquid phase together with two methods of thawing: method 1: 37–40°C for 4 min and method 2: 10 min at –80°C, 6 min at room temperature, 4 min at 40°C (Müller-Schweinitzer, 1998).

Six working groups were designated: group 1: the gas phase-thawing 1, group 2: the gas phase-thawing 2, group 3: the gas/liquid phase-thawing 1, group 4: the gas/liquid phase-thawing 2, group 5: the liquid phase-thawing 1, group 6: the liquid phase-thawing 2.

2.3. Cryopreservation

The samples of antibiotic solution were taken off. The samples were placed in a sterile GAMBRO DF 700 bag containing a solution composed of 100 ml RPMI + 10% dimethylsulfoxide

(DMSO) and this solution was equilibrated during 30 min at 4°C before cryopreservation.

Cryopreservation was carried out in the biologic refrigerator (CM25 Carburos Metalicos S.A., Madrid, Spain) with a programmed decrease of –1°C/min until –40°C was reached and faster rates thereafter until –150°C was reached.

Samples were then transferred to the liquid nitrogen tanks for their storage. Groups 1 and 2 were kept in the gas phase while Groups 5 and 6 were kept in the liquid phase in a storage tank for 2 weeks. Groups 3 and 4 were kept in gas phase for 1 week and then were submerged in liquid nitrogen for three days. After this, these groups were kept in gas phase again for 1 week.

Arterial segments were quickly thawed by means of total immersion of the bag in a water bath (37–40°C) for 4 min (the warming rate was about 130°C/min) or they were slowly thawed. Samples were first placed on dry ice for 10 min before being exposed for 6 min at room temperature and thawed within 4 min at 40°C at a mean thawing rate of approximately 15°C/min.

To reduce cell damage due to osmotic changes, the cryoprotectant was removed by sequential washing and progressive dilution at 4°C in saline solution (NaCl 0.9%). Cryoprotectant was performed by one minute immersion of the artery in successive baths of diluted DMSO (8%, 6%, 4%, 2% NaCl 0.9%) at 4°C.

2.4. Macroscopic investigation

After obtaining each artery it was examined by the naked eye for the presence of fractures. Together with the visual inspection we carried out macroscopic inspections, anatomopathologic studies are made to detect the presence of alterations of the vessel prior to cryopreservation.

After two weeks of storage in nitrogen tanks, the thawing was carried out and the results were checked by means of macroscopic inspection and anatomopathologic study.

2.5. Anatomopathologic study

In the anatomopathologic studies the tissues were fixed in 10% formol, included in paraffin, dyed with hematoxylin-eosin and orcein. The structures studied were endothelium cells, the internal elastic lamina and elastic lamelli.

2.5.1. Endothelial cells

The following are analysed:

- *Degree of partially detached,* which can be divided in different categories depending on the degree on partially detached so samples are included where the endothelium is not partially detached, is partially detached < 25%, partially detached < 50% or partially detached > 50%.
- *Degree of completely detached,* which can be divided in different categories depending on the degree of completely detached so samples are included where the endothelium is not completely detached, or on the contrary is slightly or fully detached.

2.5.2. Internal elastic lamina

The following are analysed:

- *Presence or absence* of internal elastic lamina.
- *Degree of fragmentation* of internal elastic lamina, which can be divided into different categories depending on the degree of fragmentation of internal elastic lamina so samples are included where the internal elastic lamina is not fragmented, or on the contrary is little, a lot or total fragmented.

2.5.3. Elastic lamelli

The degree of circular fragmentation is analysed. It can be divided into different categories depending on the degree of circular fragmentation of elastic lamelli so samples are included where the elastic lamelli is not fragmented, or on the contrary is little, a lot or totally fragmented.

2.6. Statistical analysis

Statistical analysis of the results was performed using the logistic regression model where overall significance ($p < 0.05$) was attained to analyse the effect of cryopreservation and thawing on endothelial cells, the internal elastic lamina and elastic lamelli (SPSS statistical package for the personal computer).

3. Results

No cases where macroscopic fractures were observed in the cryopreserved arteries which had been kept in the gas phase or the liquid phase, independently of the storage and thawing method used.

3.1. Effect of cryopreservation on the preservation of the endothelium.

Partially and completely detached endothelial cells were analysed. Observation that cryopreservation was a negative factor in the cryopreservation of the endothelium was able to be done in this study. Independently of the storage method used (the gas phase or the liquid phase), the percentages of correct endothelial cryopreservation without partial detachment were 53% and without complete detachment 89% in the 90 samples analysed [Fig. 1(a)]. Great differences in the preservation of the endothelium in the cryopreserved vessels were noticed when the quality of the control used was taken into account, it was observed that when the control showed an important degree of partial or complete detachment [Fig. 1(b)], the effect on the cryopreservation doubles this damage ($p < 0.05$) (Fig. 2). Table 1 shows the average percentages of endothelial cryopreservation in the samples that were kept in the different phases (the gas phase, the gas-liquid phase and the liquid phase).

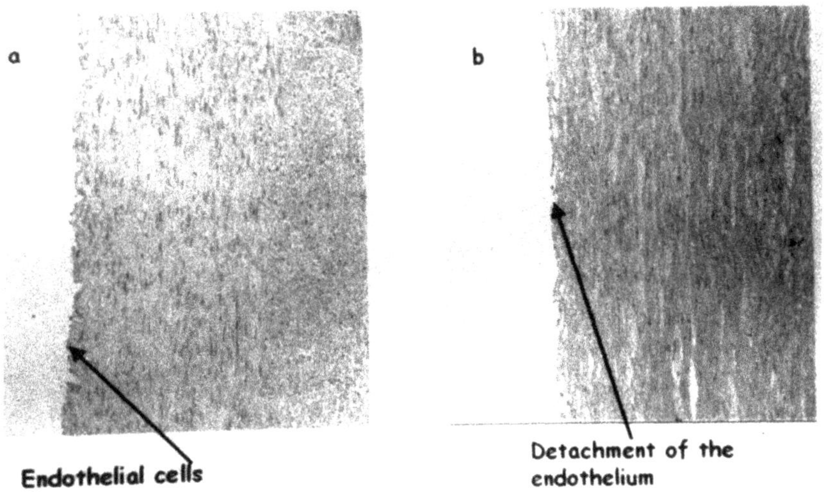

Fig. 1. Well endothelial cells (a) and partial detachment of the endothelium (b) (original magnification × 400).

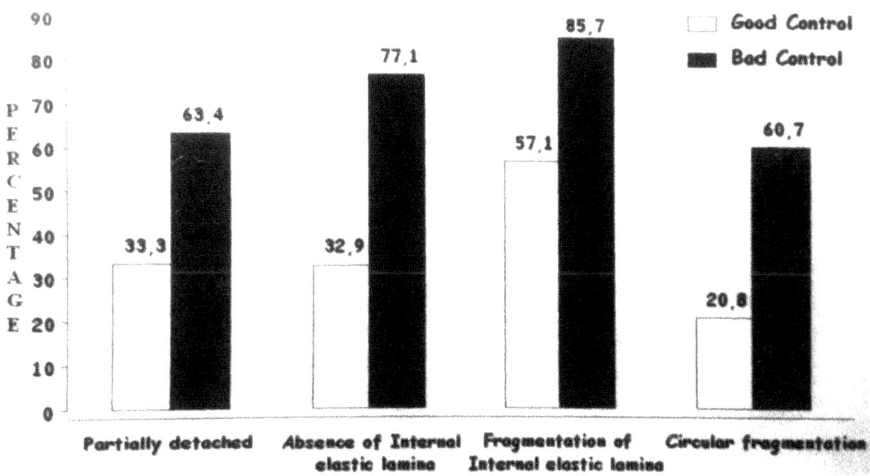

Fig. 2. Influence of the quality of the control used.

Table 1. Effect of cryopreservation on the preservation of the endothelium and internal elastic lamina and elastic lamellae.

		% Gas	% Gas/liquid	% Liquid	% Total
Degree of partial detachment of endothelium	No partially detached or partially detached < 50%	47	48	65	53
	Partially detached > 50%	53	52	35	47
Degree of complete detachment of endothelium	No completely detached or completely detached < 50%	83	90	94	89
	Completely detached > 50%	17	10	6	11
Presence of internal elastic lamina	Presence > 50%	65	55	65	50
	Presence < 50% or absence	35	45	35	50
Degree of fragmentation of internal elastic lamina	No or little fragmentation	33	19	29	27
	A lot of fragmentation	67	81	71	67
Degree of circular fragmentation	No or little fragmentation	82	73	73	67
	A lot of fragmentation	18	27	27	33

The effect of the storage method on endothelial cryopreservation was analysed by means of the logistic regression model with random effects.

When the degree of partial and complete detachment of the endothelium were analysed, it could be observed that there was no significative effect at 95%. However, a tendency was observed for the probability to have a determined level of partial or complete detachment of the endothelium decreased when the cryopreservation in liquid phase was carried out ($p < 0.1$).

3.2. Effect of cryopreservation on the preservation of the internal elastic lamina and elastic lamelli

The presence or the absence of the internal elastic lamina and its degree of fragmentation were analysed, taking into account at the same time the degree of fragmentation of the other elastic fibres.

It was seen that cryopreservation was also a negative factor for elastic preservation. Only in 50% of the 90 analysed samples was the presence of the internal elastic lamina observed. Independent of the preservation method used the total preservation without fragmentation of the internal elastic lamina was 27% and the other elastic fibres 67% [Fig. 3(a)]. Also, great differences in the preservation of the elastic fibres in the cryopreserved vessels were observed when the quality of the control used was taken into

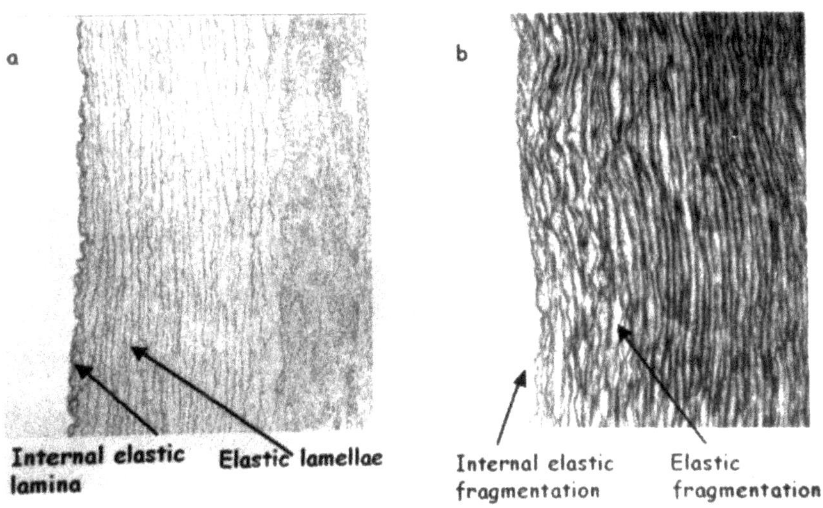

Fig. 3. Elastic preservation (a) and internal elastic lamina and circular elastic fragmentation (b) (original magnification × 400).

account ($p > 0.05$) (Fig. 2). In the same way, a similar effect of damage could be observed due to the method of cryopreservation used and depending on the quality of the control used [Fig. 3(b)]. The average percentages of the preservation of the elastic internal lamina and the elastic lamelli that were kept in the different phases (gas phase, liquid phase, gas-liquid phase) are shown in Table 1.

The effect of the preservation method on the elastic preservation was studied by means of a logistic regression model with random effects.

When the presence of elastic fibres and their degree of fragmentation were analysed, although it was observed that there were no significant effects on the 95%, it could be seen that the preservation was better in the gas phase.

When the effect of accidental immersion in liquid nitrogen was studied, it was seen that the internal elastic lamina showed a greater degree of fragmentation (Table 1) ($p < 0.1$).

3.3. Effect of the thawing method on the preservation of the endothelium and the elastic lamelli

When the influence of the thawing method used (slow thawing at 15°C/min or rapid thawing at 100°C/min) on the preservation of the endothelial cells or the elastic lamelli were analysed, we were able to observe that the preservation of the endothelium without partial detachment was 53% and without complete detachment 89% as well as the preservation without fragmentation of the intact internal elastic lamina of 27% and and the other elastic fibres of 67%, independent of the thawing method used. The average percentages observed in the cells that were thawing with each of the two methods are shown in Table 2.

The effect of the thawing method on the preservation of the endothelium and the elastic cells were analysed by means of a logistic regression model with random effects.

The statistical analysis showed that there was a better preservation of the vascular segment structure when a slow thawing method was used ($p < 0.05$).

Table 2. Effect of the thawing method used on the preservation of the endothelium and the other elastic fibres.

		% Rapid thawing	% Slow thawing	% Total
Degree of partial detachment of endothelium	No partially detached or partially detached < 50%	59	63	53
	Partially detached > 50%	41	37	47
Degree of complete detachment of endothelium	No completely detached or completely detached < 50%	87	91	89
	Completely detached > 50%	13	9	11
Presence of internal elastic lamina	Presence > 50%	60	63	49
	Presence < 50% or absence	40	37	51
Degree of fragmentation of internal elastic lamina	No or little fragmentation	33	33	27
	A lot of fragmentation	67	67	67
Degree of circular fragmentation	No or little fragmentation	70	82	67
	A lot of fragmentation	30	18	33

4. Discussion

The use of vascular grafts is of great importance in surgery. Venous autografts (saphenous veins, for example) have shown better results than allografts or synthetic grafts. Unfortunately, in many cases, this vein can not be used because of previous surgery, thromboflebitis or its small calibre. This unavailability has led to the development of other types of grafts such as synthetic grafts made with Dacron and other materials, umbilical vein grafts, etc., which are used in the cases of infections, thrombosis, formation of false aneurysms, rejection and formation of fibrosis around the graft. Due to this, alternative techniques, which are easy to use, have become necessary and these tecniques offer certain advantages (Faggioli *et al.*, 1994; Goffin *et al.*, 1998; Zapalski *et al.*, 1998).

At the beginning of the last century, Carrel introduced the preservation of grafts under cold with the objective of reducing the metabolic activity of the cells (Carrel *et al.*, 1907; 1912). Many investigators have studied the influence of two important variables in the process of cryopreservation: type of cryoprotectant and temperature of preservation. In 1975, Wewer *et al.* introduced DMSO for the cryopreservation of grafts and they indicated a success rate of 87.5% with this substance at an adequate concentration (Weber *et al.*, 1975). Although during the last years, the technique of tissue cryopreservation has been improved and shown to be effective in preserving several biochemical functions of soft muscle, the preservation of the endothelium and of the rest of the structure of the wall during the freezing/thawing is not yet satisfactory.

The use of low temperatures for the storing of allografts can produce significant damage in the cells and in the extracellular matrix, specially during the process of freezing-thawing.

In the present study, the formation of macroscopic fractures have not been observed in the cryopreserved arteries which had been kept in the gas phase or in liquid phase and thawed both slowly or rapidly. However, in previous studies carried out to –196°C approximately 40% of fractures could be seen after thawing in a bath at 37°C (Buján *et al.*, 2000). Pegg *et al.* attributed the phenomenon to the thermal stress produced by the rapid heating of the material (Hunt *et al.*, 1994; Pegg *et al.*, 1997).

The temperature and final preservation in liquid nitrogen are basic factors in the preservation of endothelial and elastic fibres (Müller-Schweinitzer *et al.*, 1997; Pascual *et al.*, 2000). This histologic study shows that the structural integrity was only partially preserved depending on the cryopreservation method with morphological changes in the endothelium and in the elastic laminas.

For Muller-Schweinitzenei *et al.* (1997) the preservation in liquid nitrogen produces 50% of the endothelial denudation. In our experience with pig aortas observed using optic microscopy with staining with hematoxilin-eosin and orcein which were kept in gas phase or in liquid phase, it can be seen that the endothelium is affected by the process of cryopreservation but nevertheless a better preservation in liquid phase with a smaller degree of partial and complete detachment could be seen. These morphologic studies show a similar endothelial preservation independent of the thawing method used. Different authors have indicated this damage to the endothelial level after the cryopreservation process. In this way, Nataf *et al.* (1994) in the carotid arteries of sheep, find that, after cryopreservation in the gas phase and rapid thawing, the endothelial cells are nearly all altered and separated from the subendothelial layer. On the other hand, Bellón *et al.* (1999) in the iliac ateries of rats that had been kept in the liquid phase and that had been rapidly thawed, observed extensive zones without endothelium and other zones where the endothelium was present. This layer was of a laminar appearance and weakly united to the subendothelial surface. Nevertheless, very few isolated areas of intact endothelium were observed.

The immunological properties of vascular endothelium are well known. The endothelium is an antigenically active layer. The chronic rejection of the endothelium, has been suggested as being the mechanism that reduces the long-term success of the cryopreserved allografts. However, there still exists controversy as to the importance of the endothelial viability in the success of the cryopreserved allografts.

On the other hand, in our work with cryopreserved pig aortas which were kept in the gas phase or in the liquid phase, the cryopreservation process also produces an alteration in the preservation of the internal elastic lamina together with the other elastic laminas that form the vessel wall. Nevertheless, the internal elastic lamina show a better preservation in the liquid phase with a fewer number of fractures. While the other elastic laminas did not show any differences in their preservation, although the gas phase was better with a smaller circular fragmentation. These studies also show an identical preservation of the internal elastic lamina and the elastic lamelli independent of the thawing method used. Nevertheless, these data are not in concordance with the data shown by some authors such as Eskew *et al.* (1997), who in cryopreserved femoral arteries of dogs, kept in the gas phase and rapidly thawed, did not observe alterations in the elastic fibres after cryopreservation. In the same way, Mesa *et al.* (1997) also did not see alterations in the preservation of the internal and external elastic lamina of femoral arteries of rabbit that had been kept at –80°C and rapidly thawed. On the contrary to these authors, Nataf *et al.* (1996), in the carotid arteries of sheep that had been cryopresered in the gas phase and rapidly thawed, observed a loss of the internal elastic lamina mainly in the external part of the media. Although the other elastic fibres, differently to what had been observed, were not modified by the cryopreservation or the thawing.

This apparent discrepancy in some results already published when compared to this work could be due to the different types of arteries used, because the structure of each one shows slight differences. This would signal the need to look for protocols of cryopreservation and storing adjusted to each arterial segment (Ku *et al.*, 1990; 1992; Stanke *et al.*, 1998).

The preservation of the vascular segments in the vapour phase of the liquid nitrogen needs frequent refilling of the storage tank either manually or automatically when the level of nitrogen drops below the pre-established level. During the refilling of the liquid nitrogen tank, the vascular segments can be submerged resulting in a rapid reduction of the temperature (Adam *et al.*, 1990). Our work reveals that this immersion of the vascular segments during three days in liquid nitrogen produced a higher grade of fragmentation of the internal elastic lamina.

In our study, a great difference in the results obtained depending on the quality of the vascular segment that is going to be cryopreserved can be observed. So, a better preservation of the endothelium and the elastic lamina can also be observed after cryopreservation when a vascular segment of good quality was used ($p < 0.05$). For this reason, it is necessary to use a good protocol of donor and vascular segment selection for the segment that was going to be cryopreserved to minimise the damage produced in the process of freezing and thawing

5. Conclusion

Preservation in the liquid phase or in the gas phase and the thawing method used seem influencing the structure of the vascular segments and this damage is produced in the whole structure of the wall, both at the endothelial level and at the level of the elastic fibres, after the cryopreservation process. Rapid thawing with respect to slow thawing produces more important damage in these arteries. The consequences of these post-cryopreservation histological changes on the success of the graft and the rejection process should be evaluated after the implantation of these arterial allografts.

6. Acknowledgments

We gratefully acknowledge grant support by the Xunta of Galicia (No. PR404A99/64-0). Also we would also like to thank the unit of experimental surgery of our hospital for their collaboration.

7. References

ADAM, M., FEI HU, J., LANGE, P. and WOLFINBARGER, L. (1990). The effect of liquid nitrogen submersion on cryopreserved human heart valves, *Cryobiology* **27**, 605–614.

BELLÓN, J.M., BUJÁN, J., GARCÍA-HONDUVILLA, N., HERNANDO, A. and NAVLET, J. (1995). Behaviour of cryopreserved endothelial cells in different phases: Their application in the seeding of vascular prostheses, *Ann. Vasc. Surg.* **9**, 266–273.

BELLÓN, J.M., GIMENO, M.J., PASCUAL, G., GARCÍA-HONDUVILLA, N., DOMÍNGUEZ, B. and BUJÁN, J. (1999). Arterial damage induced by cryopreservation is irreversible following organ culture, *Eur. J. Vasc. Endovasc. Surg.* **17**, 136–143.

BUJÁN, J., PASCUAL, G., GARCÍA-HONDUVILLA, N., GIMENO, M.J., JURADO, F., CARRERA-SAN MARTÍN, A. and BELLÓN,

J.M. (2000). Rapid thawing increases the fragility of the cryo-preserved arterial wall, *Eur. J. Vasc. Endovasc. Surg.* **20**, 13–20.

CARREL, A. (1907). Heterotransplantation of blood vessels preserved in cold storage, *J. Exp. M.* **9**, 226.

CARREL, A. (1912). The preservation of tissues and its applications in surgery, *JAMA* **59**, 523.

ESKEW, T.D., OLLERENSHAW, J.D., PHILPOTT, J.M., DENNIS, K., DAWSON, P., SUN, Y.S., CHITWOOD, W.R. and LUST, R.M. (1997). Successful small diameter arterial grafting using cryo-preserved allograft arteries, *ASAIO J.* **43**, M522–M526.

FAGGIOLI, G.L. and RICOTTA, J.J. (1994). Cryopreserved vein homografts for arterial reconstruction, *Eur. J. Vasc. Surg.* **8**, 661–669.

GOFFIN, Y.A., GRANDMOUGIN, D., WOZNIAK, G., KEPPENNE, V., NEVELSTEEN, A., VOGT, P., VAN DAMME, H., STANKOWIAK, C., DAPPER, F., DE GEEST, R., DEUVAERT, F.R. and VAN HOECK, B. (1998). Banking and distribution of large cryopreserved arterial homografts in Brussels: Assessment of 4 years of activity by the European Homograft Bank (EHB) with reference to implantation results in reconstruction of infected arterial prostheses and mycotic aneurysm, *Vasc. Surg.* **32**, 19–32.

HUNT, C.J., SONG, Y.C., BATESON, E.A.J. and PEGG, D.E. (1994). Fractures in cryopreserved arteries, *Cryobiology* **31**, 506-515.

KU, D.D., WILLIS, W.L. and CAULFIELD, J.B. (1990). Retention of endothelium-dependent vasodilatory responses in canine coronary arteries following cryopreservation, *Cryobiology* **27**, 511–520.

KU, D.D., WINN, M.J., GRIGSBY, T. and CAULFIELD, J.B. (1992). Human coronary vascular smooth muscle and endothelium-dependent responses after storage at 75°C, *Cryobiology* **29**, 199–209.

KU, D.D., LIU, Q., NORTON, P. and CAULFIELD, J.B. (1994). Cryopreservation of coronary endothelium and endothelial-mediated responses, *Cryobiology* **31**, 82–89.

LEHALLE, B., GESCHIER, C., FIEVE, G. and STOLTZ, J.F. (1997). Early rupture and degeneration of cryopreserved arterial allografts, *J. Vasc. Surg.* **25**, 751–752.

MESA, F., SERRA, J.M. and HERREROS, J. (1997). Vascular cryopreservation in microsurgery, *J. Reconst. Microsurg.* **13**, 245–250.

MÜLLER-SCHWEINITZER, E., MIHATSCH, M.J., SCHILLING, M. and HAEFELI, W.E. (1997). Functional recovery of human mesenteric and coronary arteries after cryopreservation at –196°C in a serum-free medium, *J. Vasc. Surg.* **25**, 743–750 .

MÜLLER-SCHWEINITZER, E., STULZ, P., STRIFFELER, H. and HAEFELI, W.E. (1998). Functional activity and transmembrane signaling mechanisms after cryopreservation of human internal mammary arteries, *J. Vasc. Surg.* **27**, 528–537.

MÜLLER-SCHWEINITZER, E., BRETT, W., ZERKOWSKI, H.R. and HAEFELI, W.E. (2000). The mechanism of cryoinjury: *In vitro* studies on human internal mammary arteries, *Br. J. Pharmacol.* **130**, 636–640.

NATAF, P., HADJIISKY, P., LECHAT, P., PEUCHMAURD, M., GUETTIER, C., MOUGENOT N., GANDJBAKHCH, I. and CABROL, C. (1994). Viability of cryopreserved arterial wall: Enzyme-histochemical and physiological markers, *Cell. Mol. Biol.* **40**, 803–810.

NATAF, P., GUETTIER, C., BOURBON, A., NAPPI, F., LIMA, L., DORENT, R., PAVIE, A. and GANDJBAKHCH, I. (1996). Influence of arterial allograft preparation techniques on chronic vascular rejection: A histological study, *Transplant Proc.* **28**, 2890–2892.

PASCUAL, G., GARCÍA-HONDUVILLA, N., GIMENO, M.J., JURADO, F., TURÉGANO, F., BELLÓN, J.M. and BUJÁN, J. (2000). El proceso de la descongelación lenta mantiene la viabilidad de la pared arterial criopreservada, *Angiología* **1**, 25–32.

PEGG, D.E., WUSTEMAN, M.C. and BOYLAN, S. (1997). Fractures in cryopreserved elastic arteries, *Cryobiology* **34**, 183–192.

STANKE, F., RIEBEL, D., CARMINE, S., CRACOWSKI, J.L., CARON, F., MAGNE, J.L., EGELHOFFER, H., BESSARD, G. and DEVILLIER, P. (1998). Functional assessment of human femoral arteries after cryopreservation, *J. Vasc. Surg.* **28**, 273–283.

TAKEISHI, M., HIRASE, Y. and KOJIMA, T. (1994). Experimental study of cryopreserved allogenic transfer of vessel: Preliminary report, *Microsurgery* **15**, 55–62.

WEBER, T.R., DENT, T.L., LINDENAUER, S.M., ALLEN, E., WEATHERBEE, L., SPENCER, H.H. and GLEICH, P. (1975). Viable vein graft preservation, *J. Surg. Res.* **18**, 247.

ZAPALSKI, S., PUKACKI, F. and GABRIEL, M. (1998). Allogenic arterial grafts in therapeutic management of patients with infected aortic grafts, *Polish Surg. Rev.* **70**, 488–493.

Advances in Tissue Banking Vol. 6

20 ISCHAEMIC HEART DISEASE: SEARCHING FOR THERAPEUTICAL SOLUTIONS

V. MIRABET, P. VACA and F. CARBONELL-UBEROS

Tissue Bank and Immunohaematology Service
Transfusion Centre of Valencia, Avenida del Cid
65-A, 46014 Valencia, Spain

J.A. MONTERO, M.J. DALMAU and F. HORNERO

Service of Cardiac Surgery
University General Hospital of Valencia
Avda. Tres Cruces, s/n, 46014 Valencia, Spain

J.L. LEQUERICA and S. PIQUER

Biomedicine Institute of Valencia
Spanish Council for Scientific Research
Jaime Roig 11, 46010 Valencia, Spain

1. Background

The adult heart lacks reserve cells and cannot regenerate. Unlike foetal cardiomyocytes, mammalian adult cardiomyocytes retain little (Beltrami *et al.*, 2001) or none of their developmental capacity for hyperplastic growth. Thus, individuals who suffer a large acute myocardial infarction (heart attack) in which part of the myocardium dies often develop congestive heart failure.

Satellite cells (SC, also known as myoblasts) play a primary role in the regeneration of skeletal muscle after injury. These muscle SC are small mononucleated fusiform cells, lying between the basal lamina and the sarcolemma of muscle fibres (Mauro, 1961). These myoblasts are normally mitotically quiescent, but can be activated during postnatal muscle growth or by muscle damage: they undergo extensive proliferation for a few days before beginning to fuse, thereby regenerating new muscle fibres or adding nuclei to pre-existing fibres (Moss and Lebond, 1970; Campion, 1984).

Human SC can be isolated from muscle biopsies, but due to the contamination by fibroblasts or other cell types, SC cannot be identified immediately after dissociation. Therefore, SC can only be studied after they have entered their proliferative state. To overcome this problem, different techniques have been proposed: (i) gradient centrifugation method (Yablonka-Reuveni and Nameroff, 1987), (ii) purification by flow cytometry sorting (Baroffio *et al.*, 1993), (iii) pre-plating method (Rajnoch *et al.*, 2001), and (iv) positive selection using a specific monoclonal antibody (Lequerica *et al.*, 2000).

In vitro, immature skeletal myoblasts fuse to form contracting multinucleated myotubes expressing specific skeletal muscle gene products including the transcription factor, myogenin (Blanton *et al.*, 1999). Autologous myoblasts transplantation in ischemic disease, which Taylor *et al.* (1998) called cellular cardiomyoplasty, is a method that could intervene between an acute myocardial infarction and the development of serious disease sequel such as congestive heart failure.

Approximately 4% to 8% of mammalian skeletal muscle cells are skeletal myoblasts. Because this percentage decreases with the age, a large quantity of muscle will be required to obtain sufficient numbers of cultured skeletal myoblasts to improve heart function (Tomita *et al.*, 1999).

As muscle, other adult tissues contain populations of stem cells that have the capacity for renewal after trauma, disease, or aging. These cells may be found within the tissue or in other tissues that serve as stem cell reservoirs. The adult bone marrow contains, in addition to haematopoietic stem cells, mesenchymal stem cells (MSC), which contribute to the regeneration of mesenchymal tissues such as bone, cartilage, muscle, ligament, tendon, adipose, and stroma (Conget and Minguell, 1999; Pittenger *et al.*, 1999; Bruder *et al.*, 1998; Cheng *et al.*, 1994). These MSC they may be cultured *ex vivo* for many passages without overt expression of a differentiated phenotype (Bruder *et al.*, 1997; Jaiswal *et al.*, 1997; Digirolamo *et al.*, 1999).

In addition, it must be considered that, one bone marrow aspirate (≈ 20 ml) obtained under local anaesthesia, cultured under adequate conditions can yield about 10^{13} cells, a number that approaches the total number of cells in the adult body (Colter *et al.*, 2000).

In the present report we have studied the behaviour of myoblasts and mesenchymal stem cells cultured *in vitro* as alternative sources for cellular cardiomyoplasty.

2. Myoblasts

2.1. Culture

Satellite cells were isolated and cultured as previously described (Lequerica *et al.*, 2000). Biopsies were from vastus medialis or vastus lateralis muscles of male donors ranging in age from 18 to 58 years old (mean 31.1 ± 15.7). Briefly, the muscle biopsy was rinsed in a buffered solution. The tissue was cleaned to remove fat and connective tissues and minced directly in trypsin-EDTA (0.12% and 0.02%, respectively, in Hank's balanced salt solution, HBSS). Then pieces were digested in the trypsin-EDTA solution for 45 min. To remove trypsin-EDTA, the suspension was centrifuged at 100 g for 2 min in order to pellet the majority of muscular debris. The supernatant was collected, pelleted at 600 g for 10 min and the pellet containing the cells was rinsed twice with growth medium and platted. Cell number and viability were determined using a Neubaüer chamber and trypan-blue vital staining, respectively.

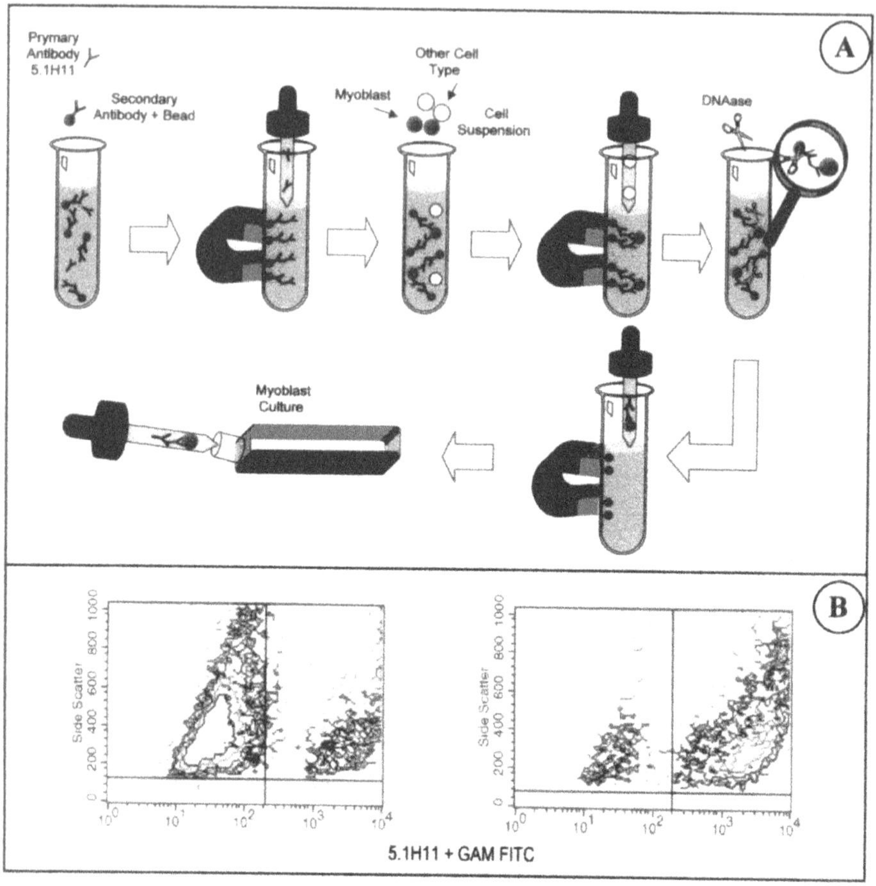

Fig. 1. (A) Procedure for direct myoblast selection. The primary antibody is 5.1H11, specific for a surface antigen in myoblasts and myotubes. The secondary antibody is an anti-mouse antibody with a paramagnetic bead attached. With the help of a magnet, the beads can be retained together with the antibodies and cells adhering to beads. The unwanted fraction is then eliminated by removing the supernatant Finally, the beads can be released from the cell by the action of a DNAase which acts at the level of a DNA fragment found between the bead and the secondary antibody. Purified myoblast population is seeded and incubated in culture conditions. (B) Populations of subcultured cells before (left) and after (right) to myoblast selection observed by flow cytometry. Positive cells for 5.1H11 are in the upper-right quadrant of each graph.

Growth medium was Dulbecco's modified Eagle medium (DMEM) to which was added: foetal bovine serum (10%), L-glutamine (2 mM), insulin (10 μg/ml), transferrin (5.5 μg/ml), sodium selenite (6.7 μg/l), penicillin (100 U/ml), streptomycin (100 μg/ml) and amphotericin B (2.5 μg/ml). Passaging of the cultures was carried out at subconfluency to preclude myogenic differentiation at high density.

2.2. Purification

Analysis of cell suspensions obtained from primary cultures revealed the presence of two populations of cells, that can be identified by their affinity to 5.1H11. The monoclonal antibody 5.1H11 (Developmental Studies Hybridoma Bank, Iowa University, Iowa) is specific for a surface antigen in myoblasts and myotubes. Special care was required to prevent fibroblasts overgrowth, 5.1H11-positive cells was decreased from 60% at primary culture to 3% at 7th passage. To avoid it, a positive selection based on an immunotechnique combined with paramagnetic polystyrene beads was used (Lequerica *et al.*, 2000) (Fig. 1).

2.3. Characterisation

In addition to the ability to fuse in myotubes, myogenic cells were identified by flow cytometry using antibodies against: myoblasts and myotubes cell adhesion molecule (5.1H11); desmin (muscle-specific intermediate filament protein); α-sarcomeric actin; and myosin heavy chain [Fig. 2(A)].

2.4. Differentiation

To induce differentiation, foetal bovine serum (FBS) was replaced in growth medium by horse serum. Confluent cultures were transferred to this differentiation medium, where satellite cells fuse to form myotubes [Fig. 2(D)].

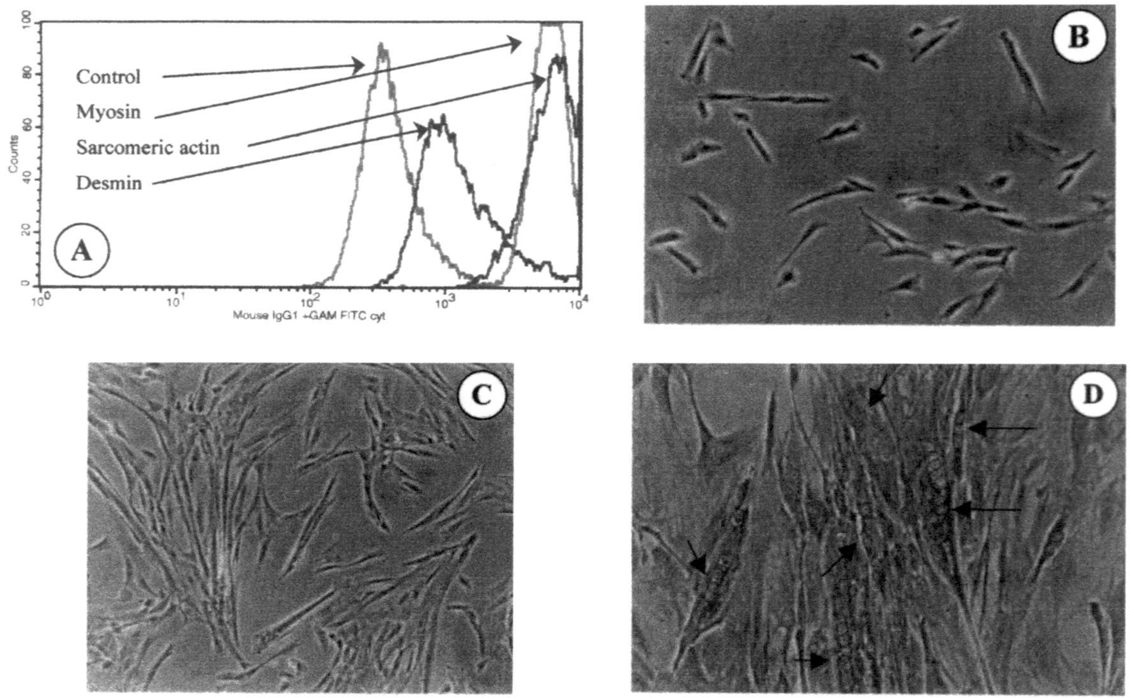

Fig. 2. (A) Cultured cells from purified suspensions showed muscle-specific markers. (B) At first of culture, myoblasts grown as small cells with dense cytoplasm. (C) Then, proliferating cells yield a population with a lower ratio nucleus/cytoplasm (these cells probably correspond to those desmin-positive/α-actin-positive cells that Baroffio *et al.* (1996) called predifferentiated myoblasts). (D) When growth medium was replaced with differentiation medium, some cells fuse to form syncitial myotubes (arrows).

3. Mesenchymal Stem Cells

3.1. Culture

Leftover material was obtained from heparinised bone marrow from multi-organ donors or femoral head (hip surgery) donors, ranging in age from 22 to 84 years old (mean 45.4 ± 16.9).

Mononuclear cells were layered over a Percoll cushion adjusted to 1.073 g/ml and separated by centrifugation (1100 g, 30 min at room temperature). The mononuclear cell layer was removed from the gradient interface and washed with Hank's balanced saline solution (HBSS). Cells were spun at 900 g for 15 min and resuspended in growth medium (α-minimum essential medium supplemented with: FBS (20%), L-glutamine (2 mM), insulin (10 μg/ml), transferrin (5.5 μg/ml), sodium selenite (6.7 μg/l), penicillin (100 U/ml), streptomycin (100 mg/ml) and amphotericin B (2.5 μg/ml). Cultures were maintained at 37°C in a humidified atmosphere containing 5% CO_2. After 24 hours, nonadherent cells were removed, and fresh growth medium was added. The medium was replaced every 3 or 4 days.

Between 10 to 14 days later, when the monolayer of adherent cells grown to 70–90% confluence, MSC were harvested by trypsinisation (0.06% trypsin and 0.02% EDTA), washed, resuspended in growth medium, and subcultured at a concentration of 800 cells/cm^2. Again grown to near confluence, and harvested with the same protocol. Adherent cells after this secondary subculture were used for the experiments described. At the end of each passage, cell number and viability were determined using a Neubaüer chamber and trypan-blue vital staining, respectively.

3.2. Characterisation

Identity and purity of cell population were estimated using a panel of antibodies raised against the expression of several MSC antigens including CD90, CD105 and CD106. Moreover, cultured cells remaining negative for haematopoeitic markers (CD34 and CD45) (Fig. 3).

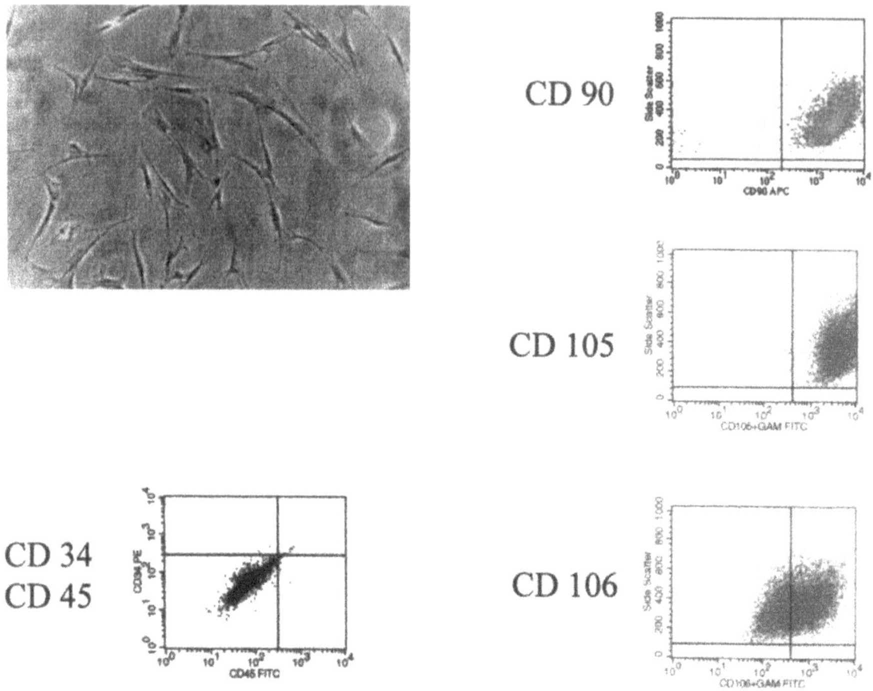

Fig. 3. Bone marrow cells grown in culture as spindle-shaped cells which expressed antigens which are known to be present in mesenchymal stem cells. No differences were observed in cell growth related to the gender or age of the donors. Graphs show that MSC cells were positive for CD90, CD105 and a great majority for CD106 (present in MSC), and negative for CD45 and CD34 (present in haematopoietic cells).

No subpopulations of marrow-derived mesenchymal cells could be discerned morphologically by microscopic observation.

3.3. Differentiation

3.3.1. 5-Azacytidine

The compound 5-azacytidine, an analogue of cytidine, causes hypomethylation of some cytosine in DNA which may be involved in activating phenotype-specific genes (Wakitani *et al.*, 1995). We have exposed human bone marrow-derived MSC to 5-azacytidine and have focused our analysis on their conversion to cardiomyogenic phenotype, analysing immunohistochemically by the presence of the cardiac-specific protein, troponin T.

Twice passaged MSC were seeded into 75 cm^2 flasks at 800 cel/cm^2 and exposed to 10 μmol/L 5-azacytidine for 24 hours. The medium containing the 5-azacytidine was removed after the 24-hours exposure period and replaced with growth medium lacking 5-azacytidine. The morphology of the cultures was screened by microscopic observation every day with a phase-contrast microscope (Fig. 4).

3.3.2. Co-culture

Myoblasts and mesenchymal stem cells were co-cultured using cell culture inserts. Myoblasts grew at the bottom surface of the dish, while MSC were grown attached to the insert polycarbonate membrane. Then, there is not direct contact between the different types of cells, but delivered citokines to the medium can cross the porous membrane. Cultures were maintained with the same medium and conditions that were used above for MSC cultures.

No signs of myogenic differentiation were observed in MSC, even when the growth medium was replaced with myoblast

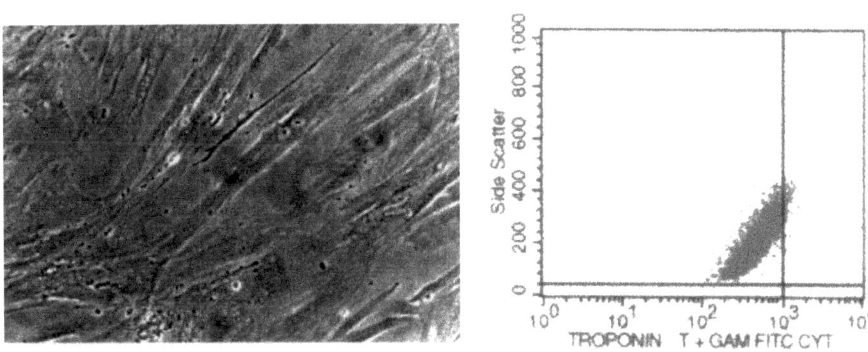

Fig. 4. 5-Azacytidine-treated mesenchymal stem cells were enlarged but not spontaneous beating or other markers of myogenic differentiation were observed. The presence of cardiomyogenic phenotype was assessed by the expression of troponin T which results negative by flow cytometry (left).

differentiation medium for several days, and myoblasts fuse to myotubes.

4. *In vitro* Implantation Model

Ventricles from heart valve multiorgan donors was kept in antibiotic solution (tobramycin, vancomycin, trimethropin/sulfamethoxazole and amphotericin B, 50 μg/ml each one, in HBSS) at 4°C for 24 hours. Then, the myocardium was minced in fragments of 0.1 cm^3. These fragments were stored either embedded in medium 199 supplemented with 4% human albumin, penicillin (100 U/ml) and streptomycin (100 mg/ml) at 4°C; or previously to fresh storage, they were exposed to a freezing/thawing cycle (immersed directly in liquid nitrogen and warming in a 42°C water bath). Anyway, tissue pieces were maintained at these conditions between 20 to 30 days, and the medium was refresh twice a week.

At the time of implantation, cultured cells were previously detached by tripsinisation and resuspended in DMEM at a density of 2×10^4 cel/ml. Aliquots of 50 ml were injected in the core of the tissue fragments. These fragments containing cells were embedded in DMEM supplemented with foetal bovine serum (10%), L-glutamine (2 mM), insulin (10 μg/ml), transferrin (5.5 μg/ml), sodium selenite (6.7 μg/l), penicillin (100 U/ml), streptomycin (100 mg/ml) and amphotericin B (2.5 μg/ml) and were incubated at 37°C, 5% CO_2 and 95% humidity for 3–5 days. Then, tissue blocks were prepared for light microscopy, embedded in paraffin, sectioned, and stained with haematoxylin and eosin (Fig. 5).

Three type of cells were implanted in separated experiments: myoblasts, non-treated mesenchymal stem cells, and 5-azacytidine-treated mesenchymal stem cells.

Tissue samples, non-injected (control) and containing cultured cells, were taken at each time period: at the reception of the ventricles, before fragments processing, after fragments processing, at storage beginning, and finish it.

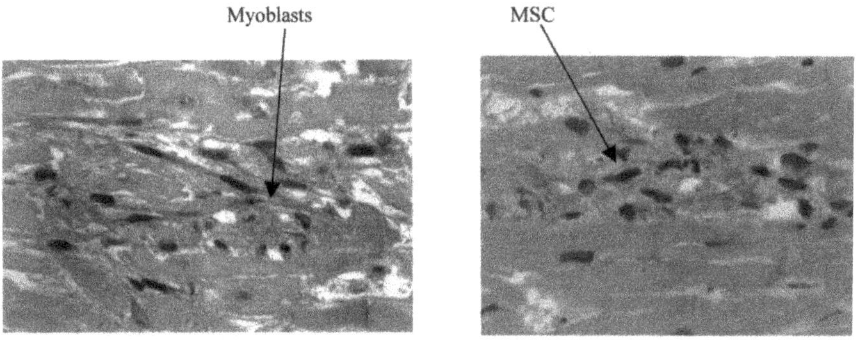

Fig. 5. In order to determine the capacity of stem cells to differentiate depending on local cues provided by specific tissue environments or in response to specific recruitment signals, an *in vitro* model was designed. Injected cells were present as clusters in the core of the myocardium fragments. No histological differences were observed between myoblasts (left) and mesenchymal stem cells (MSC) (right).

5. Comment

Cardiac transplantation, which is an efficient solution to myocardial dysfunction, is limited by a shortage of donors, complications of immunosupression, and graft failure. As an alternative, dynamic cardiomyoplasty has been purposed (Tomita *et al.*, 1999; Rajnoch *et al.*, 2000; Scorsin *et al.*, 2000; Taylor *et al.*, 1998).

As it has been demonstrated in experimental model of Matsusita *et al.* (1999), gap junctions were distinctly observed between grafted/grafted and grafted/host cardiomyocytes.

Using satellite cells for transplantation, these cells seemed to align with the native myocytes and most significantly have exerted an angiogenic influence resulting in marked increase in vascularity in the adjacent myocardium (van Meter *et al.*, 1995).

Chiu *et al.* (1995) have hypothesised that implanted satellite cells can undergo a milieu-induced differentiation that leads them to acquire a cardiac-like phenotype. This hypothesis is largely based on the observation by these authors of structures resembling intercalated disks between transplanted myoblasts.

However, the use of foetal cardiomyocytes raises several issues, including availability, immunogenicity, and ethics. By other hand, skeletal muscle biopsies may not yield sufficient cell numbers to repair the myocardium. Moreover, it must be considered the problem of fibroblasts overgrowth, reason why a myoblast purification method must be implemented. Nevertheless, it is possible that implanted fibroblasts play a role to improve ventricular contraction (Rajnoch, 2001).

These considerations have led to the search for alternate cell types among which MSC are attractive candidates because the ability to isolate, expand the culture, and direct the differentiation *in vitro* to myogenic lineages provides the opportunity to obtain a source for cellular cardiomyoplasty.

We have not observed myogenic differentiation *in vitro* with 5-azacytidine-treated or myoblast-cocultured MSC; however, cultures of non-treated MSC differentiated to adipogenic lineage when specific medium was used. The reason to this lack of myogenic differentiation is unknown. Perhaps cell-cell contact between progenitors and differentiated cells is required (Condorelli *et al.*, 2001). Wakitani *et al.* (1995) observed that nontransformed MSC cells do differentiate when exposed to hypomethylating agents, but not to the cartilage or bone phenotype. They suggested that differentiation down the endochondral pathway, which occurs when these bone marrow cells are implanted *in vivo*, may require additional regulatory mechanisms not triggered by hypomethylation of gene loci.

Nevertheless, several authors have shown that mesenchymal stem cells derived from adult bone marrow can differentiate toward a cardiac phenotype and have been induced to beat synchronously *in vitro* (Makino *et al.*, 1999). These cells communicate with the normal cardiomyocytes by intercalated disks (Soonpaa *et al.*, 1994). However, mesenchymal cells implanted in an infarct region did not beat synchronously with the heart (Tomita *et al.*, 1999).

Other aspect of MSC arousing intense interest is that they might be recruited to the site of degeneration and/or injury, to be directly involved in the repair process or to give rise to more specialised cells performing that task (van den Bos, 1997).

It is clear that cell transplantation offers the promise of restoring cardiac function for patients who have had an extensive myocardial infarction, but what type of cell is appropriate for autotransplantation? Undifferentiated myoblasts, predifferentiated myoblasts, undifferentiated MSC and differentiated MSC to myogenic lineage, are candidates to constitute the optimal source for obtaining the best yield for cellular cardiomyoplasty. Future studies will need to establish the relative efficacy of each cell type. Currently, our labs are working in the study of the *in vivo* effect of implanted MSC and myoblasts in experimentally infarcted myocardium in rat model.

6. Acknowledgements

This work was supported by grants of Instituto de Medicina Cardiovascular (Fundación Mapfre Medicina), and in part by the Generalitat Valenciana under GV98-12-5 and GR00-157 projects.

7. References

BAROFFIO, B., AUBRY, J.P., KAELIN, A., KRAUSE, R.M., HAMANN, M. and BADER, CH.R. (1993). Purification of human muscle satellite cells by flow cytometry, *Muscle Nerve* **16**(5), 498–505.

BELTRAMI, A.P., URBANEK, K., KAJSUTRA, J., YAN, S.M., FINATO, N., BUSSANI, R., NADAL-GINARD, B., SILVESTRI, F., LERI, A., BELTRAMI, C.A. and ANVERSA, P. (2001). Evidence that human cardiac myocytes divide after myocardial infarction, *N. Engl. J. Med.* **344**(23), 1785–1787.

BLANTON, J.R., GRANT, A.L., MCFARLAND, D.C., ROBINSON, J.P. and BIDWELL, CH.A. (1999). Isolation of two populations of myoblasts from porcine skeletal muscle, *Muscle Nerve* **22**, 43–50.

BRUDER, S.P., JAISWALL, N. and HAYNESWORTH, S.E. (1997). Growth kinetics, self-renewal, and the osteogenic potential of purified human mesenchymal stem cells during extensive subcultivation and following cryopreservation, *J. Cell. Biochem.* **64**, 278–294.

BRUDER, S.P., KURTH, A.A., SHEA, M., HAYES, W.C., JAISWAL, N. and KADIYALA, S. (1998). Bone regeneration by implantation of purified, culture-expanded human mesenchymal stem cells, *J. Orthop. Res.* **16**(2), 155–162.

CAMPION, D.R. (1984). The muscle satellite cell: A review, *Int. Rev. Cytol.* **87**, 225–251.

CHENG, S.L., YANG, J.W,. RIFAS, L., ZHANG, S.F. and AVIOLI, L.V. (1994). Differentiation of human bone marrow osteogenic stromal cells *in vitro*: Induction of the osteoblast phenotype by dexamethasone, *Endocrinology* **134**(1), 277–286.

CHIU, R.C.J., ZIBAITIS, A., and KAO, R.L. (1995). Cellular cardiomyoplasty: Myocardial regeneration with satellite cell implantation, *Ann. Thorac. Surg.* **60**, 12–18.

CONGET, P.A. and MINGUELL, J.J. (1999). Phenotypical and functional properties of human bone marrow mesenchymal progenitor cells, *Cell Physiol.* **181**(1), 67–73.

COLTER, D.C., CLASS, R., DIGIROLAMO, C.M. and PROCKOP, D.J. (2000). Rapid expansion of recycling stem cells in cultures of plastic-adherent cells from human bone marrow, *Proc. Natl. Acad. Sci. USA* **97**(7), 3213–3218.

CONDORELLI, G., BORELLO, U., DE ANGELIS, L., LATRONICO, M., SIRABELLA, D., COLETTA, M., GALLI, R., BALCONI, G., FOLLENZI, A., FRATI, G., CUSELLA DE ANGELIS , M.G., GIOGLIO, L., AMUCHASTEGUI, S., ADORINI, L., NALDINI, L., VESCOVI, A., DEJANA, E. and COSSU, G. (2001). Cardiomyocytes induce endothelial cells to trans-differentiate into cardiac muscle: Implications for myocardium regeneration, *Proc. Natl. Acad. Sci. USA* **98**(19), 10733–10738.

DIGIROLAMO, C.M., STOKES, D., COLTER, D., PHINNEY, D.G., CLASS, R. and PROCKOP, D.J. (1999). Propagation and senescence of human marrow stromal cells in culture: A simple colony-forming assay identifies samples with the greatest potential to propagate and differentiate, *Br. J. Haematol.* **107**(2), 275–281.

JAISWAL, N., HAYNESWORTH, S.E., CAPLAN, A.I. and BRUDER, S.P. (1997). Osteogenic differentiation of purified, culture-expanded human mesenchymal stem cells *in vitro*, *J. Cell. Biochem.* **64**, 295–312.

LEQUERICA, J., MIRABET, V., MONTERO, J.A., HURTADO, C., PIQUER, S. and CARBONELL, F. (1999). *In vitro* proliferation, differentiation and immuno-magnetic bead purification of human mioblasts, *Ann. Transplant.* **4**(3–4), 103–108.

MAKINO, S., FUKUDA, K., MIYOSHI, S., KONISHI, F., KODAMA, H., PAN, J., SANO, M., TAKAHASHI, T., HORI, S., ABE, H., HATA, J., UMEZAWA, A. and OGAWA, S. (1999). Cardiomyocytes can be generated from marrow stromal cells *in vitro*, *J. Clin. Invest.* **103**(5), 697–705.

MAURO, A. (1961). Satellite cell of skeletal muscle fibres, *J. Biophys. Biochem. Cytol.* **9**, 493–495.

MOSS, F.P. and LEBOND, C.P. (1970). Satellite cells as the source of nuclei in muscles of grwing rats, *J. Cell Biol.* **44**, 459–462.

PITTENGER, M.F., MACKAY, A.M., BECK, S.C., JAISWAL, R.K., DOUGLAS, R., MOSCA, J.D., MOORMAN, M.A., SIMONETTI, D.W., CRAIG, S. and MARSHAK, D.R. (1999). Multilineage potential of adult human mesenchymal stem cells, *Science* **284**, 143–147.

MATSUSHITA, T., OYAMADA, M., KURATA, H., MASUDA, S., TAKAHASHI, A., EMMOTO, T., SHIRAISHI, I., WADA, Y., OKA, T. and TAKAMATSU, T. (1999). Formation of cell junctions between grafted and host cardiomyocytes at the border zone of rat myocardial infarction, *Circulation* **100**(Suppl. II), 262–268.

RAJNOCH, C., CHACHQUES, J.C., BERREBI, A., BRUNEVAL, P., BENOIT, M.O. and CARPENTIER, A. (2001). Cellular therapy reverses myocardial dysfunction, *J. Thorac. Cardiovasc. Surg.* **121**, 871–878.

SCORSIN, M., HAGÉGE, A., VILQUIN, J.T., FISZMAN, M., SAMUEL, J.L., RAPPAPORT, L., SCHWARTZ, K. and MENASCHÉ, P.H. (2000). Comparison of the effects of fetal cardiomyocyte and skeletal myoblast transplantation on postinfarction left ventricular function, *J. Thorac. Cardiovasc. Surg.* **119**(6), 1169–1175.

SOONPAA, M.H., KOH, G.Y., KLUG, M.G. and FIELD, L.J. (1994). Formation of nascent intercalated disks between grafted fetal cardiomyocytes and host myocardium, *Science* **264**(5155), 98–101.

TAYLOR, D.A., ATKINS, B.Z., HUNGSPREUGS, P., JONES, T.R., REEDY, M.C., HUTCHESON, K.A., GLOWERZ, D.D. and KRAUS, W.E. (1998). Regenerating functional myocardium: Improved performance after skeletal myoblast transplantation, *Nature Med.* **4**(8), 929–933.

TOMITA, S., LI, R.K., WEISEL, R.D., MICKLE, D.A.G., KIM, E.J., SAKAI, T. and JIA Z.Q. (1999). Autologous transplantation of bone marrow cells improves damaged heart function, *Circulation* **100**(Suppl. II), 47–56.

VAN DEN BOS, C.H., MOSCA, J.D., WINKLES, J., KERRIGAN, L., BURGESS, W.H. and MARSHAK, D.R. (1997). Human mesenchymal stem cells respond to fibroblast growth factors, *Hum. Cell* **10**(1), 45–50.

VAN METER, C.H. Jr., CLAYCOMB, W.C., DELCARPIO, J.B., SMITH, D.M., DEGRUITER, H., SMART, F. and OCHSNER, J.L. (1995). Myoblast transplantation in the porcine model: A potential technique for myocardial repair, *J. Thorac. Cardiovasc. Surg.* **110**(5), 1442–1448.

WAKITANI, S., SAITO, T. and CAPLAN, A.I. (1995). Myogenic cells derived from rat bone marrow mesenchymal stem cells exposed to 5-azacytidine, *Muscle Nerve* **18**(12), 1417–1426.

YABLONKA-REUVENI, Z. and NAMEROFF, M. (1987). Skeletal muscle cell populations. Separation and partial characterization of fibroblast-like cells from embryonic tissue using density centrifugation, *Histochemistry* **87**, 27–38.

Advances in Tissue Banking Vol. 6

21 PROCUREMENT, PROCESSING AND CLINICAL ASPECTS OF CRYOPRESERVED VASCULAR HOMOGRAFTS

JACINTO SÁNCHEZ-IBÁÑEZ

Human Tissue Transplant Programme Juan Canalejo Hospital, As Xubias, 84
15006 La Coruña, Spain

1. Introduction

Historically, one of the main purposes in the field of vascular surgery was to find an "ideal conduit" when necessary to restore the normal blood flow to a limb or an organ.

The "ideal conduit" has to meet several requirements, such us, to maintain the physiological blood flow without turbulence that could otherwise produce thrombosis phenomenons; the graft should not be identified as foreign; a long durability should prevail as well as a high grade of resistance to infections.

Despite the advances in the field of vascular anastomosis in the beginning of the century by Alexis Carrel, it was necessary to wait until the 1950s when, vascular prostheses began to be investigated and developed. The arterial vascular allografts began to be studied as a possible substitute of vascular prostheses at the end of 1940s through the investigations of Charles Hufnagel (1947) and Robert Gross (1948).

The first cases of allografts using abdominal aortic surgery were published in the beginning of the 1950s (Oudet, 1953). After that, the first vascular allografts tissue banks were established using lyophilization ("freeze drying method") as the principal preservation method. A few years later a degeneration and failure of these grafts were recorded and communicated (Szylagyi, 1957; Crawford, 1960). In the 70s, the glutaraldehyde-stabilised human umbilical cord vein started to be used as a vascular prosthesis in limb revascularisation (Darking, 1973). Nowadays, there is a follow-up of over 20 years with human umbilical cord vein (Darking, 1995).

The first clinical use of cryopreserved vascular grafts was communicated at the same time (Tice, 1972). The use of dymethyl sulfoxide (DMSO) as a cytoprotective substance has bestowed the cryopreservation method as an eligible way to preserve vascular allografts .

In this chapter, we will review the principal aspects concerning the procurement, processing and the different clinical aspects of human vascular allografts.

2. Procurement

The donor criteria used by almost all human vascular tissue banks are based on the standards of the European Association of Tissue Banks and the American Association of Tissue Banks.

The donors of our Vascular Tissue Bank are all originally "organ donors". The upper age limit is 50 years old. The atherosclerosis criteria is the principal cause of refusal in approximately 90% of our donors between the ages of 45 and 50 years old. Due to this reason, a very careful examination at retrieval is performed to identify if there is an important atherosclerosis component or not.

The harvesting has to be aseptic to avoid any cross contamination. The most frequent harvested allografts are: thoracic aorta, aortoiliac bifurcation and femoral arteries. If possible, we also try to perform a musculoskeletal retrieval without increasing the percentage of allografts contamination.

The extraction must to be very meticulous, without any vascular pulling whatsoever in order to avoid the collateral tears. We cut them distally without ligatures. Some Tissue Banks prefer to suture the collaterals at the same moment of the processing, but in our case, it is fulfilled in the operation theatre by vascular surgeons that like to review, inspect and suture the allografts by themselves.

At this stage, a ring for a histological study is obtained. We take another one just before the processing. It allows us to know the exact damage incurred during the cool ischaemia period. If the allografts is implanted, after thawing we obtain a ring to know the exact cryopreservation damage.

The vascular allografts are previously washed in a saline solution. Afterwards, they are kept at 4°C in the same saline solution and duly transported to the Tissue Bank. The maximum cool ischaemia period should not exceed 12 hours. In our Tissue Banks the average is only 8 hours.

3. Processing

The cryopreservation technique gives us the possibility to store the allografts for a long time (perhaps years). The introduction of DMSO as a cytoprotector in 1975 (Weber, 1975) made it possible to apply the cryopreservation method and prevented the formation of ice crystals inside the cell.

The technique is carried out in a biological refrigerator with a programmed temperature decrease from –1°C/min until –40°C is attained and at even faster rates until –150°C is reached. Samples are transferred to liquid nitrogen tanks. We could either store samples in the vapour phase with a temperature of around –150°C or directly in liquid nitrogen (–196°C).

There is not consensus regarding which storage method is better (Adam, 1990; Stanke, 1998). The liquid nitrogen method assures a constant temperature. One of the main problems with the vapour phase is that the graft may be accidentally immersed in liquid nitrogen during the refilling stage. The different results achieved in literature could be due to the different types of arteries that were used (Rendal, in press). It is nonetheless necessary to fulfill more studies to determine which storage method is the best; although the different storage method applied seems to depend on the type of artery we use.

One important factor in the presence of fractures is the optimum thawing method used. It seems that there is a direct relationship between the heating rate and the presence of fractures (Hunt, 1994; Pegg, 1997). There is however an agreement to use a slow thawing method to avoid fractures in vascular and valvular homografts.

Several investigations could not demonstrate that the cryopreservation method produced a reduction in the antigenicity of the cryopreserved vascular allografts (Shuji, 2001; Marois, 1999), probably because in these studies, the storage in liquid nitrogen was very short. Even more, some authors (Ono, 1999) found that the immunogenicity of the cryopreserved endothelial cell was stronger that the fresh endothelial cell due to the greater production of interleukin 6 and interleukin 8.

4. Clinical Aspects

In the past years, various papers that demonstrate the clinical use of vascular homografts, both fresh and cryopreserved, can we found. Review of the principal indications of these kinds of allografts will be presented.

The interest is mainly based on the good results achieved with the cryopreserved human aortic valves in the treatment of the destructive aortic endocarditis. This endocarditis makes it necessary to replace the proximal part of the ascending aorta. The use of the cryopreserved human aortic valves reduce the reinfection rates in comparison with the use of mechanical heart valves prostheses.

This property of the vascular allografts has been demonstrated in experimental studies. They were more resistant to reinfection than synthetic grafts after *in situ* replacement of an infected prosthetic graft (Knosalla, 1998). The authors think that it is probably due to the antibiotic loading of the cryopreserved vessels. This anti-infectivity property seems to be maintained over a long time (Ohkado, 1999).

The incidence of prosthetic infection continues to be around 2% despite the use of antibiotics. The risk depends on several factors, principally surgical indications as well as proper patient factors. The treatment can be conservative, although feasible in very few cases, with proper drainage of the infection. In extranatomic or *in situ* revascularisation, the complete excision is the choice alternative. The first treatment has some disadvantages such us, a low middle-term patency with a relatively high amputation rate (Hepp, 1988).

With the *in situ* revascularisation technique, we can use either synthetic graft materials, with a high risk of reinfection or allografts (both fresh or cryopreserved). Good results had been published using fresh allografts (Ruotolo, 1997). A recent study (Vogt, 1998) compares the synthetic prosthetic material in an extranatomic reconstruction and cryopreserved allografts with *in situ* repair in the treatment of mycotic aneurysms and infected vascular prostheses in the aorta. In the first group the post-operative mortality was 18% vs. 6% with the allografts. In total, the disease-related death was 32% with the classical technique and only 12% using cryopreserved allografts.

One of the first multi-centre study published (Chiesa, 1998) comparing fresh and cryopreserved arterial allografts in the treatment of prosthetic graft infections gave good results, independent of the fact that the allograft used was either fresh or cryopreserved, with a hospital mortality rate of 13.6% (6 cases, where 4 of them due to allograft ruptures). They could not find any differences between fresh or cryopreserved arterial allografts.

Another multi-centre study (Verhlest, 2000) with 90 patients, also gave promising results using cryopreserved allografts after a mean follow-up of 36 months, with a 30-day mortality rate of 17% (16 cases), in six cases the cause of death was related to an early rupture of the anastomosis between the aorta and the homografts, 5 of them had enteroprosthetic fistula (EPF). ABO compatibility between the donor and recipient was not a prerequisite. During the follow up, a graft rupture occurred in 3 patients with EPF. They recommended special care when using allografts in patients with EPF. Similar results were obtained by others authors (Nevelsteen, 1998; Vogt, 1999).

Other authors found more problems in patient without non-ABO compatible allografts (Desgranes, 1998; Alonso, 1999) and tried to find a direct relationship between cellular viability and infection resistance.

Another clinical situation is below-knee revascularisation in patients with limb-threatening conditions. The most suitable conduit in these cases is the autogenous saphenous vein, but not all patients have an adequate vein. Bypass with synthetic prostheses, such us expanded polytetrafluoroethylene (PTFE) grafts have given disappointing results (such us patency rate) in the below-knee revascularisation. Another option is to use human umbilical cord vein (Darking, 1973; 1995), or cryopreserved venous allografts.

The results with cryopreserved venous allografts seem to be better than human umbilical cord vein (Lesèche, 1997). They achieved a secondary patency rate of 52% after one year with a relatively low incidence of early post-operative complications, but with a high incidence of graft thrombosis during the first post-operative year, probably due to immunological problems.

More recently, several investigations gave good results using cryopreserved arterial allografts (Alonso, 1999; Castier, 1999). The secondary patency rate is between 59% and 74% at 18 months, with a limb salvage rate higher than 70%. Castier and co-workers reviewed the patency rates reported using cryopreserved venous allografts and these were low. They recommended the use of cryopreserved arterial grafts in below-knee revascularisation for limb salvage.

A recent retrospective multi-centre study (Albertini, 2000) used fresh (45 allografts) and cryopreserved (120 allografts) arterial grafts. The primary patency rate at 5 years was 16.1% and the secondary patency 25.9% with a limb salvage rate of 74.2% at 5 years. The primary failure due to allograft degradation was 3% and the incidence of graft rupture was 1.2%. They could not find any difference in the patency rates according to the ABO matching. In the explanted graft they found a lymphocyte infiltration of the media and the adventitia.

Comparing the patencies between PTFE grafts and arterial allografts, there are differences but depending on which reports are used. This is probably due to the surgical technique, severity of the disease and co-morbid factors. A randomised trial seems to be necessary to answer this question.

Cryopreserved vascular allografts are also clinically used in vascular accesses in haemodyalisis. Matsuura (2000) compared cryopreserved femoral vein grafts with prosthetic arteriovenous grafts (PTFE). The resulting patency rates were similar. They recommended the use of cryopreserved vascular allografts in an infected area. Similar results were found using arterial allografts (Takamoto, 1998).

We tried to maintain cells viable by using cryopreservation; cells such us endothelial and smooth muscle cells. These cells can express Class II antigens and theoretically produce an immunological rejection that seems to be the cause of vascular degeneration. For this reason, some authors recommend the use of immunosuppression treatment. The use of cyclosporine A (Vischjager, 1996), in an experimental study, increased the repopulation of host smooth-muscle cells and gave better functional responses when compared to a group of non cyclosporine A. The use of arterial allografts in patients taking immunosuppression treatment, such as, in kidney transplant patients achieved good results (da Gama, 1994).

This phenomenon can be observed with fresh homografts (Mirelli, 1999). Early post-operative immunological studies showed a significant increase in total T lymphocytes (CD3), helper T lymphocytes (CD4) and cellular immune response. Afterwards, there is an increase in CD8 T lymphocytes. Apparently, fresh allografts seem to induce a strong anti-HLA antibody response similar to a chronic rejection.

The problem is that immunosuppression treatment can not be given to patients with prosthetic graft infections and the benefits in patients with a severe ischaemic limb seem to be out of proportion (Callow, 1996).

5. Conclusions

A careful procurement and a good cryopreservation method, with a slow thawing protocol, will assure good results in the treatment of prosthetic graft infections and below-knee revascularisation in patients with limb-threatening conditions; but it is still necessary to develop further comparative studies between human allografts and synthetic prostheses.

6. References

ADAM, M., FEI HU, J., LANGE, P. and WOLFINBARGER, L. (1990). The effect of liquid nitrogen submersion on cryopreserved human heart valves, *Cryobiology* **27**, 605–614.

ALBERTINI, J.N., BARRAL, X., BRANCHEREAU, A., FAVRE, J.P., GUIDICELLI, H., MAGNE, J. and MAGNE (2000). Long-term results of arterial allograft below-knee bypass grafts for limb salvage: A retrospective multicenter study, *J. Vasc. Surg.* **31**(3), 426–435.

ALONSO, M., SEGURA, R., PRADA, C., CAEIRO, S., CACHALDORA, J.A., DIAZ, E., LUJAN, S., CAL, L. and VIDAL, J. (1999). Cryopreserved arterial homografts: Preliminary results in infrageniculate arterial reconstructions, *Ann. Vasc. Surg.* **13**(3), 261–267.

CALLOW, A.D. (1996). Arterial homografts, *Eur. J. Vasc. Endovasc. Surg.* **12**, 272–281.

CASTIER, Y., LESÈCHE, G., PALOMBI, T., PETIT, M.D. and CERCEAU, O. (1999). Early experience with cryopreserved arterial allografts in below-knee revascularisation for limb salvage, *Am. J. Surg.* **117**, 197–202.

CHIESA, R., ASTORE, D., PICCOLO, G., MELISSANO, G., JANNELLO, A., FRIGERIO, D., AGRIFOGLIO G. *et al.* (1998). Fresh and cryopreserved arterial homografts in the treatment of prosthetic graft infections: Experience of the Italian Collaborative Vascular Homograft Group, *Ann. Vasc. Surg.* **12**(5), 457–462.

CRAWFORD, E.S., DeBAKEY, M.E., MORRIS, G.C. *et al.* (1960). Evaluation of late failures after reconstructive operations for occlusive lesions of the aorta and iliac, femoral and popliteal arteries, *Surgery* **47**, 79–104.

DAGAMA, A.D., SARMENTO, C. and DOCARMO X. (1994). The use of arterial allografts for vascular reconstruction in patients receiving immunosuppression for organ transplantation, *J. Vasc. Surg.* **20**, 217–278.

DARDIK, H. (1995). The second decade of experience with the umbilical vein graft for lower-limb revascularisation, *Cardiovasc. Surg.* **3**, 265–269.

DARDIK, I. and DARDIK, H. (1973). Vascular heterograft: Human umbilical cord vein as an aortic substitute in baboon. A preliminary report, *J. Med. Primatol.* **2**, 269–301.

DESGRANGES, P., BEAUJAN, F., BRUNET, S., CAVILLON, A., QVARFORDT, P., MELLIÈRE, D. and BECQUEMIN, J.P. (1998). Cryopreserved arterial allografts used for the treatment of infected vascular grafts, *Ann. Vasc. Surg.* **12**(6), 583–588.

GROSS, R.E., BILL, A.H. Jr. and PEIRCE, E.C.II (1949). Method for preservation and transplantation of arterial grafts: Observations on arterial grafts in dog: Report on transplantation of preserved arterial grafts in nine human cases, *Surg. Gynecol. Obstet.* **88**, 689–701.

GROSS, R.E., HURWITT, E.S., BILL, A.H. Jr. *et al.* (1948). Preliminary observation on the use of the human arterial grafts in the treatment of certain defects, *N. Engl. J. Med.* **239**, 578–591.

HEPP, W., de JONGHE, K. and PALLUA, N. (1988). Late results following extranatomic bypass procedures for chronic aortoiliac occlusive disease, *J. Cardiovasc. Surg.* **29**, 181–185.

HUFNAGEL, C.A. (1947). Preserved homologous arterial transplant, *Bull. Am. Coll. Surg.* **32**, 231.

HUNT, C.J., SONG, Y.C., BATESON, E.A.J. and PEGG, D.E. (1994). Fractures in cryopreserved arteries, *Cryobiology* **31**, 506–515.

KNOSALLA, C., GOËAU-BRISSONNIÈRE, O., LEFLON, V., BRUNEVAL, P., EUGÈNE, M., PECHÈRE, J.C., KOSKAS, F., NICOLAS, M., LESCHI, J.P., GEROTA, J. and KIEFFER, E. (1998). Treatment of vascular graft infection by *in situ* replacement with cryopreserved aortic allografts: An experimental study, *J. Vasc. Surg.* **27**, 689–698.

LESÈCHE, G., PENNA, C., BOUTTIER, S., JOUBERT, S. and ANDRÉASSIAN, B. (1997). Femorodistal bypass using cryopreserved venous allografts for limb salvage, *Ann. Vasc. Surg.* **11**(3), 230–236.

MAROIS, Y., WAGNER, E., PARIS, E., ROY, RAYNALD, R., DOUVILLE, Y. and GUIDOIN, R. (1999). Comparison of healing in fresh and preserved arterial allografts in the dog, *Ann. Vasc. Surg.* **13**(2), 130–140.

MATSUURA, J., JOHANSEN, K., ROSENTHAL, D., CLARK, M., CLARKE, K. and KIRBY, L. (2000). Cryopreserved femoral vein grafts for difficult haemodyalisis access, *Ann. Vasc. Surg.* **14**(1), 50–55.

MIRELLI, M., STELLA, A., FAGGIOLI, G., SCOLARI, M., IANNELLI, S., FREYRIE, A., BUSCAROLI, A., DE SANTIS, L., RESTA, F. BONOMINI, V. and D'ADDATO, M. (1999). Immune response following fresh arterial homograft replacement for aortoiliac graft infection, *Eur. J. Vasc. Endovasc. Surg.* **18**, 424–429.

MORIYAMA, S., UTOH, J., SUN, L., TAGAMI, H., OKAMOTO, K., KUNITOMO, R. and KITAMURA, N. (2001). Antigenicity of cryopreserved arterial allografts: Comparison with fresh and glutaraldehyde treated grafts, *Am. Soc. Artificial Int. Organ J.* **47**(3), 202–205.

NEVELSTEEN, A., FERYN, T., LACROIX, R., SUY, R. and GOFFIN, Y. (1998). Experience with cryopreserved arterial allografts in the treatment of prosthetic graft infections, *Cardiovasc. Surg.* **6**(4), 378–383.

OHKADO, A., HACHIDA, M., HOSHI, H., NONOYAMA, M., SAITOU, S., BONKOHARA, Y., HANAYAMA, N., FURUKAWA, H. and KOYANAGI, H. (1999). Problems in preservation of allogenic heart valves and vessels, *Transplant. Proc.* **31**, 2047–2048.

ONO, M., NAKAJIMA, J., LEE, M. and TAKAMOTO S. (1999). Influence of cryopreservation on antigen expression and immunogenicity of human vascular endothelial cells, *Transplant. Proc.* **31**, 985–987.

OUDET, J. and BEACONSFIELD, P. (1953). Thromboses of the aortic bifurcation treated by resection and homograft replacement, *Arch. Surg.* **66**, 365–374.

PEGG, D.E., WUSTEMAN, C.M. and BOYLAN, S. (1997). Fractures in cryopreserved elastic arteries, *Cryobiology* **34**, 183–192.

RENDAL, E., FERNÁNDEZ, R.O., RODRÍGUEZ, M., SÁNCHEZ, J., SEGURA, R., MATÉU, G., FILGUEIRA, P., PERTEGA, S., RODRÍGUEZ, C. and ANDION, C. Possible factors that influence the structure of vascular segment: Cryopreservation and thawing. Importance of the quality control used. *Cryobiology*, in press.

RUOTOLO, C., PLISSIONNIER, D., BAHNINI, A., KOSKAS, F. and KIEFFER, E. (1997). *In situ* arterial allografts: A new treatment for aortic prosthetic infection, *Eur. J. Vasc. Endovasc. Surg.* **14**(Suppl. A), 103–107.

STANKE, F., RIEBEL, D., CARMINE, S., CRACOWSKI, J.L., CARON, F., MAGNE, J.L., EGELHOFFER, H., BESSAR, G. and DEVILLIER, P. (1998). Functional assessment of human femoral arteries after cryopreservation, *J. Vasc. Surg.* **28**, 273–283.

SZYLAGYI, D.E., McDONALD, R.T., SMITH R.F. *et al.* (1957). Biologic fate of human arterial homografts, *Arch. Surg.* **75**, 506–529.

TAKAMOTO, S., NAKAYIMA, S., OKITA, Y., YUASA, M., SADA, M., IMAKITA, M. and TSUJI, T. (1998). Cryopreserved femoral arterial allografts for vascular access in hemodyálisis, *Transplant. Proc.* **30**, 3917–3919.

TICE, D.A. and ZERBINO, V.R. (1972). Clinical experience with preserved human allografts for vascular reconstruction, *Surgery* **72**, 260–267.

VISCHJAGER, M., VAN GULIK, M., VAN MARLE, J., PFAFFENDORF, M. and JACOBS, M. (1996). Function of cryopreserved arterial allografts under immunosuppressive protection with cyclosporine A, *J. Vasc. Surg.* **24**(5), 876–882.

VOGT, P.R., BRUNNER-LA ROCCA, H.P., CARREL, T., VON SEGESSER, L., RUEF, C., DEBATIN, J., SEIFERT, B., KIOWSKI, W. and TURINA M.I. (1998). Cryopreserved arterial allografts in the treatment of major vascular infection: A comparison with conventional surgical techniques, *J. Thoracic Cardiovasc. Surg.* **116**(6), 965–971.

VOGT, P.R. and TURINA, M.I. (1999). Management of infected aortic grafts: Development of less invasive surgery using cryopreserved homografts, *Ann. Thorac. Surg.* **67**, 1986–1989.

WEBER, T.R., DENT, T.L., LINDENAUER, S.M., ALLEN, E., WEATHERBEE, L., SPENCER, H.H. and GLEICH, P. (1975). Viable vein graft preservation, *J. Surg. Res.* **18**, 247–255.

Advances in Tissue Banking Vol. 6

22 FACTORS THAT INFLUENCE THE VIABILITY AND THE HISTOLOGIC STRUCTURE OF THE CRYOPRESERVED PIG CARDIAC VALVES: CRYOPRESERVATION AND THAWING

Mª ESTHER RENDAL VÁZQUEZ,[1] M. RODRÍGUEZ CABARCOS,[1] R.O. FERNÁNDEZ MALLO,[1] J. SÁNCHEZ IBÁÑEZ,[2] N. DOMENECH CARCÍA,[3] T. DÍAZ ROMÁN,[3] C. ZAVANELLA BOTTA,[4] M. GONZÁLEZ CUESTA,[5] P. FILGUEIRA FERNÁNDEZ,[5] S. PÉRTEGA DÍAZ,[6] C. RODRÍGUEZ FERNÁNDEZ[1] and C. ANDIÓN NÚÑEZ[1]

[1]Cryobiology Unit, [2]Transplant Coordination Office, [3]Unit of Experimental Surgery, [4]Department of Cardiac Surgery, [5]Pathology Department, [6]Statistics Department, Complejo Hospitalario Juan Canalejo, La Coruña, Spain

1. Introduction

The use of mechanical or biological valves often produces serious clinical problems but the use of valve homografts has several advantages such as freedom from coagulation, low incidence of the thromboembolism, relative resistance to endocarditis. However, it could possibly have less durability (Courtman *et al.*, 1995; Pidisheim *et al.*, 1996).

Cryopreservation of tissues from –70°C to –196°C offers the prospect of virtually indefinite storage. Improvements in the preservation techniques, such as the introduction of cryoprotectant agents and a controlled rate of freezing, the greater number of multiorganic donors in developed countries and major demand for cryopreserved valves, have allowed the development of human cardiac valve banks.

It has been suggested that the possible factors that influence the long-term durability of cardiac valves are immunological compatibility and cellular viability (Angell *et al.*, 1989; Fischlein *et al.*, 1995). Nevertheless, the impact of allograft valve viability on valve durability remains controversial. There is no strong substantial evidence that viable cells are still present, still produce collagen and repair damaged matrix. Their presence, before implantation, indicates optimal tissue preservation overall. This at least may be the strongest argument for aiming for and maintaining allograft valve viability in clinical implantation of valves (Gall *et al.*, 1998). The fibroblasts, which are paramount to the preservation of the matrix of the valve, are an ideal cell for this analysis. The fibroblasts are essential cells for the functioning of the valves because they produce the principal constituents of the extracellular matrix: collagen, elastin, reticulin and mucopolisacarides (Van der Kamp *et al.*, 1979; Hu *et al.*, 1989; 1990). A variety of techniques for harvesting, disinfecting, and storing the valves have been used, each of which may play an important role in affecting fibroblast viability. Tissue-processing steps, such as warm (WIT) and cold ischaemic time (CIT), antibiotic disinfection, cryopreservation, and thawing all affect viability (McNally *et al.*, 1992).

Although during the last few years the technique of cryopreservation of isolated tissues in nitrogen has been improved, significant morphological and metabolical changes are produced. Changes in the structure may also lead to implant failure. The occurrence of fractures in the frozen and thawed grafts is another significant problem. The fractures occurrence in the media layer of the arterial wall and in the leaflet where microfractures can be seen on both surfaces of the tissue, penetrating into the collagen/proteoglycan matrix. This has been observed in valves which were kept in liquid nitrogen (Adam *et al.*, 1990). These fractures can produce serious problems in a planned operation. There is no clear consensus as to which of the two methods of storage is preferable.

The experiments described in this paper explore the influence of CIT, sterilisation with or without antifungicides and accumulation on the viability of the valve leaflets and in the histological structure of the arterial wall, valve leaflet and myocardium and the influence of the storage method used on the presence of fractures.

2. Material and Methods

2.1. Tissue extraction

Hearts from 40 pigs (80 kg) in an aseptic procedure are used.

2.2. Tissue preparation

Samples were carefully harvested from 40 pigs. The hearts were immediately placed in saline solution, with 1 hour of WIT, at 4°C and transported to the laboratory where dissecction and obtainment of aortic and pulmonary valves was carried out with 2 or 24 hours of CIT. The incubation in antibiotic solution was carried out (amikacyn 50 μgr/ml, metronidazol 50 μgr/ml, vancomycin 50 μgr/ml), for 20 hours at 4°C, with or without antifungus (amphotericin B 5 μgr/ml).

2.3. Experimental design

Each valve was randomly assigned to each study group.

Eight working groups were designated and were cryopreserved for one week:

Group 1: 2 hours of CIT, with amphotericin, gas phase
Group 2: 24 hours of CIT, with amphotericin, gas phase
Group 3: 2 hours of CIT, with amphotericin, liquid phase
Group 4: 24 hours of CIT, with amphotericin, liquid phase
Group 5: 2 hours of CIT, without amphotericin, gas phase
Group 6: 24 hours of CIT, without amphotericin, gas phase

Group 7: 2 hours of CIT, without amphotericin, liquid phase
Group 8: 24 hours of CIT, without amphotericin, liquid phase.

2.4. Cryopreservation

The samples were taken out of antibiotic solution. The samples were placed in a sterile GAMBRO DF 700 bag containing a solution composed of 100 ml RPMI 1640 medium +10% dimethylsulfoxide (DMSO) and this solution was equilibrated during 30 min at 4°C before cryopreservation.

Cryopreservation was carried out in the biologic refrigerator (CM25 Carburos Metalicos S.A., Madrid, Spain) with a programmed decrease of –1°C/min until –40°C was reached and faster rates thereafter until –150°C was reached.

Samples were then transferred to the liquid nitrogen tanks for their storage, in gas phase or in liquid phase, for one week.

Cardiac valves were slowly thawed. The valves that were stored in liquid phase were passed to gas phase for 24 hours, afterwards they were kept for 10 min at room temperature and then at 37°C in a bath until their thawing. The valves that were stored in gas phase were placed 10 min at room temperature and then at 37°C in a bath until their thawing.

To minimise toxic effects of DMSO, the cryoprotectant was removed by sequential washing and progressive dilution at 4°C in saline solution.

2.5. Controls

After obtaining each valve it was examined by the naked eye for the presence of fractures.

After one week of storage in nitrogen tanks, the thawing was carried out and the results were checked by means of macroscopic inspection, anatomopathologic study, and studies with vital dyes (trypan blue or citometry) were performed looking for the presence of fractures, presence of alterations in the myocardium, wall and valve leaflet and checking viability of the valve leaflets.

2.6. Anatomopathologic study

In the anatomopathologic studies the tissues were fixed in 10% formol, included in paraffin, dyed with hematoxylin-eosin and orcein. The structures studied were wall, leaflet and myocardium.

2.6.1. Wall

The endothelial cells were subdivided into 2 groups:

- Partially detached, which can be attached, detached < 25%, detached < 50% or detached > 50%.
- Completely detached which can be attached, slightly or greatly detached.

Elastic lamelli where circular fragmentation also ranges from values of none, a little, a lot or total.

2.6.2. Leaflet

The endothelial cells were subdivided into 2 groups:

- Partially detached, which can be attached, detached < 25%, detached < 50% or detached > 50%.
- Completely detached which can be attached, slightly or greatly detached.

Elastic lamelli where circular fragmentation also ranges from values of none, a little, a lot or total.

Fibroblasts where it could be analysed:

- Number of fibroblasts could be few, normal or a lot.
- Distribution which was regular or irregular.
- Inflammation subdivided into two groups: acute or chronic.

2.6.3. Myocardium

- *The ischaemia* were subdivided into 3 groups: focal, multifocal or confluent.
- *The inflammation* subdivided into two groups: acute or chronic.

2.7. Cellular viability

To analyse the cellular viability, valve fibroblasts. To do this, the leaflet was removed from the valve and washed in phosphate buffered saline solution (PBS). It was then minced and transferred into a digestion buffer [M199 medium, 15% heat-inactivated bovine fetum serum, penicillin (100 U/ml), streptomycin (100 μgr/ml), amphotericin B (5 μgr/ml) and 0.08% collagenase]. Afterwards, it was incubated on a shaker at 37°C until the fragments were digested. Then the cells were washed twice in culture medium [M199 medium, 15% heat-inactivated bovine fetum serum, penicillin (100 U/ml), streptomycin (100 μgr/ml), amphotericin B (5 μgr/ml)] before being used to determine the viability with trypan blue and with citometry using propidium iodure (PI) or fluorescent diacetate (FDA).

Collagen type I and platelet derived growth factor (PDGFR) receptor α (fibroblasts) and von Willebrand's antigen (endothelial cells) expression was analysed by immunocytochemistry, in fibroblasts of cardiac valves from each one of the cryopreservation groups to see whether the cells analysed were fibroblasts and not other types of cells such as endothelial cells.

2.8. Statistical analysis

Statistical analysis of the results was performed using the logistic regression model where overall significance ($p < 0.05$) was attained to analyse the effect of cryopreservation and thawing on the viability of the valve leaflets and on the histological structure of the arterial wall, valve leaflet and myocardium (SPSS statistical package for the personal computer).

3. Results

No case of macroscopic fractures were observed in the cryopreserved cardiac valves which were stored in gas phase or in liquid phase, independently of the storage method used.

3.1. Cell viability

When the methods to determine the fibroblast viability were observed, a discrepancy between the two methods employed (trypan blue and citometry) could be observed ($p < 0.05$) (Fig. 1). With citometry, a greater number of dead cells were observed. The trypan blue test by itself is not a perfect viability assay and it may be susceptible to some errors. For this reason, citometry is employed to determine cellular viability.

After cryopreservation, the viability was reduced to 56% in all groups studied, independently of CIT, treatment with amphotericin and storage method used.

No differences between the number of viable cells between 2 and 24 hours of CIT were seen (Fig. 2). In the same way, no influence of treatment with or without amphotericin on cellular viability was observed (Fig. 3). On the contrary, a greater viability in liquid phase could be observed ($p < 0.05$) (Fig. 4).

When we independently analysed aortic and pulmonary valves, we were able to observe differences in cellular viability between

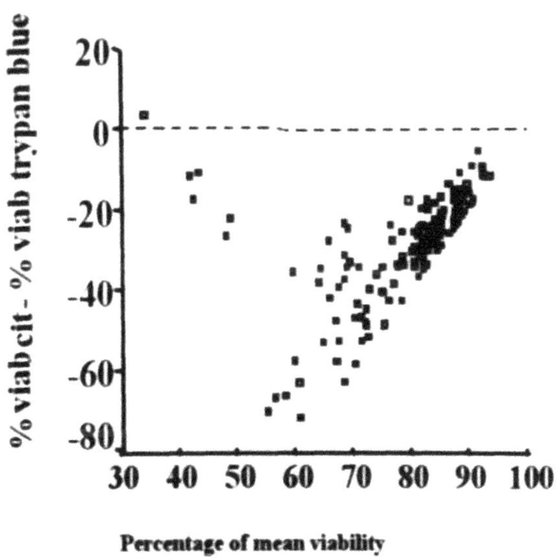

Fig. 1. Comparison between trypan blue test and citometry to determine the cellular viability.

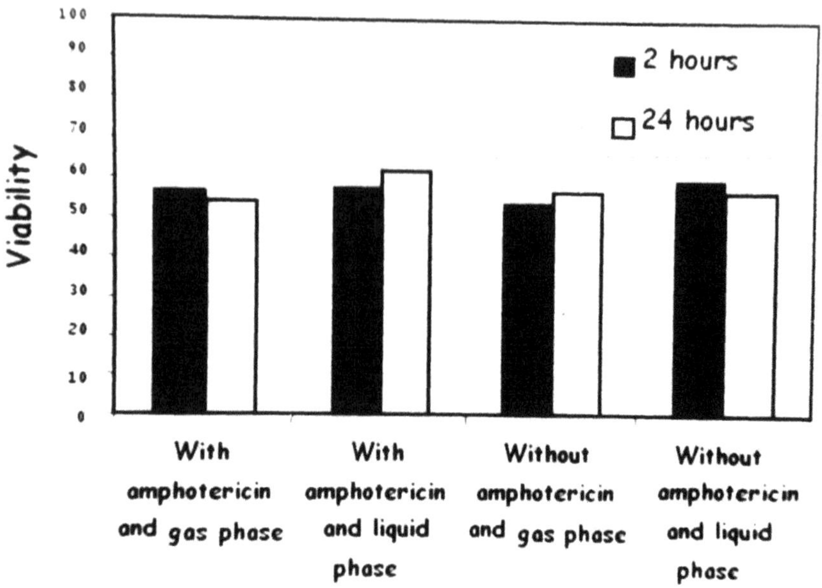

Fig. 2. Influence of cold ischaemic time on cellular viability.

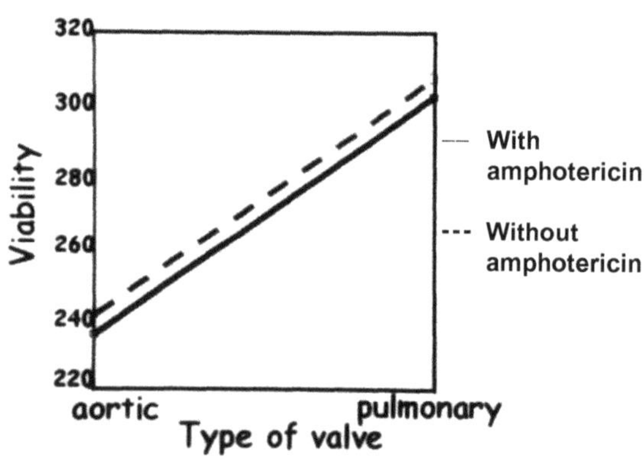

Fig. 3. Influence of treatment with or without amphotericin on cellular viability.

these two valves ($p < 0.05$). The pulmonary valve showed a greater viability than the aortic valve ($p < 0.05$), but the differences in their responses to CIT, treatment with amphotericin and storage method used were not significantly different between them (Figs. 3–4).

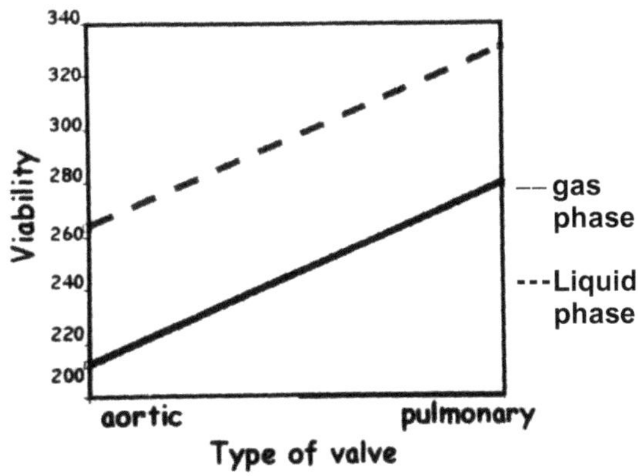

Fig. 4. Influence of storage method used on cellular viability.

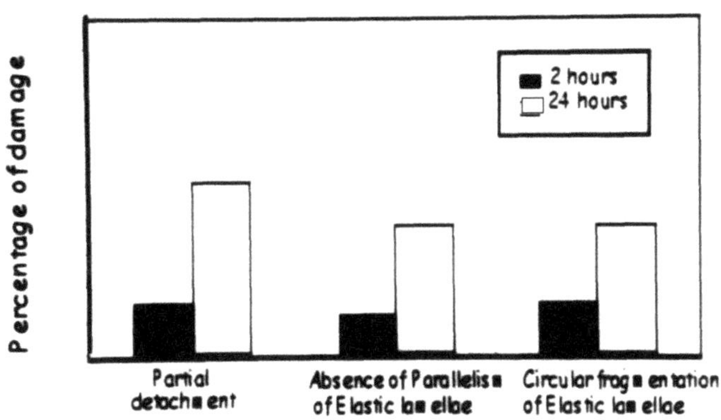

Fig. 5. Influence of cold ischaemic time on the histology of wall valve.

3.2. Histology

3.2.1. In the wall of the valve

A better preservation of endothelium cells and elastic lamellae with 2 hours of CIT could be observed (Fig. 5). The valves with 24 hours of CIT had a greater level of partial detachment of endothelium and a greater level of absence of parallelism of elastic lamellae with a greater number of valves with circular fragmentation (Fig. 6). On the other hand, any influence of treatment with antifungus and of the storage method used could not be observed.

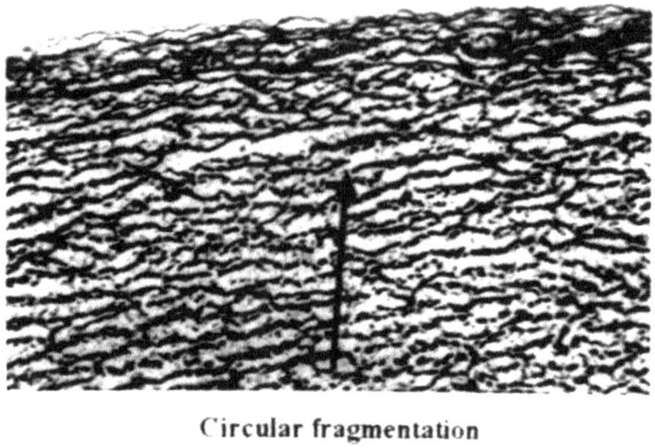

Fig. 6. Circular fragmentation of elastic lamellae in the wall of the valves (original magnification × 400).

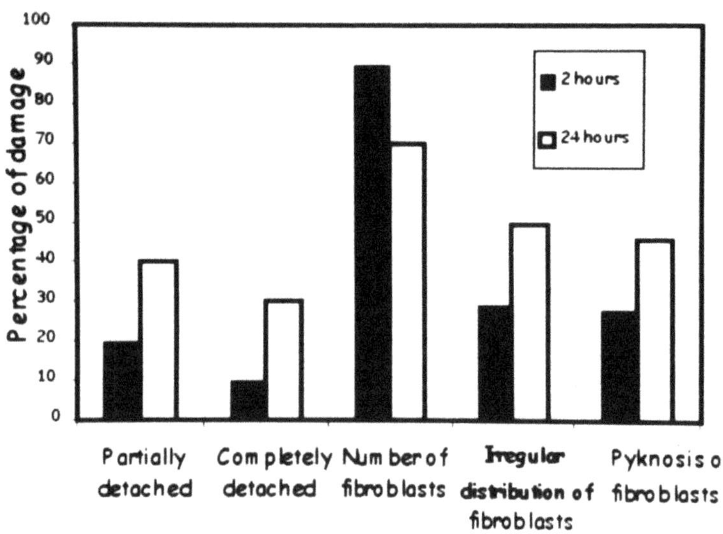

Fig. 7. Influence of cold ischaemic time on the histology of leaflet valve.

3.2.2. In the leaflet

A better preservation of endothelium cells and fibroblasts with 2 hours of CIT could be observed (Fig. 7). In the valves with 24 hours of CIT, a greater degree of partial and complete detachment in the endothelium were observed and the fibroblasts

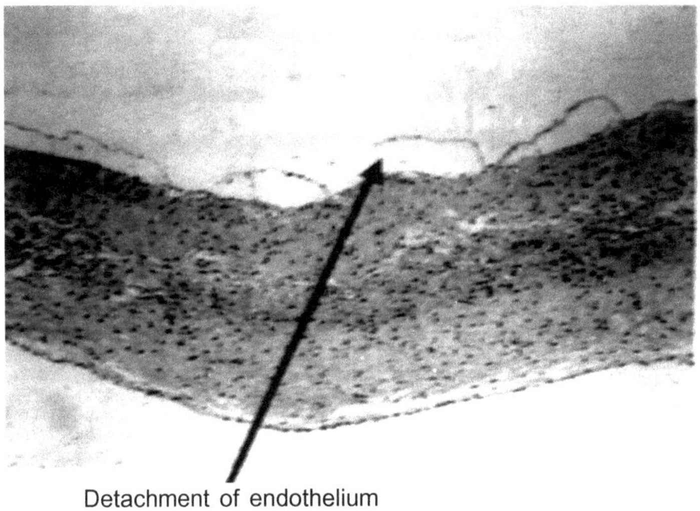

Fig. 8. Partial detachment of endothelium in leaflets of aortic valves (original magnification ×400).

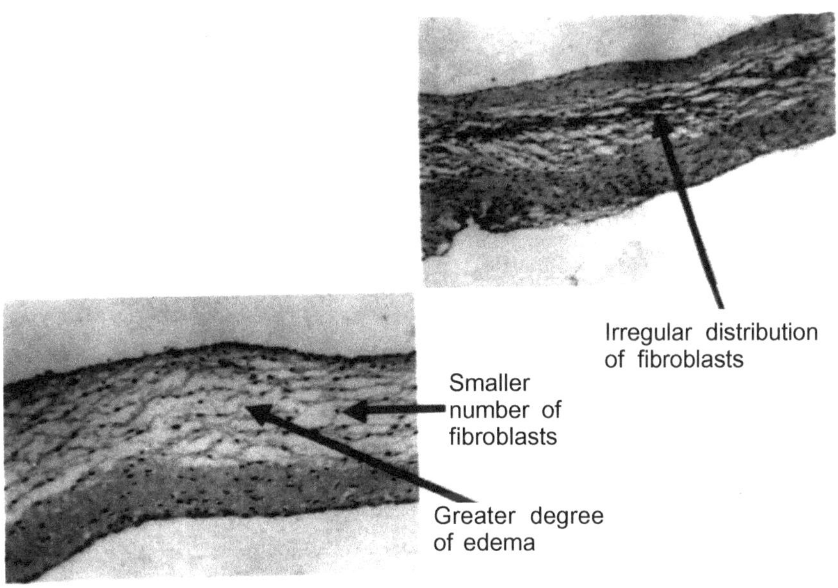

Fig. 9. Greater degree of edema, smaller number of fibroblasts and irregular distribution of fibroblasts in the leaflets of aortic valves (original magnification × 400).

appeared in smaller numbers, with an irregular distribution and a greater degree of pyknosis could be observed (Figs. 8 and 9). On the other hand, no influence of treatment with antifungus and the storage method used on the histologic structure of the leaflet could

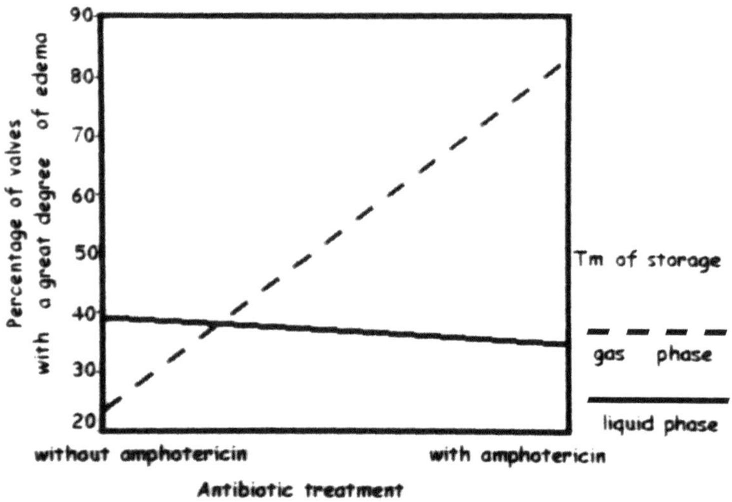

Fig. 10. Influence of treatment with or without antifungus and storage method used on the degree of edema of valve leaflet.

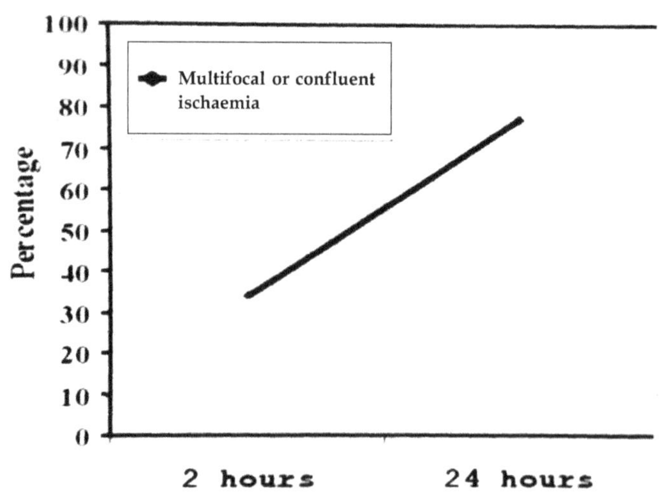

Fig. 11. Effect of cold ischaemic time in the myocardium.

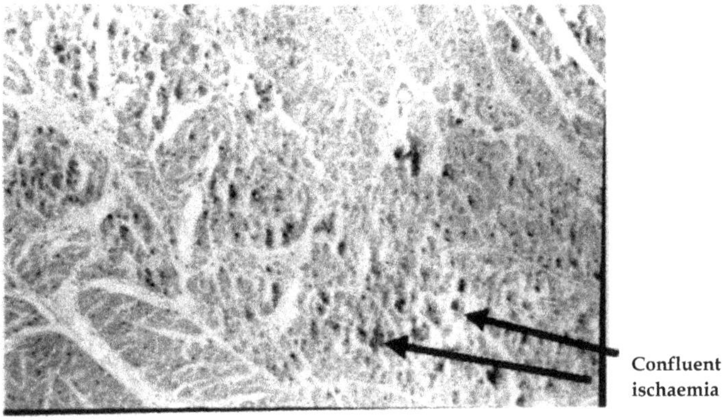

Fig. 12. Confluent ischaemia on myocardium of aortic valves (original magnification × 400).

be observed, except on the degree of edema where it could be observed that in gas phase the degree of edema was similar with or without amphotericin but in liquid phase the degree of edema was greater with amphotericin (Fig. 10).

3.2.3. In the myocardium

An important effect of CIT could be observed (Fig. 11). In the valves with 24 hours of CIT, the ischaemia was principally multifocal or confluent (Fig. 12). On the other hand, no effect of treatment with or without amphotericin and storage method used could be observed.

4. Discussion

The use of prosthetic valves or bioprosthetic tissue valves often produces serious clinical problems such as thrombogenesis and the consequent clinical effects can occur (Pidisheim *et al.*, 1996).

Courtman *et al.* (1995) indicated that a tissue valve for homograft or xenograft transplant has several advantages such as freedom from coagulation, low incidence of thromboembolism and relative resistance to endocarditis, however it could be possible less durability than a mechanical valve (Borkon *et al.*, 1987). With the development of cryopreservation techniques, their use has become more widespread (Niwaya *et al.*, 1995). The cryopreserved homografts have shown favourable clinical results after implantation and the reason seems to be attributable to the viable fibroblasts in the valve (Angell *et al.*, 1987; O'Brien *et al.*, 1991). Viable fibroblasts synthesise the main constituents of extracellular matrix: collagen, elastin, reticulin and mucopolysaccharides (Van der Kamp *et al.*, 1979), and therefore, the viability of fibroblasts in the implanted valves is related to the longevity of the implantation (Hu *et al.*, 1990).

Nevertheless, there are several factors that have influence on the viability of the valvular leaflets such as ischaemic time (WIT or CIT), sterilisation technique (with or without antifungus), cryopreservation and thawing (Wassenaar *et al.*, 1997).

Controversy has surrounded and continues to surround the importance of the presence of viable fibroblasts within the allograft leaflet matrix at the time of implantation. There is evidence that cryopreserved valves, viable at the time of cryopreservation, have a much lower level of structural deterioration than non-viable valves (Grunkemeier *et al.*, 1994; Gall *et al.*, 1998). In addition, Yacoub *et al.* (1995) observed that valves with some or major fibroblast viability produced improved clinical long-term durability. Fibroblasts may be responsible for remodelling the extracellular matrix to maintain the durability of the implanted allograft valves, or fibroblast viability may simply be an indirect marker for overall structural integrity of the valve. In any event, assessment of cell viability may prove to be an important determinant of the overall quality of cryopreservation of allograft valves (Gall *et al.*, 1998).

Several methods for determining valve viability have been proposed. Methods using marked aminoacids, enzymatic using a tetrazolium salt and other methods such as assessment of collagen synthesis, protein synthesis, dye uptake, and dye exclusion (Lu *et al.*, 1997; Gall *et al.*, 1998). In this work, the use of trypan blue and citometry to determine the viability were compared. A discrepancy could be observed between these two methods. The trypan blue test by itself is not a perfect viability assay because it cannot predict future viability or indicate origin and mitotic potential of cells, and it may be susceptible to denominator errors (Messier *et al.*, 1999). The citometry is more exact than trypan blue to determine the viability. Double staining with fluorescein diacetate-propidium iodure (FDA-PI) is reported to be a rapid, convenient, and reliable method for assessing cell viability compared with the trypan blue dye exclusion method (Wilczek *et al.*, 1996). For this reason, we believe that FDA-PI staining is useful and convenient for identifying damaged or dead cells. Cytometry offers rapid, precise and objective estimation of the examined material (Wilczek *et al.*, 1996).

Verghese *et al.* (1999) said that viable valves are those in which at least 50% of the fibroblasts were preserved. In our study after cryopreservation, the viability was reduced to 56% in all groups studied independently of cold ischaemic time, treatment with amphotericin and storage method used.

To obtain a sterile graft for its implantation, it is necessary to identify and eliminate any potential contaminate. Multiple combinations of antibiotics have been used, with or without antifungus, with different results as to cellular viability. Nevertheless, it has been observed that cryopreservation after a period of antibiotic sterilisation significantly reduces viability and with the use of antifungus (amphotericin B) this viability is reduced even more (Gall *et al.*, 1998; Kashima *et al.*, 1999; Suh *et al.*, 1999). Gall *et al.* (1998) said that whereas the use of amphotericin B at 37°C for 6 and 24 hours had no significant effect on viability before cryopreservation, the combination of exposure to amphotericin B and cryopreservation was very deleterious to viability. Amphotericin B should not be used in conjunction with cryopreservation. Most investigators use amphotericin B at a concentration of 25–100 μgr/ml (Niwaya *et al.*, 1995; Armiger, 1998; Verghese *et al.*, 1999). In the present work, it could be observed that the viability was reduced after cryopreservation but independently of the treatment with or without amphotericin. Perhaps, this could be due the use of a low concentration of amphotericin B (5 μgr/ml) which was considered sufficient to cover the necessary antibiotic level.

All harvested valves have an unavoilable period of ischaemic time. The ischaemic time is divided into WIT, which is defined as the interval from heart cessation until procurement of the heart, and CIT, which is the interval between procurement and dissection. This ischaemic time has been shown to affect cell viability (Suh *et al.*, 1999). Shetty *et al.* (1996) showed that warm ischaemic time played an important role with respect to the viability of valves. Their study indicated that under their circumstances, WIT of more than 11 hours showed no viability. Niwaya *et al.* (1995), using flow citometry, showed that that cell viability of the cryopreserved allograft valves was well preserved with a WIT < 520 min (8.7 hours). Crescenzo *et al.* (1993), in an electron microscopic study, observed that heart valves harvested within a warm ischaemic time of 12 hours contain a population of matrix cells, a majority of which are morphologically unaltered and thus presumably will be viable after transplantation.

In most of the studies, the influence of WIT only was analysed, but Suh *et al.* (Suh *et al.*, 1999) also analysed the influence of cold ischaemic time. They showed that cryopreservation did not significantly affect the viability of fibroblasts and there was a significantly negative correlation between WIT and cell viability, and they demonstrate that extending WIT decreased the viability of fibroblasts. Cold ischaemia for 24 hours did not produce a significant change of enzymatic activities after 2 hours of WIT. Although the cellular mechanism involved in cold ischaemia is unclear, it has been reported that cold ischaemia for 24 hours with less than 12 hours of WIT did not have a significant effect on the cellular metabolism.

In our work, the fibroblasts of the cryopreserved pig cardiac valves were well preserved with 2 hours of WIT and 2 or 24 hours of CIT. No differences between the number of viable cells between 2 and 24 hours of CIT were seen. We also think that the WIT has a greater influence on cellular viability.

In the different studies, the cardiac valves have been stored in gas or in liquid phase. Although there are no reasons to think that this could have an influence on cellular viability, in the present study, a greater viability in liquid phase that in gas phase was observed.

Even although the correlation between pulmonary and aortic valve viability had not been assessed, there was no obvious reason to suspect that there would be a significant difference (Gall *et al.,* 1998). Nevertheless in this study, differences could be observed in cellular viability between aortic and pulmonary valves. The pulmonary valve showed a greater viability than the aortic valve, but the differences in their responses to cold ischaemic time, treatment with amphotericin and storage method used were not significantly different between them. This outcome indicated that not all types of valve had the same biological features in cryopreservation.

The controversy regarding valve graft viability has been due to a great extent to the fact that preparation procedures vary from institution to institution, so that accurate comparisons are difficult to make (Armiger, 1998).

When the histological structure of cardiac valves was analysed, the different authors were able to observe changes after cryopreservation due to different factors. Mitchell *et al.* (1995) and Vogt *et al.* (1999) showed that cryopreserved aortic valves presented a loss of surface endothelium but with near-complete preservation of the elastic lamina. Armiger *et al.* (1998) in turn expressed that in most cryopreserved valves the leaflet fibroblasts showed a diminished cellularity and irregular distribution. Crescenzo *et al.* (1993) noticed a significant association between the morphological changes that were produced in the structure of cardiac valves and the WIT.

In the present study, when the histological structure of the valve was analysed, no influence was observed in either the treatment with amphotericin or the storage method used. On the contrary, it was necessary to reduce the CIT to minimise the damage in the histological structure because in this study a short period of WIT was already used. The valves with 24 hours of CIT showed a greater degree of partial detachment and complete detachment of endothelium in the wall and in the leaflet. The fibroblasts were present in smaller numbers and irregular distribution and they were more pyknotic. Finally, the ischaemia in the myocardium was mainly multifocal nature.

On the other hand, a very important problem that has been seen in cryopreserved valvular homografts is the presence of macroscopic fractures of the arterial wall produced after the thawing and which mainly appear in the media layer of the arterial wall. Also, numerous microfractures in the leaflet had been observed that can be present over both surfaces of the tissue penetrating into the collagen/proteoglycan matrix. This has been observed in valves that were kept in the liquid phase (Adam *et al.*, 1990). These fractures can produce serious problems in a programmed operation. In the present study, no macroscopic fractures could be observed in the valves that were kept in liquid phase and slowly thawed.

From this study, we can conclude that the use of correct protocols of selection (donor, ischaemia) or cryopreservation decreases the effects of the same in the viability of the fibroblasts and in the histological structure of the cardiac valves. The consequences of the reduction in the post-cryopreservation viability of the fibroblasts and the histological changes for the success of the graft and the process of rejection should be evaluated after the implantation of these valvular grafts.

5. Acknowledgements

The authors gratefully acknowledge grant support by the Xunta of Galicia (N° PR404A99/64-0). The authors also would like to thank the unit of experimental surgery of Juan Canalejo Hospital for their collaboration.

6. References

ADAM, M., FEI HU, J., LANGE. P. and WOLFINBARGER, L. (1990). The effect of liquid nitrogen submersion on cryopreserved human heart valves, *Cryobiology* **27**, 605–614.

ANGELL, W.W., ANGELL, J.D., OURY, J.H., LAMBERTI, J.J. and GREHL, T.M. (1987). Long-term follow-up of viable frozen aortic homografts, *J. Thorac. Cardiovasc. Surg.* **93**, 815–822.

ANGELL, W.W., OURY, J.H., LAMBERTI, J.J. and KOZIOL, J. (1989). Durability of the viable aortic allograft, *J. Thorac. Cardiovasc. Surg.* **98**, 48–56.

ARMIGER, L.C. (1998). Postimplantation leaflet cellularity of valve allografts: Are donor cells beneficial or detrimental? *Ann. Thorac. Surg.* **66**, S233–S235.

BORKON, A.M., SOULE, L.M. and BANGHMAN, K.L. (1987). Comparative analysis of mechanical and bioprosthetic valves after aortic valve replacement, *J. Thorac. Cardiovasc. Surg.* **94**, 20–23.

COURTMAN, D.W., PEREIRA, C.A., OMAR, S., LANGDON, S.E., LEE, J.M. and WILSON, G.J. (1995). Biomechanical and ultrastructural comparison of cryopreservation and a novel cellular extraction of porcine aortic valve leaflets, *J. Biomed. Mater. Res.* **29**, 1507–1516.

CRESCENZO, D.G., HILBERT, S.L., MESSIER, R.H. Jr., DOMKPWSKI, P.W., BARRICK, M.K. and LANGE, P.L. (1993). Human cryopreserved homografts: Electron microscopic analysis of cellular injury, *Am. Thorac. Surg.* **55**, 25–31.

FISCHLEIN, T., SCUTZ, A., HAUSHOFER, M., FREY, R., UHLIG, A. and DETTER, C. (1995). Immunologic reaction and viability of cryopreserved homografts, *Ann. Thorac. Surg.* **60**, S122–S126.

GALL, K.L., SMITH, S.E., WILLMETTE, C.A. and O'BRIEN, M.F. (1998). Allograft heart valve viability and valve processing variables, *Ann. Thorac. Surg.* **65**, 1032–1038.

GRUNKEMEIER, G.L. and BODNAR, E. (1994). Comparison of structural valve failure among different models of homograft valves, *J. Heart. Valve. Dis.* **3**, 556–560.

HU, J.F., GILMER, L., HOPKINS, R. and WOLFINBARGER, L. Jr. (1989). Effect of antibiotics on cellular viability in porcine heart valve tissue, *Cardiovasc. Res.* **23**, 960–964.

HU, J.F., GILMER, L., HOPKINS, R. and WOLFINBARGER, L. Jr. (1990). Assessment of cellular viability in cardiovascular tissue as studied with 3H proline and 3H insulin, *Cardiovasc. Res.* **24**, 528–531.

KASHIMA, I., YOZU, R., SHIN, H., YAMADA, T., HATA, J. and KAWADA, S. (1999). Effect of storage temperature on cell viability in cryopreserved canine aortic, pulmonic, mitral, and tricuspid valve homografts, *Jap. J. Thorac. Cardiovasc. Surg.* **47**, 153–157.

LU, J.H., CHIU, Y.T., SUNG, H.W., HWANG, B., CHONG, C.K., CHEN, S.P., MAO, S.J., YANG, P.Z. and CHANG, Y. (1997). XTT-colorimetric assay as a marker of viability in cryoprocessed cardiac valve, *J. Mol. Cell. Cardiol.* **29**, 1189–1194.

MC NALLY, R.T. and BROCKBANK, K.G.M. (1992). Issues surronding the preservation of viable allografts heart valves, *J. Med. Eng. Technol.* **16**, 34–38.

MESSIER, R., BASS, B., DOMKOWSKI, P. and HOPKINS, R. (1999). Interstitial cellular and matrix restoration of cardiac valves after cryopreservation, *J. Thorac. Cardiovasc. Surg.* **118**, 36–49

MITCHELL, R.N., JONAS, R.A. and SCHOEN, F.J. (1995). Structure-function correlations in cryopreserved allofgraft cardiac valves, *Ann. Thorac. Surg.* **60**, S108–S113

NIWAYA, K., SAKAGUCHI, H., KAWACHI, K. and KITAMURA, S. (1995). Effect of warm ischaemia and cryopreservation on cell viability of human allograft valves, *Ann. Thorac. Surg.* **60**, S114–S117.

O'BRIEN, M.F., MEGRIFFIN, D.C. and STAFFORD, E.G. (1991). Allograft aortic valve replacement: Long-term comparative clinical analysis of the viable cryopreserved and antibiotic 4°C stored valves, *J. Cardiac. Surg.* **32**, 534–543.

PIDISHEIM, D. and WATSON, J.T. (1996). Cardiovascular applications. In: *Biomaterials Science*, B.D. RATNER, A.S. HOFFMAN, F.J. SCHOEN and J.E. LEMONS, eds., Academic Press, San Diego, pp. 283–297.

SHETTY, S., KAUSHIK, S., MOJAMDAR, M., GOGATE, A. and CHAUKAR, A.P. (1996). Viability testing of homograft valves using methyl thiazol tetrazolium assay, *J. Postgrad. Med.* **42**, 72–75.

SUH, H., LEE, J.L., PARK, J.C., HAN, D.W., YOON, C.S., PARK, Y.H. and CHO, B.K. (1999). Viability and enzymatic activity of cryopreserved porcine heart valve, *Yonsei Mediacal J.* **40**, 184–190.

VAN DER KAMP, A.W.M. and NAVITA, J. (1979). Fibroblast function and the maintenance of the aortic-valve matrix, *Cardiovasc. Res.* **13**, 167–172.

VERGHESE, S., SUDHA, P., PADMAJA, P., MATHEW, T., PRABHAKAR, P., ARUMUGAN, S., MURTHY, K.S. and CHERLAN, K.M. (1999). Cryopreservation of cardiac homografts, *Indian Heart J.* **51**, 301–306.

VOGT, P.R., STALLMACH, T., NIEDERHÄUSER, U., SCHNEIDER, J., ZÜND, G., LACHAT, M., KÜNZLI, A. and TURINA, M.I. (1999). Explanted cryopreserved allografts: A morphological and immunohistochemical comparison between arterial allografts and allograft heart valves from infants and adults, *Eur. J. Cardio-Thorac. Surg.* **15**, 639–645.

WASSENAAR, C., BAX, W.A., SUYLEN, R.J., VUZEVSKI, V.D. and BOS, E. (1997). Effect of cryopreservation on contractile properties of porcine isolated aortic valve leaflets and aortic wall, *J. Thorac. Cardiovasc. Surg.* **113**, 165–172.

WILCZEK, P., SZYDLOWSKA, I., LOTYSZ, D. and RELIGA, Z. (1996). The use of flow citometry for the investigation of viability of heart valve-derived fibroblasts, *Folia histochemica et cryobiologica* **34**, 41–42.

YACOUB, M., RASMI, N.R.H. and SUNDT, T.M. (1995). Fourteen-year experience with homovital homografts for aortic valve replacement, *J. Thorac. Cardiovasc. Surg.* **110**, 186–194.

Advances in Tissue Banking Vol. 6

23 THE CRYOPRESERVED THORACIC AORTA AS A VASCULAR GRAFT

CARLOS-A. MESTRES, IVÁN GARCÍA, MIGUEL A. CASTRO, ELBA AGUSTÍ, VICENTE RIAMBAU, RAMÓN CARTAÑÁ, MIGUEL JOSA, JOSÉ L. POMAR, JAIME MULET

Department of Cardiovascular Surgery
and Cardiovascular Tissue Bank, Hospital Clinic
University of Barcelona, Barcelona, Spain

1. Introduction

The last decade of the 20th century has seen the renewal of interest among surgeons for the use of homologous vascular tissue, namely vascular homografts or vascular allografts. The initial attempts to use vascular homografts are very old and date back to early the past century when Alexis Carrel (Carrel, 1902) started to work on organ transplantation and vascular sutures for he was awarded the Nobel Prize for Medicine 1912. Robert Gross was credited to be among the first surgeons to use vascular homografts after an extensive experimental work (Gross and Hufnagel, 1945). In the following years, vascular transplantation became popular and Dubost was the first who operated on an abdominal aortic aneurysm, replacing the patient's aorta with a vascular homograft, a segment of the thoracic aorta (Dubost, Allary and Oeconomos, 1951; Dubost, 1952). Previously there were some other examples with regards to the use of vascular tissue (Lam, 1951; Oudot, 1952).

The disappointing results due to a high degeneration rate of the homograft in form of aneurysm or rupture led the surgeons to abandon this biological graft in favour of the synthetic fabric grafts developed and first implanted in the 1950s (Voorhees, Jaretzki and Blakemore, 1954). What it seemed clear since the beginning is that the method of preservation had some influence in the performance of the transplanted vessel.

In the beginning of the clinical experience, the vascular homografts were harvested in sterile fashion, placed in culture medium and stored in the fridge at 4°C. As this was working for short periods of time, some other methods of preservation were used including chemicals like formaldehyde, deep freezing immersion in dry alcohol or physical methods like gamma radiation. The combination of shortage of tissues, late infections or degeneration mostly related to an improper processing methodology prompted the surgeons to move towards the fabric vascular prostheses (Dennis, 1987).

The experience with replacement of the cardiac valves with homograft valves that always include a segment of ascending aorta showed that very good results could be achieved with proper harvesting, handling and processing. This included a serious decontamination protocol in an attempt to make sure that all tissues were free of bacteria at the time of implantation. The initial method of preservation at 4°C was later replaced by the new cryopreservation techniques, especially after the work of O'Brien and co-workers who started using cryopreserved heart valves around 1975 (O'Brien *et al.*, 1990). They have reported excellent graft survival after more than 20 years (O'Brien *et al.*, 2001).

The renewed interest in the use of vascular tissue was a consequence of the abovementioned experiences with heart valves. By the late 1980s and the beginning of 1990s, some groups reported early experiences with vascular transplantation with different graft preservation methodologies. The group of La Pitié from Paris advocated the use of fresh homografts as the turnover for organ transplantation at their institution was very fast (Koskas *et al.*, 1996). In Zürich, the group of the University Hospital used cryopreserved homografts that were later stored at –196°C in liquid nitrogen following the advice of O'Brien's group (Vogt P *et al.*, 1996). This was the same policy at our institution since we started using heart valves in 1988 (Mestres, Miró and Pomar, 1994) and later vascular tissue in 1992 (Mestres, Mulet and Pomar, 1995).

By now, it can be said that vascular transplantation has become a useful tool among practising cardiovascular surgeons and that almost every artery is transplanted for a variety of vascular conditions. The ascending aorta, the descending thoracic aorta, the aortic bifurcation and the femoropopliteal axis are all individually transplanted. A number of references can be found in the international literature that report results with these different homografts. Especially popular in peripheral vascular reconstruction are the aortic bifurcation and the femoropopliteal axis. They are usually implanted for primary or secondary vascular infection or because of occlusive disease.

It is interesting to notice that despite the current popularity of vascular homografts, most of the information available refers to aortoiliac or femoropopliteal reconstruction. In these areas, the specific grafts to be used are the aortic bifurcation and the iliac or femoral arteries. The ascending aorta is usually implanted as part of the aortic root complex in cases of replacement of the aortic valve and the ascending aorta, most frequently for endocarditis or vascular graft infection. There is not much information accumulated regarding the descending thoracic aorta as a vascular graft. There is a belief that this is a poor quality graft due to early degeneration. However there have been scanty reports reporting good long-term results (Cornelissen, Hamerlinjck and Vermeulen, 1994).

The current chapter will, deal with the descending thoracic aorta as vascular graft. It has been used in Hospital Clinic University of Barcesince since 1992 and the information available now confirms that this is a useful vascular graft that can be implanted in different conditions.

2. The Anatomy and Physiology

2.1. The normal aorta

The normal aorta changes its size and shape after it takes off the left ventricle at the level of the aortic root. It is interesting to notice the complexity of the aorta when attention is paid at its gross anatomy. The region of the aortic root, that includes the left ventricular outflow tract, the aortic valve and the sinuses of valsalva, which is not the subject of this document, is interesting and complex enough to deserve specific literature. After the aorta separates from the aortic root, its three main segments, ascending, arch and descending are clearly different. The aortic valve is considered to be an specialised region of the aorta and it deserves a separate consideration. One of the best descriptions of the whole of the aorta from anatomy to pathology and function is that of Lindsay and Hurst (1979) that can be considered as a classic monography.

The descending aorta is subdivided in two portions, thoracic and abdominal. Considering its anatomy they are two clearly differentiated vascular segments. The thoracic aorta extends from the body fourth thoracic vertebra to the lower part of the body of the 12th thoracic vertebra. In normal adult individuals of an average of 70 kg is 20 cm long and lies within the posterior mediastinum. Its average diameter changes from 23 mm proximal to 20 mm distal at the beginning of the abdominal segment (Morse, 1979). Its main anatomical features are a number of parietal arteries of which the intercostals are usually the most important. There is a variety of esophageal and pericardial branches from individual to individual. On the average there are ten pairs of intercostal arteries. Once dissected, this results in a long arterial homograft with an important number of large and small collateral vessels that need careful examination at the time of processing and implantation as they need to be secured.

The aorta is an elastic artery with the typical distribution in three layers, the tunica intima, the media and the adventitia, being the media the thickest of them. The tunica intima is a thin layer that consists of collagen, smooth muscle and endothelial cells. The tunica media contains collagen, elastin, smooth muscle and ground substance with the specific distribution in circumferential elements of elastic fibers that are thought to be the foundation functional structure of the aorta (Golinsky and Glagov, 1967). Ground substance has a complex chemical structure based on mucopolysaccharides (Hirst and Gore, 1983). The adventitia is loose, based on a large fibroblast population and provides external support.

From the functional point of view, extensive research has shown that the structure of the aorta allows for the transmission of the pulse wave. There is no doubt that anatomy plays a fundamental role in this transmission (Kot and Rose, 1979). The velocity of the pressure wave generated by the ejection of blood by the left ventricle depends on the structure and characteristics of the wall of the aorta. As the aorta is an elastic vessel it will transmit the wave at a lower velocity. The ascending aorta and the thoracic aorta have a larger content of elastic tissue and therefore pulse wave transmission is faster than in peripheral arteries. The thoracic aorta is a large diameter structure and therefore pulse wave transmission under normal circumstances is fast. Blood flow is not turbulent and therefore thrombus formation is unlikely unless heavy endothelial damage or significant dilation can alter this pattern of flow.

3. The Thoracic Aortic Homograft

Once the basic anatomy and physiology has briefly been reviewed, we are in the position of getting a better understanding of the homograft. It has to be said that conceptually there are no differences among all types of large-diameter vascular homografts rather than those strictly anatomical. The length of the homograft will change according to the type of donation. The type of donor and the presence of arterial disease will strongly influence on the type of homograft that will be processed. Typically, a good quality thoracic aortic homograft will extend from the left subclavian artery down to the abdominal aorta at the level of the celiac axis. Sometimes it will include the neck vessels (Fig. 1).

It is to be noticed that extreme care should be taken at the time to secure the thoracic collateral vessels arising from the homograft. Upon gross inspection at the Tissue Bank where it will be performed at the laminar flow hood, collaterals should be double ligated. It is important not to place ligatures too close to the body of the homograft. If during the banking process there is any doubt

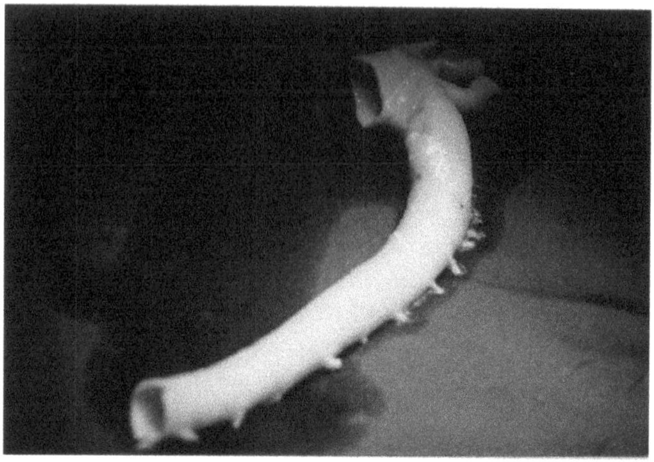

Fig. 1. A view of a thoracic aorta after harvesting. Notice that the aortic arch has been preserved. The neck vessels and the intercostal arteries are clearly seen. The arch can be used if required. The intercostals arteries should be doubly ligated during processing but final checkup is performed in the operating room.

regarding any of the collaterals, the implanting surgeon will take care of them at the time of implantation. A good quality homograft is user friendly and easy to handle.

4. Banking Process

Regulations on banking process have been published elsewhere and at the Tissue Bank, we follow the standards for tissue banking of the European Association of Tissue Banks (EATB) that were published in 1995 and those of the Spanish Association of Tissue Banks (AEBT) published early in 1999.

4.1. Dissection and measurements

As with any other type of homograft, valvular or vascular, dissection is carried in a laminar flow hood using aseptic techniques. Visual inspection leads to discard unsuitable arteries showing calcification or gross atheroma. The biologist and the consultant cardiovascular surgeons will be very strict as morphology represents a major criteria to discard tissues. Calibrated Hegar dilators are used to measure the internal diameter of the aorta. Length is also recorded.

4.2. Decontamination

Several antibiotic-antifungal regimens have been tested. Since January 1996, a protocol including streptomycin 50 μg/ml, penicillin 50 IU/ml and amphotericin B 10 μg/ml for 24 hours at 4°C is used. This protocol guarantees complete sterilisation.

4.3. Cryopreservation and storage

The homografts are placed in a cryoprotective solution that consists in 10% dimethlysulfoxide (DMSO) and 10% fetal calf serum and placed in a double bag. The homografts are freezed at a rate of 1°C/min for the first 20°C. The overall duration of the freezing process of 80 min. A Cryoson BV-10® system (Carburos Metálicos SA, Barcelona, Spain) that incorporates a specifically developed software for the different tissues is used. The bags are finally immersed in liquid nitrogen at –196°C. The cryopreservation protocol has been previously described by us (Mestres *et al.*, 2000).

4.4. Quality control

Full screening is performed before and after decontamination. All donors are tested against aerobic and anaerobic bacteria and fungus. In addition, and according to the Spanish law, the following studies are performed: HIV1 & HIV2, HbsAg and anti-HBc, anti-HCV, VDRL, Q fever and Cytomegalovirus.

4.5. Effects of cryopreservation

Since the inception of cryobiology in clinical practice, it seems clear that cryopreservation has a number of effects that may influence in the performance of the tissue (Bank and Brockbank, 1987). Cryopreservation eliminates a significant amount of antigenicity, decreases cell viability, reduces the contents of connective tissue of the homograft and alters the terminal nerve function. In addition it decreases smooth muscle contractility in up to 50%. Recent work (Adham *et al.*, 1996) shows that mechanical characteristics of the homograft and especially the thoracic aorta, may be influence by the freezing process as well.

5. Clinical Experience

Cardiovascular tissues have been implanted by the members of the Department of Cardiovascular Surgery, Hospital Clinic, University of Barcelona, since early 1989. The initial work included just cardiac valves but after October 1992, large-caliber vascular homografts have been regularly implanted (Mestres, Pomar and Mulet, 1995). Between October 1992 and December 2000, 51 vascular homografts were implanted in 49 patients.

5.1. Patient profile

In 13 of them the implanted homograft was the thoracic aorta. All these patients were male ages ranging from 44 to 80 with a mean age of 64.8. The clinical picture of the patients included ischaemic heart diseases (7), gastrointestinal cancer (5), previous vascular operations (4), chronic renal failure (2) and mitral valve endocarditis and Stanford type A aortic dissection in one case each.

Pre-operative diagnosis included abdominal aortic aneurysm in 7 cases, prosthetic vascular infection in 2, mycotic aneurysm in 2 and thrombosis of a previous vascular reconstruction in 2. The cases with vascular infection were related to a bifurcation graft and an ascending aortic prosthesis.

5.2. Operations

The operations performed varied regarding the location of the pathology. As this is a good graft for a straight tube repair, it was

Fig. 2. Intra-operative view of a thoracic aorta that has been used to construct an aortofemoral bifurcation homograft. Note the suture line with an extension bifurcation.

used in this fashion in 9 cases in the abdominal aorta, two of them with associated reimplantation of visceral vessels. In four additional operations, the graft was used for reconstruction of the ascending aorta, aortofemoral bifurcation grafts and aortouniiliac repair. It is interesting to notice that in 10 cases only one homograft was used and in 3 cases a composite homograft was required (Fig. 2). In these latter cases, 3 thoracic aortas were extended using iliac arteries and a bifurcation homograft as well. The average homograft diameter was 16 mm and the average length 10.6 mm. A number of associated procedures were performed, including 3 cases of gastrectomy, 2 cases of colectomy, 1 splenectomy and repair of a mycotic popliteal aneurysm each.

5.3. Results

Three patients died in the hospital, due to acute myocardial infarction, disruption of anatomotic line and sepsis with multiorgan failure. One patient died at home 6 months after surgery because of myocardial infarction. The survivors are leading an active life with no complications related to their grafts.

Mean intensive care time was 3 days. Mean postoperative length of stay was 23 ± 4 days. This reflects that the average population receiving a vascular homograft is usually a population with severe pathology and comorbidites.

Mean follow-up is 42 ± 5 months (3–75). There has been no recurrence of infection in those patients operated upon for active primary or secondary vascular infection. All homografts are patent. Patency has been assessed with computed axial tomography and angiography. Computed tomography has been performed 16 times (2/pt) and angiography has been performed 12 times (1.5/pt). Up to now, no case of homograft-related complication has been detected.

6. Comment

After a decade of personal experience, we have been able to collect a series of 49 complex patients undergoing 51 operations in which a cryopreserved vascular homograft was implanted for a variety of indications, where 13 of them were implanted with a cryopreserved thoracic aorta. This type of vascular homograft has been used since the beginning of vascular replacement surgery. In 2001. we have celebrated the 50th anniversary of the first implantation of such a homograft (Friedman, 2001).

It is agreed that in the early phases homografts were popular but after some time they were abandoned due to degeneration, in form of dilatation or rupture. In a similar way, some years later the homograft valves were found to behave in such a way that they also required early explantation in a number of cases (Barratt-Boyes *et al.*, 1967). What actually seemed true is that manipulation and preservation were considered the most important influencing factor regarding tissue performance on the mid- and long-term. This is why cryobiology certainly modified the approach to human tissues (May and Baust, 1987).

The review of the literature confirms the renewed interest in the use of vascular homografts over the past decade and a number of lessons has been learnt. First, the preservation methodology plays an important role in tissue behaviour as stated earlier. This was confirmed by the data from Paris (Koskas *et al.*, 1996) when showing a high homograft attrition rate in less than five years, compared to the data from Zurich (Vogt *et al.*, 1996). Secondly, that homografts undergoing cryopreservation tend to last longer, a fact that has been particularly demonstrated when analysing the performance of cryopreserved aortic valves as the follow-up is longer (O'Brien, 2001).

In our own experience, we have been able to found that the cryopreserved thoracic aortic homograft shows an adequate behaviour during a follow-up of 7 years, as we started with this type of homograft in late 1993. Conversely to previous opinions, we found that the cryopreserved thoracic aorta works well as a replacement biological prosthesis. No cases of dilatation nor rupture during the follow-up have been found.

Vascular homografts are an extremely good material to work with. This is a biological user-friendly material which adapts perfectly to the needs of the surgeon and the requirements of the patient. It also seems true that according to published material and our own experience, the thoracic aorta is also a material resistant to infection and, very important from the technical point of view, it allows for anatomic and extra-anatomic reconstructions alone or in combination with additional homografts of different anatomic locations (iliacs, bifurcations). The operation is usually carried out with less fear of intraoperative bleeding through the suture lines as the homograft is hemostatic itself.

Patients suffering from a variety of conditions, including reoperations, active infections and, also interesting, patients with intra-abdominal neoplasms can be operated. Homografts have helped to perform combined operations, thus reducing the need for second look operations and to allow for a complete and combined cure of simultaneous conditions. At the end, we have found that the thoracic aorta is a versatile homograft, suitable for implantation in different locations and that benefits from the currently available methods of preservation.

7. References

ADHAM, M., GOURNIER, J.P., FAVRE, J.P., ROCHE, E. DE LA, DUCERF, C., BAULIEUX, J., BARRAL, X. and POUYET, M. (1996). Mechanical characteristics of fresh and frozen human descending thoracic aorta, *J. Surg. Res.* **64**, 32–24.

BANK, H.L. and BROCKBANK, K.G.M. (1987). Basic principles of cryobiology, *J. Cardiovasc. Surg.* **3**(Suppl. I), 137–143.

BARRATT-BOYES, B.G., ROCHE, A., BRANDT, P., SMITH, J. and LOWE, J. (1969). Aortic homograft valve replacement. A long-term follow-up of an initial series of 101 patients, *Circulation* **XL**, 763.

CARREL, A. (1902). La technique opératorie des anastomoses vasculaires et la transplantation des viscères, *Lyon. Méd.* **98**, 859.

CORNELISSEN, P.H.J., HAMERLINJCK, R.P. and VERMEULEN, F.E. (1994). Aneurysmal dilatation of an aortic homograft more than 30 years after implantation into the thoracic aorta, *Eur. J. Cardio-Thorac. Surg.* **8**, 447–448.

DENNIS, C. (1987). Brief history of development of vascular grafts. In: *Modern Vascular Grafts*, P.N. Sawyer, ed., McGraw-Hill, New York, pp. 1–25.

DUBOST, C., ALLARY, M. and OECONOMOS, N. (1951). A propos du traitement des anevrismes de l'aorte. Ablation de l'anevrisme, rétablissement de la continuité par greffe d'aorte humaine conservée, *Mem. Acad. Chir.* **77**, 38.

DUBOST, C. (1952). Resection of an aneurysm of the abdominal aorta, *Arch. Surg.* **64**, 405–408.

ESTÁNDARES DE LA ASOCIACIÓN ESPAÑOLA DE BANCOS DE TEJIDOS: AEBT (1999). AEBT, Spain.

EUROPEAN ASSOCIATION FOR TISSUE BANKING. (1995). *General Standards For Tissue Banking.* Viena.

FRIEDMAN, S.G. (2001). The 50th anniversary of abdominal aortic reconstruction, *J. Vasc. Surg.* **33**, 895–898.

GROSS, R.E. and HUFNAGEL, C. (1945). Coarctation of the aorta: Experimental studies regarding its surgical correction, *N. Engl. J. Med.* **233**, 287.

HIRST, A.E. and GORE, I. (1983). The etiology and pathology of aortic dissection. In: *Aortic Dissection*, Doroghazi RM, E.E. Slater, eds., New York, McGraw-Hill Book Company, pp. 13–53.

KIEFFER, E., SABATIER, J., PLISSONNIER, D. and KNOSALLA, C. (2001). Prosthetic graft infection after descending thoracic/thoracoabdominal aortic aneurysmectomy: Management with in situ arterial allografts, *J. Vasc. Surg.* **33**, 671–678.

KOSKAS, F., PLISSONIER, D., BAHNINI, A., RUOTOLO, C. and KIEFFER, E. (1996). *In situ* arterial allografts for aortoiliac graft infection: A 6-year experience, *Cardiovasc. Surg.* **4**, 454–459.

KOT, P.A. and ROSE, J.C. (1979). Physiology of the aorta. In: *The Aorta*, J. Jr. Lindsay and J.W. Hurst, eds., New York, Grune & Stratton, pp. 39–47.

LAM, C.R. and ARAM, H.H. (1951). Resection of the descending thoracic aorta for aneurysm: Report of the use of a homograft in a patient and an experimental study, *Ann. Surg.* **134**, 743–752.

LINDSAY, J. Jr. and HURST, J.W. (1979). *The Aorta.* New York, Grune & Stratton.

MAY, S.R. and BAUST, J.G. (1987). Cryobiology of tissues. In: *Cardiac Valve Allografts (1962–1987)*, A.C. Yankah, R. Hetzer, D.C. Miller, D.N. Ross, J. Somerville and M.H. Yacoub, eds., Steinkpoff Verlag Darmstadt, pp. 27–34.

MESTRES, C.A., MIRO, J.M. and POMAR, J.L. (1994). Cryopreserved homografts in the treatment of relapsing fungal cardiovascular infections, *J. Thorac. Cardiovasc. Surg.* **108**, 990–991.

MESTRES, C.A., MULET, J. and POMAR, J.L. (1995). Large-caliber cryopreserved arterial allografts in vascular reconstructive operations: Early experience, *Ann. Thorac. Surg.* **60**, S105–S107.

MESTRES, C.A., AGUSTI, E., MARTINEZ, A., CABRER, C., MANYALICH, M., CARTAÑA, R. and POMAR, J.L. (2000). Cardiovascular tissue banking in the non-cadaveric setting: Ten-year experience of a University Hospital-based bank with active organ donation program, *J. Heart Val. Dis.* **9**, 523–529.

MORSE, D.E. (1979). Embryology, anatomy and histology of the aorta. In: *The Aorta,* J. Jr. Lindsay, J.W. Hurst, eds., New York, Grune & Stratton, pp. 15–37.

O'BRIEN, M.F. and McGIFFIN, D.C. (1990). Aortic and pulmonary allografts in contemporary cardiac surgery. In: *Advances in Cardiac Surgery,* R.B. Karp, N.T. Kouchoukos, H. Laks and A. Wechsler, eds., Year Book Medical Publishers, Inc. Chicago, pp. 1–24.

O'BRIEN, M.F., HARROCKS, S., STAFFORD, E.G., GARDNER, M.A., POHLNER, P.G., TESAR, P.J. and STEPHENS, F. (2001). The homograft aortic valve: A 29-year, 99.3% follow up of 1022 valve replacements, *J. Heart Val. Dis.* **10**, 334–344

OUDOT, J. (1951). La greffe vasculaire dans les thromboses du carrefour aortique, *Presse Med.* **59**, 234–236.

VOGT, P.R., Von SEGESSER, L.K. and GOFFIN, Y. (1996). Eradication of aortic infections with the use of cryopreserved homografts, *Ann. Thorac. Surg.* **62**, 640–645.

VOGT, P.R., BRUNNER-LA ROCCA, H.P., CARREL, T., Von SEGESSER, L.K., RUEF, C., DEBATIN, J., SEIFERT, B., KIOWSKI, W. and TURINA, M. (1998). Cryopreserved arterial allografts in the treatment of major vascular infection: A comparison with conventional surgical techniques, *J. Thorac. Cardiovasc. Surg.* **116**, 965–972.

VOORHEES, A.B. Jr., JAREZTKI, A. III and BLAKEMORE, A.H. (1952). The use of tubes constructed from Vynion "N" cloth in bridging arterial defects, *Ann. Surg.* **135**, 332.

WOLINSKY, H. and GLAGOV, S. (1967). Nature of species differences in the medial distribution of aortic vasa vasorum in mammals, *Cir. Res.* **20**, 409–421.

SECTION V:
COLLAGEN AND SKIN

Advances in Tissue Banking Vol. 6

24 CRITICAL REVIEW OF SKIN SUBSTITUTION AND BANKING IN THE THIRD MILLENNIUM

DUINSLAEGER LUC

Military Hospital Queen Astrid, Brussels, Belgium
Head of Tissue Bank Burn Centre of Brussels

1. Introduction and History

First use of heterograft (lizard skin) described by Cannady back in 1682. Heterografts from chicken, pigeon, dogs, guinea pig, rats, pigs, frogs, rabbits, were applied between 1850 and 1900. Dr. Reverdin from Paris used his own skin (being thus the first homograft) on the granulating wound of a patient in 1869. Foster and Smith who used cutaneous homografts from amputated limbs in 1871 soon followed this. Ten years later first cadaver homografts were used. Loeb and Padgett demonstrated the rejection of hetero/homografts in 1930, but it took until 1943 to elucidate the mechanism of rejection (Medawar). Miller advocated in 1967 the immediate coverage of partial thickness burns with homografts, but his infection ratio was very high. From 1970 on increasing evidence arose to confirm the usefulness of early excision. From this period on an important research effort was driven to skin substitution to be able to close the large excised burn wounds.

Rheinwald and Green described the cultivation of human keratinocytes in 1979 and soon autologous cultured keratinocytes (O'Connor, 1981) and allogeneic cultured keratinocytes (Madden,

1986) were introduced in clinical practice. Burke and Yannas developed the first bilayered collagen based synthetic skin substitute in 1981 (now commercially available as Integra®).

In the last decade of the century several new collagen-based and tissue or culture-derived skin or dermal substitutes became available to the burn surgeon: Alloderm®, Transcyte®, Apligraf®, Transderm®, Beverderm®, Ortec CCS®. The development of skin substitutes is still ongoing. Boyce has published the combination of a cultured skin substitute with a synthetic collagen-based skin substitute (Integra®) in 1999.

2. Goal of a Skin Substitute

Early excision convincingly improved burn wound treatment by avoiding invasive burn wound infection, providing a good wound surface and diminishing the immune response of the burned tissue as well as the concentration of several circulating cytotoxins. After excision, however, prompt closure is of the utmost importance, preferably using epidermal as well as dermal substitution.

A good skin substitute should be able to:

- Prevent/control infection
- Have a good adherence
- Preserve the underlying tissue
- (Timely) close the open wound
- Maintain the skin function
- Have an acceptable cosmetic and functional long-term result
- Be readily available, with a low cost and high biosafety and easy to use.

3. Types of Skin Substitutes

3.1. Types of skin substitutes according to its *function*

3.1.1. Temporary substitutes

This temporary substitute will provide a transient wound closure of fresh wounds or donor sites until healing or a physiologic closure of excised and/or (widely) meshed autografted full thickness burn wounds, e.g. Biobrane®, allogeneic keratinocyte cultures (Transderm®), pigskin, Dermagraft TC®.

3.1.2. Semi-permanent substitutes

Some material of the skin substitute will remain or will be gradually replaced, e.g. Integra®.

3.1.3. Permanent substitutes

The substitute will be permanently incorporated, e.g. Alloderm®, Beverderm®.

3.2. Types of skin substitutes according to its *source* or *production method*

These three types can be combined (composite substitutes):

Natural tissues: allografts, xenografts, amniotic membranes, Alloderm®, deepithelised human dermis (Beverderm®).

Culture-derived: Apligraf®, cultured keratinocytes, Dermagraft®, Ortec CCSÆ or other fibroblast or epithelial seeded dermal analogues

Collagen based composites: Biobrane®, TransCyte®, Integra®

4. Natural Tissue Derived Skin Substitutes

4.1. Epidermal/dermal substitution: Xenografts

- Fresh or lyophilised? This will depend on the expertise and experience of the burn centre and availability
- Acceptable results in superficial partial thickness (e.g. scalds)
- Biosafety problem
- Procurement mostly easy
- Temporary substitution/no blood supply/quick sloughing off
- Low cost, readily available.

4.2. Epidermal/dermal substitution: Homografts

- Cryopreserved or glycerolised? This question is still unanswered. Clinical data on this subject are inadequate and inconclusive Glycerolisation is cheaper in production and storage. Cryopreservation offers the possibility of using viable skin.
- Good results in superficial partial thickness (e.g. scalds).
- Biosafety problem! Some data exist to support the hypothesis that glycerolisation is safer. When accurate screening protocols are used by the skin bank the transmission risk is very low.
- Procurement is difficult in some countries due to social or religious reasons.
- Temporary/semi-permanent (when Cuono-method is used).
- Still the gold standard for temporary skin substitution especially in superficial partial thickness burn or in the treatment of large full thickness burns where early and aggressive excision is mandatory.
- Need of meticulous donor screening methods.

4.3. Dermal substitution: Alloderm®, Beverderm®

- Useful as dermal template/no epidermal "barrier"
- Promising reports about "one step" grafting (Alloderm®)
- Combination with cultured keratinocytes? No clinical data available
- No comparative studies as permanent skin substitute.

4.4. (Epidermal)/dermal substitute: Amniotic membrane

- Less adherent than homograft
- Easy to apply, transparent
- Difficult to obtain and prepare in large quantities
- Biosafety problem
- Quick disintegration, less barrier function.

5. Culture Derived Skin Substitutes

5.1. Epidermal substitution: Cultured keratinocytes

- Drawbacks of autologous (timing, take problems, blistering) and allogeneic cultures (need of autologous cells)
- High cost and meticulous techniques needed

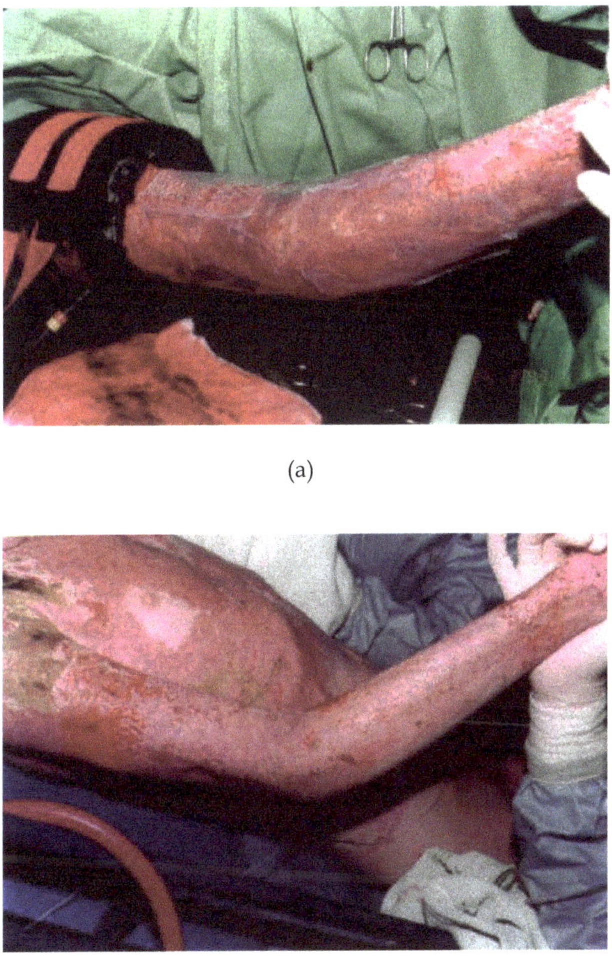

(a)

(b)

Fig. 1. Use of cultures keratinocytes on deep partial thickness burn. (a) Day 0. (b) Day 5 after operation.

- Need of dermal matrix to obtain good results:
  - — partial thickness burns: autologous dermal matrix is still present (Fig. 1)
  - — dermal substitution/(Cuono method (allogeneic dermis is left in place after homografting), Integra®, etc.
  - — modified sandwich method (see Sec. 6, Figs. 4 and 5).

6. Modified Sandwich Technique Using Allogeneic Keratinocyte Cultures

6.1. Introduction

The use of cultured epithelium autografting (CEA) is an established tool in the management of extensive burns. Since the first introduction of keratinocyte culturing techniques by Rheinwald and Green in 1975, multiple reports have been published on the use of CEA as a permanent wound coverage technique and a life-saving procedure in specific cases (De Luca *et al.*, 1989; Cuono, Langdon and McGuire, 1986; Gallico *et al.*, 1984; O'Connor *et al.*, 1981; 1990; Hefton *et al.*, 1983; Braye *et al.*, 2000; Foyatier, Faure and Latarjet, 1989; Teepe *et al.*, 1990; Herzog *et al.*, 1988; Munster, Weiner and Spence, 1990; Eldad *et al.*, 1986).

However, the use of autologous cultured epithelium has several important drawbacks: one has to wait for three weeks to 30 days before a CEA consisting of 5 to 6 cell layers thick is available, even with state-of-the-art techniques. Secondly, the surface of the CEA can be limited by technical problems during culturing (contamination, low growth potential, etc.). Thirdly, the CEA have to be used at a certain moment that does not necessarily correspond with the ideal timing for the patient, and considerable logistics and technique are required to prepare the CEA immediately before grafting, leading to high costs for personnel and infrastructure. Finally, average take rates of CEA are definitely lower than conventional split-skin autografting. Due to the mentioned drawbacks, an increasing tendency arised to use allogeneic cultured keratinocytes (De Luca *et al.*, 1989; McGuire, Birchall and Langdon, 1987;

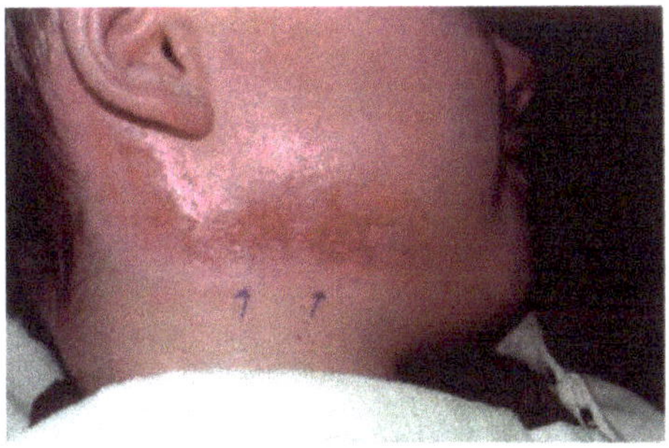

(a)

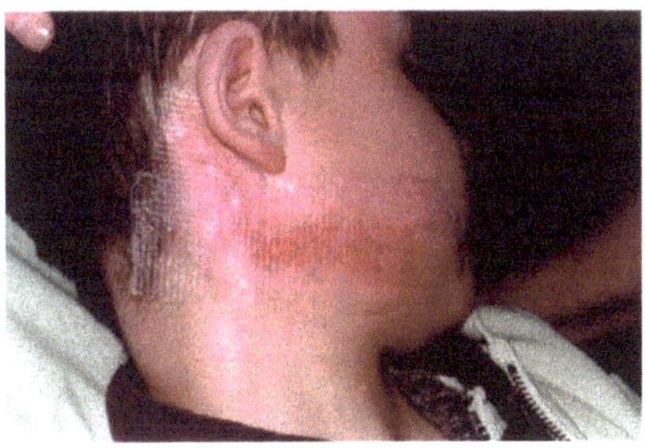

(b)

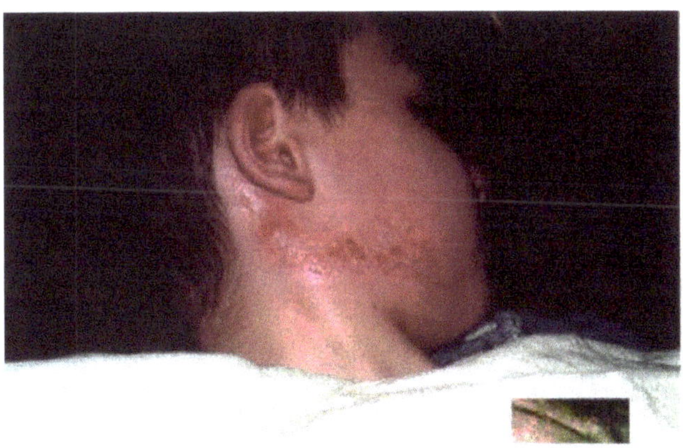

(c)

Fig. 2. Use of cultures on residual (chronic) wound. (a) Day 0. (b) + Culture (c) 5 days post-application.

Foyatier, Faure and Latarjet, 1989; Specht, Strittmatter and Bohm, 1988; Faure *et al.*, 1988; Bolivar-Flores *et al.*, 1990; Eldad *et al.*, 1987; Teepe, 1988; Phillips *et al.*, 1990).

Although convincing evidence exists that these transplanted cultured cells do not survive very long at the graft site (Brain *et al.*, 1989; Better-Galland *et al.*, 1988; Phillips and Gilchrest, 1989; Deglize, Benathan and Krupp, 1987; Roseeuw and De Coninck, 1989; Phillips *et al.*, 1990), it is assumed that they enhance wound re-epithelialisation by producing factors that stimulate the remaining keratinocytes (e.g. at the wound edge or from residual epidermal appendages in deep dermal burns) to proliferate and/or migrate. In order to combine the reliable and good take of conventional split-skin grafting with the stimulatory effect on re-epithelialisation of allogeneic cultured keratinocytes, an alternative sandwich-technique is used currently at the Brussels burn centre of the Military Hospital Queen Astrid.

6.2. Modified "sandwich" technique

Early excision is of the utmost importance when treating extensive burns: it will provide a good wound surface, ready to accept a skin graft, it will lower infection risks and diminish inflammatory cascade and circulating cytokines. This results in better mortality rates. The problem however, when donor sites are lacking, is to close the excised wound in a permanent way as quickly as possible. Conventionally, the excised burn wounds are closed temporarily with split thickness human allograft from skin banks. In the modified "sandwich" technique, the excised wounds are closed by using widely meshed split thickness autografts covered with fresh or cryopreserved sheets of cultured allogeneic keratinocytes.

Standard treatment regimen of burns over 60% TBSA in adult or 45% TBSA in children or elderly patients (over 60 years) is performed as following:

- Biopsy for autologous keratinocyte culturing on admission.
- Tangential excision starting from day 3 to 5 after admission twice a week until total surface of burn is excised.

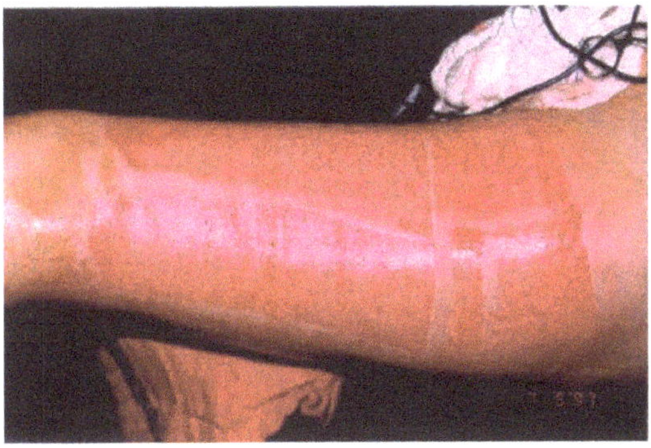

(a)

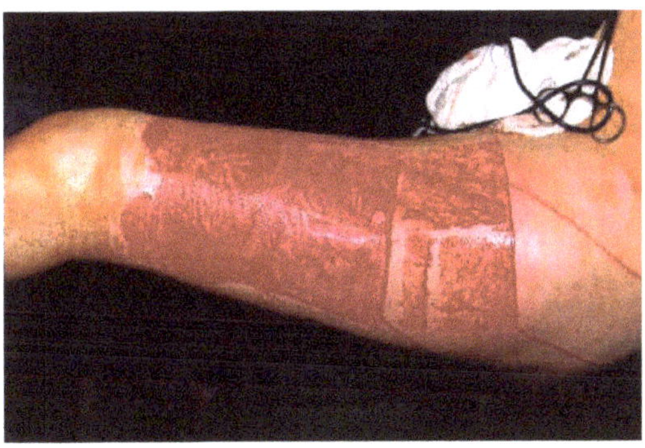

(b)

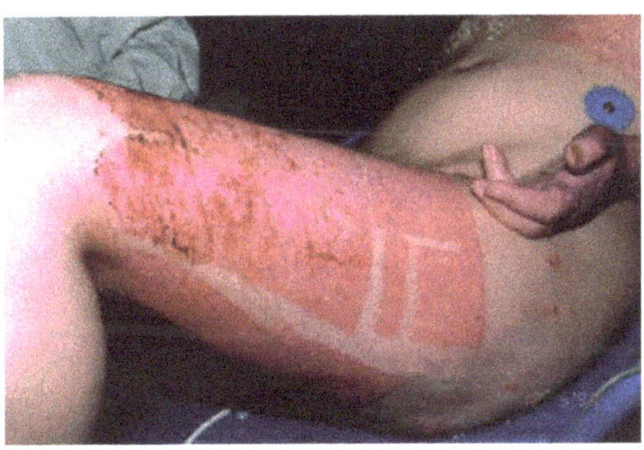

(c)

Fig. 3. Use of cryopreserved keratinocyte cultures on donor site. (a) Day 0. (b) Day 0 + culture. (c) Day 5: healed donor site (normally 8–12 days).

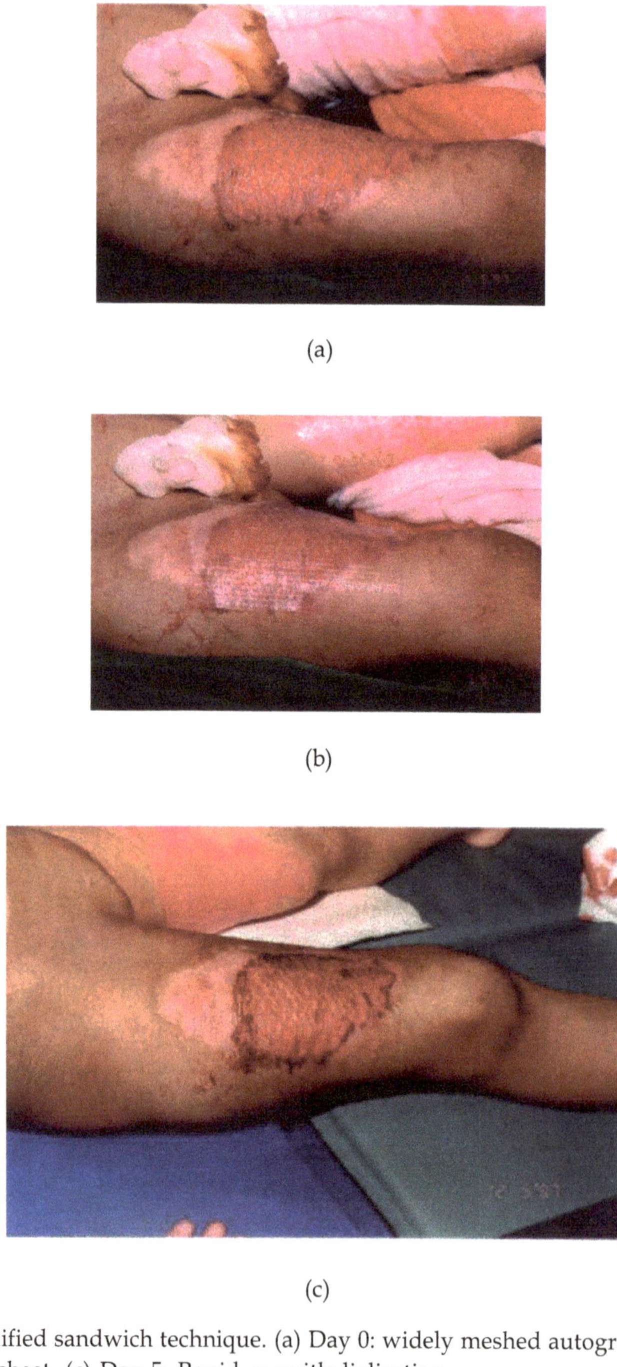

Fig. 4. Modified sandwich technique. (a) Day 0: widely meshed autograft. (b) Day 0: + cultured sheet. (c) Day 5: Rapid re-epithelialisation.

- Immediate definitive wound closure using large meshed autograft from available donor site covered with allogeneic cultured sheets.
- Treatment of donor site with allogeneic cultured keratinocyte sheets (Fig. 3).
- Temporary closure of the excised burn wounds with cryopreserved human allografts if donor site is lacking.
- Cuono's method or modified sandwich technique using autologous cultures as soon as autocultures are available (3 to 4 weeks after admission, see Figs. 4 and 5).
- Treatment of small residual defects with cultured keratinocyte sheets (Fig. 2).

From 1989 to 2000, 656 graft procedures on 622 patients were performed using this modified sandwich technique in the burn centre of Brussels. Clinical results showed similar take rates and infection risks compared to conventional split thickness grafting, but significantly better wound closure on day 5 post-grafting (Duinslaeger *et al.*, 1998). A quantification study of the effect on re-epithelialisation on 1 to 3 meshed autografted wounds showed enhancement of re-epithelialisation by 80% on day 5 (Duinslaeger *et al.*, 1996). Healing of donor site was significantly reduced compared to poly-urethane dressing treatment (Duinslaeger *et al.*, 1997).

6.3. Conclusion

The modified sandwich technique allows the surgeon to excise earlier and more aggressively extensive burns with a safety that is comparable to conventional grafting and with the advantage of a quicker wound closure. It combines the possibilities offered by biotechnology with conventional surgical techniques. In view of the related costs and needed infrastructure, this technique is only indicated for the treatment of extensive burns or in the case of high mortality risk patients.

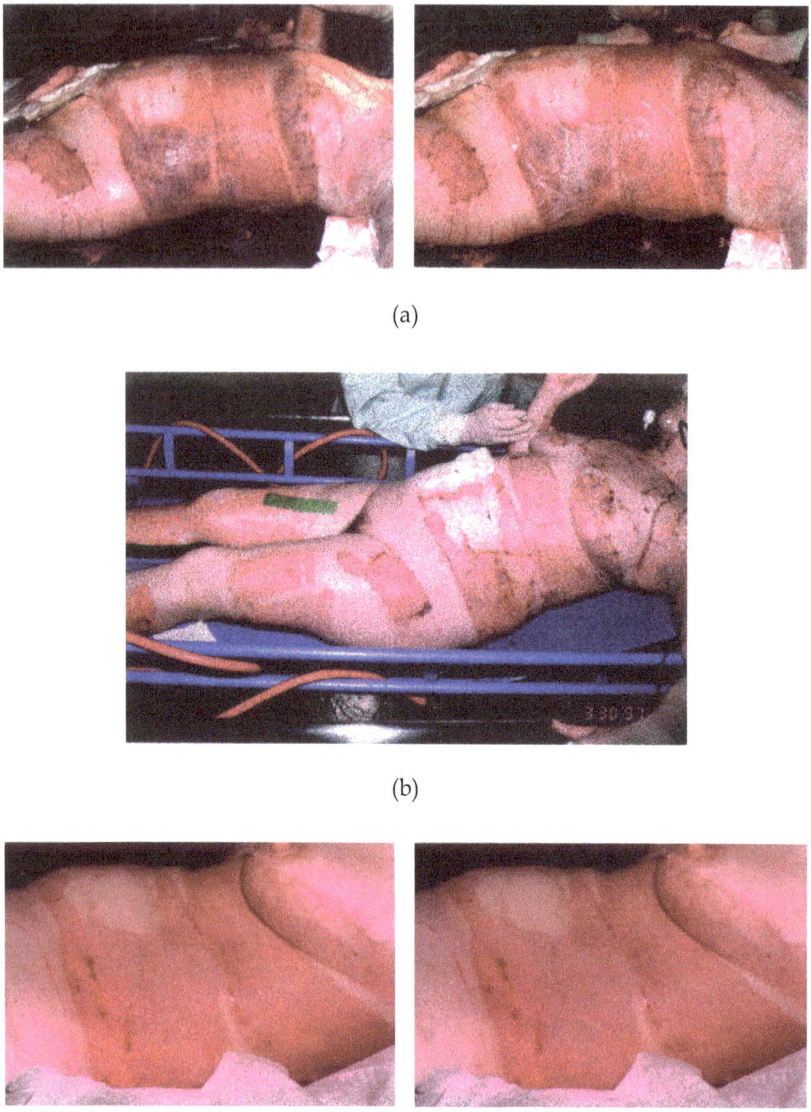

Fig. 5. Modified sandwich technique. (a) Large surface 3rd degree (full thickness) excised in one step + autograft on an elderly patient. (b) Day 5. (c) Day 10: healing.

7. Collagen-Based Synthetic Skin Substitutes

7.1. Epidermal/dermal substitution: Apligraf®

- Living epidermal substrate (stratified = barrier)
- Living dermal matrix (fibroblasts)
- Satisfying results in chronic ulcers, burn data not yet available
- Semi-permanent function.

7.2. (epidermal)/dermal substitution: Ortec CCS®

- Keratinocyte/fibroblast co-culture in collagen sponge
- No immediate barrier
- Multicentre clinical trial underway in burns
- Semi-permanent.

7.3. Temporary epi/derm substitution: Biobrane®

- Good adherence
- Removed after healing or prior to autografting
- Problem of infection/fluid accumulation
- Temporary function.

7.4. Semi-permanent epi/derm substitution: TransCyte®

- Epidermis = silicone which is removed after reepithelialisation or before autografting
- Dermis = non living, human fibroblast derived products
- Must be kept frozen until use
- Expensive technology and storage.

7.5. Temporary epidermal substitution, permanent dermal incorporation: Integra®

- Need to replace epidermis with epidermal autografts (or cultures)
- Problems of higher infection rate and lack of ingrowth

- Interesting development in progress (incorporation of epidermal elements), but no clinical data yet
- Use in reconstruction promising, though sometimes variable shrinkage.

8. General Conclusion

- Indications and clinical usefulness of temporary skin substitutes are well defined, according to availability and transmission risk assessment: homografts or synthetic composite bilayer structure skin substitute.
- Allogeneic cultures are an alternative treatment in (deep) partial thickness burns.
- Modified sandwich technique combining autografts with allogeneic cultures has a proven efficacy and safety.
- Optimising of the newer skin substitutes is necessary to lower infection rates, shorten duration of neodermis formation and improve epidermal substitution technique.

9. Future Prospects

- Growth factor (bFGF?) application may replace temporary skin substitution in partial thickness burns or used in combination with skin substitutes.
- Incorporation of antibiotics in skin substitutes.
- Introduction of "one-step" total skin substitution.
- Genetically modified cultures.

10. References

BETTEX-GALLAND, M., SLONGO, T., HUNZIKER, T., WIESMANN, U. and BETTEX, M. (1988). Use of cultured keratinocytes in the treatment of severe burns, *Z. Kinderchir* **43**, 224–228.

BOLIVAR-FLORES, J., POUMIAN, E., MARCHE-MORENO, M., MONTES DE OCA, G. and KURI-HARCUCH, W. (1990). The use of cultured human epidermal keratinocytes for allografting burns and conditions for temporary banking of cultured allografts, *Burns* **16**, 3–8.

BRAIN, A., PURKIS, P., COATES, P., HACKETT, M., NAVSARIA, H. and LEIGH, I. (1989). Survival of cultured allogeneic keratinocytes transplanted to deep dermal bed assessed with probe specific for Y chromosome, *Br. Med. J.* **298**, 917–919.

BRAYE, F., ODDOU, L., BERTIN-MAGHIT, M. *et al.* (2000). Widely meshed autograft associated with cultured autologous epithelium for the treatment of major burns in children: Report of 12 cases, *Eur. J. Pediatr. Surg.* **10**, 35–40.

CUONO, C., LANGDON, R. and MCGUIRE, J. (1986). Use of cultured autografts and dermal autografts as skin replacement after burn injury, *Lancet* **II** , 1123–1124.

DEGLIZE, B., BENATHAN ,M. and KRUPP, S. (1987). Resultats préliminaires du traitement des brûlés d'épiderme de culture, *Schweiz Med. Wochenschr* **117**, 1380–1393.

DE LUCA, M., ALBANESE, E., BONDANZA, S. *et al.* (1989). Multicentre experience in the treatment of burns with autologous and allogeneic cultured epithelium, fresh or preserved in a frozen state, *Burns* **15**, 303–309.

DUINSLAEGER, L., VERBEKEN, G., REPER, P. *et al.* (1996). Lyophilised keratinocyte lysates contain multiple mitogenic activities and stimulate closure of meshed autograft-covered burn wounds with similar efficiency as fresh keratinocyte cultures, *Plast. Reconstr. Surg.* **98**(1), 110–117.

DUINSLAEGER, L., VERBEKEN, G., VAN HALLE, S. *et al.* (1997). Cultured allogeneic keratinocyte sheets accelerate healing compared to Op-Site treatment of donor sites in burns, *J. Burn. Care Rehabil.* **18**(6) , 545–551.

DUINSLAEGER, L., DELAEY, B. and VANDERKELEN, A. (1998). Short- and long-term results of application of allogeneic cultured keratinocytes on burn wounds and burn scar, *Eur. J. Plast. Surg.* **21**, 14–18.

ELDAD, A., BURT, A., CLARKE, J.A. and GUSTERSON, B. (1987). Cultured epithelium as a skin substitute, *Burns* **13** , 173–180.

FAURE, M., DEMIDEM, A., MAUDIT, G., CHABANON, M. and THIVOLET, J. (1988). Immunological aspects of epidermal allografts. In: *Clinical Use of Cultured Epithelium in Surgery and Dermatology*, Medical and Scientific Conferences Ltd., Wheathamstead, Herts, UK, pp. 53–57.

FOYATIER, J.L., FAURE, L. and LATARJET, J. (1989). Cultured epithelium: Clinical use in burned patients. *3rd Congress of European Burn Association*, Prague, Czechoslovakia, p. 54.

GALLICO, G.G., O'CONNOR, N.E., COMPTON, C.C., KEHINDE, O. and GREEN, H. (1984). Permanent coverage of large burn wounds with autologous cultured human epithelium, *N. Engl. J. Med.* **311**, 448–451.

HEFTON, J.M., MADDEN, M.R., FINKELSTEIN, J.L. and SHIRE, G.T. (1983). Grafting of burn patients with autografts of cultured epidermal cells, *Lancet* **II**, 428–430.

HERZOG, S.R., MEYER, A., WOODLEY, D. and PETERSON, H.D. (1988). Wound coverage with cultured autologous keratinocytes: Use after burn wound excision, including biopsy follow-up, *J. Trauma* **28**, 195–198.

MCGUIRE, J., BIRCHALL, N. and LANGDON, R. (1987). Successful engraftment of allogeneic keratinocytes, *Clin. Res.* **35**, 720a.

MUNSTER, A.M., WEINER, S.H. and SPENCE, R.J. (1990). Cultured epidermis for the coverage of massive burn wounds, *Ann. Surg.* **211**, 676–680.

O'CONNOR, N.E., MULLIKEN, J.B., BANKS-SCHLEGEL, S., KEHIDE, O. and GREEN, H. (1981). Grafting of burns with cultured epithelium prepared from autologous epidermal cells, *Lancet* **I**, 75–78.

O'CONNOR, N.E., GALLICO, G.G., COMPTON, C.C., BRIGGS, S. and REMENSNYDER, J. (1990). Long-term results in major paediatric burns treated with cultured epithelial autografts, *Proc. Am. Burn Assoc.* Meet, p. 174.

PHILLIPS, T.J. and GILCHREST, B.A. (1989). Cultured allogeneic keratinocytes grafts in the management of wound healing: Prognostic factors, *J. Dermatol. Surg. Oncol.* **15**, 1169–1176.

PHILLIPS, T.J., BHAWAN, L., LEIGH, I.M,. BAUM, H.J. and GILCHREST, B.A. (1990). Cultured epidermal autografts and allografts: A study of differentiation and allograft survival, *Am. Acad. Dermatol.* **23**, 189–198.

ROSEEUW, D. and DE CONINCK, A. (1989). Epidermal DNA analysis of chronic ulcers grafted with allogeneic cultured epidermis, *J. Invest. Dermatol.* **92**, 509A.

SPECHT, B.U., STRITTMATTER, B. and BOHM, N. (1988). Allogeneic cultured epidermals for wound grafting, *J. Invest. Dermatol.* **191**, 379A.

TEEPE, R.G. (ed.) (1988). *Clinical Use of Cultured Epithelium in Surgery and Dermatology* (Leiden, March 1987). Medical and Scientific Conferences Ltd., Birmingham, UK.

TEEPE, R.G., KREIS, R.W., KOEBRUGGE, E.J. *et al.* (1990). The use of cultured autologous epidermis in the treatment of extensive burn wounds, *J. Trauma* **30**, 269–275.

Advances in Tissue Banking Vol. 6

25 COLLAGEN BIOCHEMISTRY: AN OVERVIEW

MITSUO YAMAUCHI

CB# 7455, Dental Research Center
University of North Carolina, Chapel Hill
NC 27599-7455, USA

1. Introduction: What is "Collagen"?

"Collagens" are a large family of structurally related proteins that form a supramolecular assembly in the extracellular matrix and contain one or more domain(s) of unique triple helices ("collagenous domain"). This collagen domain, the hallmark of these proteins, is a coiled-coil right-handed triple helix composed of three polypeptide chains, called α chains. In this chapter, the triple-helical and non-triple helical domains are often referred to as collagenous and non-collagenous domains. Depending on the collagen type, all three α chains can be identical (homotrimer) or different (heterotrimer). In the latter case, the molecule can be composed of 2 or 3 different α chains. Each α chain in the molecule is coiled into an extended left-handed polyproline II-type helix and then the three left-handed helical α chains are intertwined to one another and folded into a ropelike right-handed triple helix structure (Bateman *et al.*, 1996). The triple helical structure is stabilised by the high content of imino acids, i.e. proline (Pro) and hydroxyproline (Hyp), and the presence of Hyp is essential for interchain hydrogen bonds that further stabilise the triple helical structure. In order for the α chains to form the collagen triple helix, the repetitive sequences of amino acids [Gly (glycine)-X-Y]n (X and Y can be any amino acid but often X is Pro and Y Hyp) are required. Every third amino acid is situated in the centre of the triple helix in a very restricted space where only Gly, the smallest amino acid, can fit. Thus, mutations of collagen that lead to substitution of Gly residues by more bulky amino acid residues in the triple helical domain cause severe disruption of collagen structure leading to connective tissue abnormalities seen, for instance, in many cases of osteogenesis imperfecta (Byers, 2000; Forlino and Marini, 2000; Prockop and Kivirikko, 1995).

A number of excellent reviews on collagen superfamily, collagen genes and collagen-related diseases are available (Bateman *et al.*, 1996; Byers, 2000; Myllyharju and Kivirikko, 2001; Prockop and Kivirikko, 1995; Ricard-Blum, 2000), so the readers are referred to these review articles and book chapters for more extensive and specific information. In this chapter, only basic and recent findings on the collagen superfamily, type I collagen biosynthesis and some details of collagen post-translational modifications, in particular cross-linking, are described.

2. Collagen Superfamily

Collagen is the most abundant protein in vertebrates accounting for about 30% of the body's total proteins and is present in essentially all tissues and organs of the body. It is now obvious that collagen is encoded by a family of at least 38 distinct genes and that more than 21 different genetic types of collagen exist (Byers, 2000; Myllyharju and Kivirikko, 2001; Prockop and Kivirikko, 1995). These distinct types of collagen show marked diversity and complexity in the structure (e.g. length and numbers of collagenous and non-collagenous domains, their assembly mode, the extent of post-translational modifications), quantities, their splice variants, biological functions and tissue distribution. Based on the assembly mode and some structural features, this growing collagen superfamily can be classified into 3 major groups and many subgroups (Fig. 1).

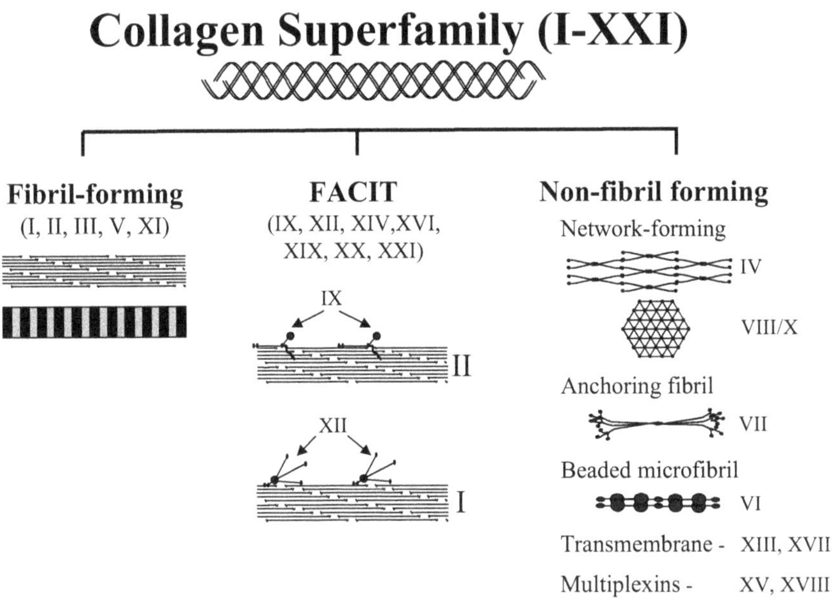

Fig. 1. Collagen superfamily classified by assembly modes and domain structures. For abbreviations and some details of the structures, see the text. Relative scale is not taken into consideration in this illustration. Some of them also might form different assembly structures due to the alternative splicing.

The first is a classical "fibril-forming collagen" group including types I, II, III, V and XI. Those molecules represent the major members of collagen superfamily in terms of their quantities and are capable of forming fibrils in the extracellular matrix. In the fibrils, the neighbouring collagen molecules are packed into a quarter-staggered array. The similarities of the nine genes that encode α chains of these collagens suggest that they arose by multiple duplications from a single ancestral gene (Vuorio and de Crombrugghe, 1990). All fibril-forming collagens consist of three domains, a short amino-terminal non-triple helical (N-telopeptide) domain, a long central uninterrupted triple helical domain and a short carboxy-terminal non-triple helical domain (C-telopeptide) (Fig. 2).

It has become clear that those collagen fibrils are composed of more than one collagen type, thus forming "heterotypic collagen fibrils". For instance, collagen type I fibrils may contain 5–10% type V and type II collagen fibrils 5–10% type XI. Those minor fibrillar collagens (i.e. V and XI) often retain a large portion of the amino propeptide (N-propeptide) due to the incomplete extracellular processing. When the molecules are incorporated into the fibrils, these extended arms appear to be exposed at the fibril surface. They could then function as a regulator of the lateral aspect of fibril growth by exerting a steric hindrance to accretion of collagen molecules to the surface of the fibrils (Birk, 2001). It has been suggested that the ratio of these minor fibrillar collagens, i.e. V and XI, to the major ones, i.e. I and II, may be one of the determinants of collagen fibril diameter. This speculation was supported by *in vitro* (Birk *et al.*, 1990) and mutant/transgenic mice studies (Garofalo *et al.*, 1993; Li *et al.*, 1995).

The major molecular form of type V collagen is composed of two identical α chains and one different chain, $[\alpha1(V)]_2\alpha2(V)$, and as a minor form, α1(V)α2(V)α3(V), where three different α chains are involved. Recently, it has been shown that Schwann cells express a heterotrimeric type V collagen composed of α1(V) and α4(V) chains which may play a role in nerve development and regeneration (Chernousov *et al.*, 2000). In some tissues, types V and XI are present as a hybrid molecule containing an α1(XI) and an α2(V) chain, for instance, an α1(XI)α2(V)α3(XI) form *in vitreous* humor (Byers, 2000; Prockop and Kivirikko, 1995).

The second group is "**F**ibril-**A**ssociated **C**ollagen with **I**nterrupted **T**riple helices" (FACIT) including types IX, XII, XIV, XVI, XIX, XX and XXI. They do not form fibrils themselves but attach to the surfaces of preexisting collagen fibrils (e.g. types I, II). The members of this group contain two or more short triple helical domains that are interrupted by short non-collagenous regions (Olsen *et al.*, 1995). By projecting their amino terminal globular domains out from the surface of the fibrils into the matrix, they are thought to function as bridging molecules between fibrils or between fibrils and other matrix molecules or cells. In this group, type IX is the most extensively characterised molecule. It is a heterotrimeric molecule consisting of α1(IX), α2(IX) and α3(IX) chains, with three collagenous and four non-collagenous globular domains. The molecules are located periodically in an anti-parallel fashion along the surface of type II collagen fibrils and this interaction is stabilised by lysyl-oxidase catalysed covalent cross-link (Eyre *et al.*, 1987; van der Rest and Mayne, 1988). Interestingly, this collagen often occurs as a proteoglycan in which a single chondroitin sulfate chain is covalently linked to a serine residue in non-collagenous domain 3 of an α2 (IX) chain. The size of this glycosaminoglycan chain is highly variable from non-existing to small seen in cartilage to very large *in vitreous* humor (Brewton *et al.*, 1991).

Types XII and XIV are homotrimeric molecules with sequence homologous to type IX and are found in connective tissues containing type I collagen fibrils except mineralised tissues. A fragment of type XII released after pepsin digestion was first discovered in periodontal ligament (Yamauchi *et al.*, 1986a). Both type XII and XIV molecules contain a central globular domain with large three finger-like extensions at the amino terminal end projecting out into perifibrillar space (Keene *et al.*, 1991). Two forms of type XII that differ in the size of this N-terminal extension domain are generated by alternative splicing of RNA transcripts. The larger form of type XII was found to contain a chondroitin sulfate glycosaminoglycan chain while the smaller did not. Type XIV collagen is also found in many type I collagen-containing connective tissues, such as skin, tendon, lung, placenta and vessel walls (Walchli *et al.*, 1994). However, it does not appear to bind to type I collagen fibrils directly but rather the interaction is via the dermatan sulfate chain of decorin, a collagen-binding small leucine-rich proteoglycan (Ehnis *et al.*, 1997).

Chick cDNA clones for a new member of FACIT subfamily have been recently isolated and characterised (Koch *et al.*, 2001). The domain structure of this new FACIT collagen is similar to those of types XII and XIV and is expressed highly in corneal epithelium.

More recently, an additional FACIT member, type XXI, has been identified and partially characterised (Tuckwell, 2002).

The third group is "non-fibrillar" collagens that do not form fibrils nor are closely associated with them. The members of this group can be divided into many subgroups: 1. network-forming collagens including basement membrane collagens (IV) and hexagonal network-forming collagens (VIII and X), 2. anchoring fibril collagen (VII), 3. microfibrillar collagen (VI), 4. multiplexins (**Multiple** triple heli**x** domains with **in**terruptions) including XV and XVIII, and 5. transmembrane collagens (XIII, XVII).

The type IV collagen is the major component of basement membrane which is critical in the separation of cells/cell sheets from underlying or surrounding connective tissues or from different types of cell sheets. This specialised connective tissue functions not only as a mechanical support for cells, but also plays a number of roles in filtration, regulation of cell migration, growth and differentiation. Although the major molecular form of this type IV collagen is a $[\alpha1(IV)]_2\alpha2(IV)$ heterotrimer, there are six type IV collagen genes identified in mammals and their products form at least three different heterotrimeric type IV collagen molecules. These isoforms may have a specific tissue distribution, e.g. in the glomerular basement membrane, a heterotrimeric form of $\alpha3(IV)\alpha4(IV)\alpha5(IV)$ exists (Boutaud *et al.*, 2000). The type IV collagen monomers associate at the C-termini to form dimers and at the N-termini to form tetramers. The tetramers then form a network-like structure with large openings (Yurchenco and O'Rear, 1994).

Both types VIII and X are similar in molecular structure and their assembly forms. The former, a heterotrimer of $[\alpha1(VIII)]_2\alpha2(VIII)$, is produced by endothelial cells and assembles into a hexagonal lattice in the Descemet's membrane which separates the endothelial cells from the stroma in cornea. Type X, a homotrimer of $[\alpha1(X)]_3$, is primarily synthesised by hypertrophic chondrocytes and forms a hexagonal network structure similar to type VIII. This collagen is present almost exclusively in the hypertrophic cartilage in the healthy organism.

Type VII is the major component of anchoring fibrils at the dermal-epidermal junction linking the basement membrane to anchoring plaques in the underlying dermis. This is a homotrimeric molecule, $[\alpha1(VII)]_3$, containing a long triple helical central domain (almost 1.5 times longer than that of type I collagen) flanked by a N- and a C-terminal non-collagenous domain (NC1 and NC2, respectively). The central triple helical structure is interrupted by 19 short non-collagenous domains. The type VII collagen is secreted as a procollagen form and, during the fibrillogenesis, the molecules assemble into antiparallel dimers through the interaction at the C-terminal regions with the large N-terminal globular domains pointing outwards. Prior to or during the assembly, the C-terminal propeptide domains (NC2) are cleaved off likely by procollagen C-proteinase/BMP-1, and disulfide bonds stabilise the dimers. The dimers then associate laterally and in register to form the main constituents of anchoring fibrils. Dystrophic epidermolysis bullosa is caused by mutations in the gene for type VII collagen (see a recent review by Bruckner-Tuderman *et al.*, 1999).

Type VI collagen is a highly glycosylated, cysteine-rich molecule with a short triple helical domain and very large terminal non-collagenous globular domains which account for more than two-thirds of the molecule. It is composed of three different chains, two short α chains, α1(VI) and α2(VI) of about 1000 amino acid residues, and a longer α3(VI) chain with about 3000 amino acid residues. The monomers are first assembled into antiparallel dimers with a 75-nm overlap between the triple helical domains and the two dimers generate a tetramer by the parallel alignment. Apparently, this process is taken place intracellularly. The secreted tetramers are the building blocks for the microfibril network present in the extracellular matrix in many soft connective tissues (Kielty and Shuttleworth, 1997; Timpl and Chu, 1997). The type VI collagen is an adhesive protein and interacts with cell surface receptors of the α1 integrin family, fibrillar collagens, proteoglycans and glycosaminoglycans, and the loss of type VI collagen expression has been suggested to contribute to the metastasis of mesenchymal tumour cells (Trueb and Odermatt, 2000).

Multiplexins and transmembrane collagens are two subgroups which recently received much attention. It has now become clear that the non-collagenous domains of many of the collagen molecules have important biological roles. Recently, the C-terminal fragments of types XVIII and XV collagen that are proteolytically released from the respective parent molecules were identified as "endostatin" or its analog, a potent inhibitor of angiogenesis and tumour growth (Felbor *et al.*, 2000; O'Reilly *et al.*, 1997; Sasaki *et al.*, 2000). Type XVIII collagen contains 10 triple helical collagen domains interrupted by 11 non-collagenous domains (Oh *et al.*, 1994). Immunohistochemical localisation of this collagen demonstrated that this is a component of vascular and epithelial basement membranes (Muragaki *et al.*, 1995; Saarela *et al.*, 1998). In addition, this collagen has been shown to be a heparan sulfate proteoglycan in basement membrane (Halfter *et al.*, 1998). Type XV, another member of multiplexin family, is structurally similar to type XVIII and the genomic structure of their genes shows a considerable conservation indicating a common ancestor. It contains several interruptions in the central collagenous domain and large non-collagenous domains at both ends. In contrast to type XVIII, it is a disulfide bonded chondroitin sulfate proteoglycan (Li *et al.*, 2000). Thus, both type XVIII and XV collagens are also "collagen-proteoglycan" hybrid molecules. Type XV collagen is immunohistochemically localised to basement membrane of capillaries and skeletal muscle cells (Hagg *et al.*, 1997). A homologous domain fragment of type XV collagen (an endostatin analog) to endostatin of type XVIII also shows anti-angiogenesis activity and, furthermore, it has been shown that the noncollagenous domain of type IV collagen also inhibits vessel growth (Marneros and Olsen, 2001). This indicates the regulatory roles of these basement membrane collagen/proteoglycan fragments in cell migration and proliferation that affect angiogenesis.

Although a vast majority of collagens are present in the extracellular matrix, some of the collagen types, such as types XIII and XVII, are found to be transmembrane molecules (thus, only a partially extracellular molecule). The former has been localised on the surface of fibroblasts and other cells possessing extensive alternative splicing variants (Hagg *et al.*, 1998; Peltonen *et al.*, 1997; Pihlajaniemi and Rehn, 1995). The latter is localised in hemidesmosomes at epithelial-mesenchymal interfaces and carries the Bullous pemphigoid autoantigen, BPAG2 (Li *et al.*, 1993). It has been reported that mutations in this collagen lead to benign forms of epidermolysis bullosa (Chavanas *et al.*, 1997; Gatalica *et al.*, 1997). Both types XIII and XVII contain a single transmembrane amino terminal domain and the extracellular C-terminal domain, and these molecules, unlike other collagens (see below), appear to fold from the N- to the C-terminal end to form the triple helical molecule (Areida *et al.*, 2001).

3. Type I Collagen Biosynthesis

As described above, collagens are vastly diverse in quantity, structure although related, size, assembly mode, tissue distribution and function. Among all types of collagen, however, type I collagen is the predominant genetic product and it is the essential molecule to provide tissues and organs with tensile strength, form and cohesiveness. It is the major component of most connective tissues including skin, tendon, ligament, cornea, blood vessels, bone, dentin and others. Type I collagen is a heterotrimeric molecule composed of two $\alpha 1$ chains and one $\alpha 2$ chain, $[\alpha 1(I)]_2\alpha 2(I)$, although a homotrimeric form of $\alpha 1$ chains, $[\alpha 1(I)]_3$ does exist as a minor form. As briefly described above, the type I collagen molecule consists of three domains: the NH_2-terminal nontriple helical (N-telopeptide), the central triple helical and the COOH-terminal nontriple helical (C-telopeptide) domains. The single (uninterrupted) triple helical domain represents more than 96% of the molecule. The parent molecule is secreted outside the cell as its precursor form, procollagen, and the collagen molecule is generated by the enzymatic cleavages at both the C- and the N-terminal ends of the procollagen molecule (Fig. 2). The essential role of type I collagen in our body is clearly exemplified by various inherited diseases caused by mutations in the genes encoding pro $\alpha 1$ and $\alpha 2$ chains of this protein. The major phenotypes caused by such mutations include various types of osteogenesis imperfecta (OI), Ehlers-Danlos syndromes (EDS) and osteoporosis (Byers, 2000; Myllyharju and Kivirikko, 2001). It has been shown that homozygosity for COL1A1 null mutation in mice is lethal. While that for COL1A2 null mutation is not lethal, connective tissues in these mice are impaired (Byers, 2000; Hata *et al.*, 1988; Schnieke *et al.*, 1983).

The biosynthesis of type I collagen from gene transcription to secretion into extracellular matrix, self-assembly and the formation

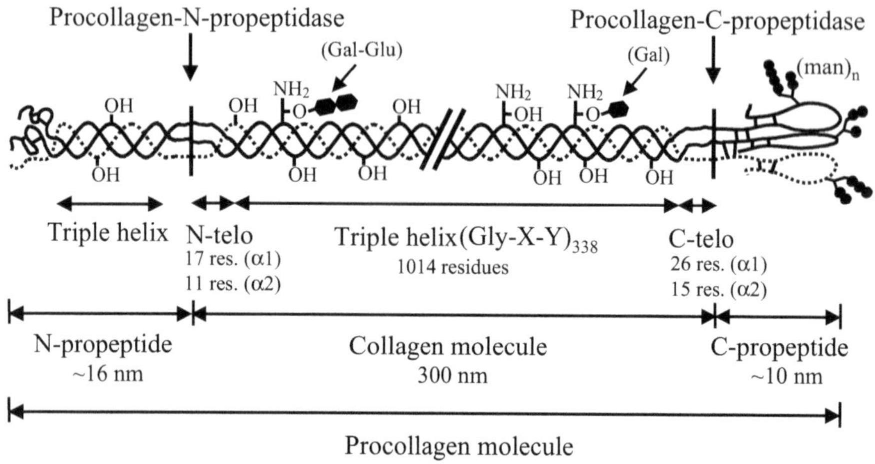

Fig. 2. The structure of a type I procollagen molecule. It is composed of two pro-α1(I) chains (solid lines) and one pro-α2(I) chain (dotted line) forming a unique triple helical collagen structure. The N-propeptide contains a short triple helical domain and intrachain disulfide bonds. Each chain of the globular C-propeptide is glycosylated by high-mannose N-linked oligosaccharide $(man)_n$ and this domain is stabilised by interchain disulfide bonds (lines). Specific proline and lysine residues are hydroxylated indicated by -OH and -OH-NH_2. Some of the hydroxylysine residues are glycosylated by O-linked galactose (Gal) or -galactose-glucose (Gal-Glu) indicated by closed hexagon. The procollagen molecule is cleaved at specific sites by procollagen-N- and C-propeptidases indicated by arrows generating a collagen molecule that consists of the N-telopeptide, uninterrupted triple helix and the C-telopeptide.

of functional stable fibrils is a very complicated, multi-step process and it requires meticulous coordination of a large number of biochemical events in and outside the cell (Bateman *et al.*, 1996) (see its simplified illustration in Fig. 3).

The process of biosynthesis includes the following four major steps:

1. Gene transcription and mRNA processing.
2. Transportation of the mRNA to endoplasmic reticulum (ER) lumen where translation of mRNA into pre-procollagen α chains, cleavage of signal peptides, a number of post-translational modifications, three chain association and folding into a triple helix procollagen molecule take place.

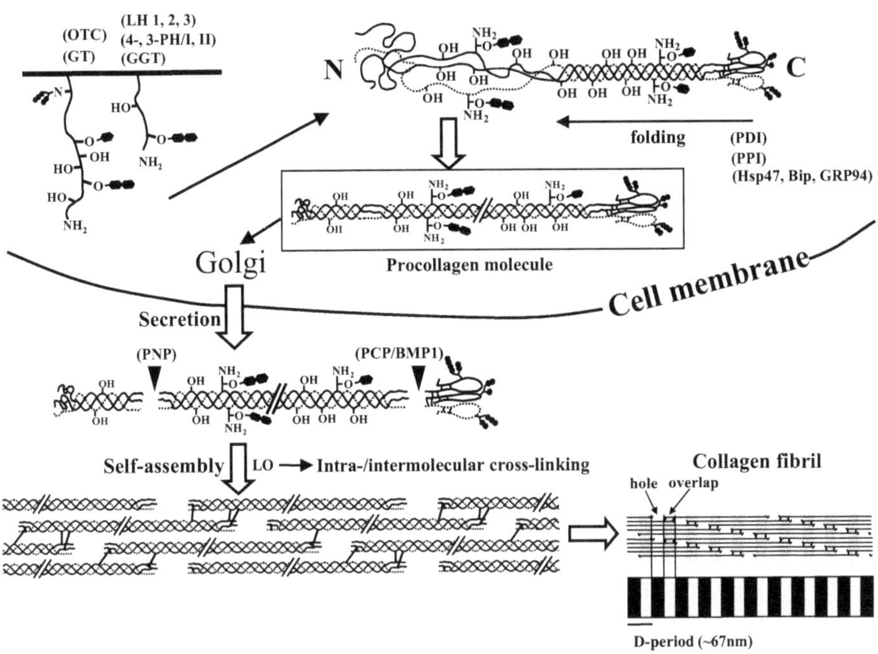

Fig. 3. Schematic for the biosynthesis of type I collagen. The upper part (above the cell membrane) summarises the intracellular and the lower (below the cell membrane) the extracellular events. The intracellular events include extensive post-translational modifications such as hydroxylation, glycosylation (both O- and N-linked), association of pro α chains and folding into a triple helical molecule from the C- to N-terminus. The extracellular events involve the removal of both N- and C-propeptide extensions, self-assembly of collagen molecules into a fibril, enzymatic oxidative deamination of lysine and hydroxylysine residues by LO and subsequent intra- and intermolecular covalent cross-linking. The collagen molecules are packed in parallel and are longitudinally staggered with respect to one another by some multiple of axial repeat distance, D (~ 67 nm). This packing arrangement creates repeated regions of high and low packing density, i.e. overlap and hole regions, respectively, showing a characteristic banding pattern of collagen fibril seen at ultrastructural level. See details of these events in the text and some explanations for illustration of the molecule in Fig. 2. LH: lysyl hydroxylase, PH: prolyl hydroxylase, GGT: galactosylhydroxylysyl glucosyltransferase, GT: hydroxylsyl galactosyltransferase, OTC: oligosaccharyl transferase complex, PDI: protein disulfide isomerase, PPI: peptidyl-prolyl *cis-trans* isomerase, Hsp: heat shock protein, Bip: binding proteins, GRP: glucose-regulated protein, PNP: procollagen-N-proteinase, PCP: procollagen-C-proteinase, BMP:bone morphogenetic protein, LO: lysyl oxidase, Solid line: pro α1 chain, Dotted line: pro α 2 chain.

3. Transportation to Golgi, secretion into the extracellular matrix and extracellular processing to form a collagen molecule.
4. Self-assembly into a fibril and its stabilisation via further modifications of telopeptidyl Lys/Hyl by lysyl oxidase and subsequent covalent intra- and intermolecular cross-linking.

A number of enzymes (more than 15) many of which are collagen-specific and several molecular chaperones including collagen-specific Hsp47 are involved in this highly complex process. Here, very basic and some new information of these events and some details on covalent cross-linking are described.

3.1. Gene transcription and mRNA processing

Like other secreted proteins, the entire gene including introns is transcribed and the transcript modified at the 5′ end by capping and at the 3′ end by polyadenylation. Some heterogeneity of collagen mRNA species is introduced by variation of polyadenylation at the 3′ end. Introns are then spliced out and the mature mRNA is transported into cytoplasm. The fibril-forming collagen genes contain more than 51–53 exons, thus the RNA transcripts undergo extensive splicing events before they are transported into cytoplasm. The genes for human pro α1(I) and pro α2(I) are localised in the chromosomes 17q21.3–q22 and 7q21.3–q22. The collagen triple helical domain is encoded by 44 exons, the majority of which are 54 bp thus coding 6 (Gly-X-Y) triplets, and to a lesser extent 45 bp, 99 bp, 108 bp and 162 bp. This highly conserved gene structure suggests that the ancestral progenitor of fibrillar collagen (possibly all collagen genes) was a 54-bp or a 45-bp Gly-X-Y coding sequence, and its tandem duplications and crossovers resulted in a multi-exon collagen gene (Sandell and Boyd, 1990).

3.2. mRNA translation, post-translational modifications and procollagen assembly in endoplasmic reticulum

After translation of mRNA, the nascent prepro α chains (translation products) are initially directed into the endoplasmic reticulum

(ER) by the N-terminal signal peptides. The signal peptide is then cleaved off from the nascent prepro α chains by signal peptidase and the procollagen α chains are extensively modified by a number of enzymes most of which are collagen-specific. The modifications include: hydroxylation of specific proline (Pro) and lysine (Lys), O-linked glycosylation of hydroxylysine (Hyl) and N-linked oligosaccharide glycosylation of asparagine (Asn) residues.

The content of Pro/Hyp in type I collagen is relatively high representing about 25% of the total amino acids and 40–45% of Pro is hydroxylated (mostly 4-Hyp). The hydroxylation of Pro is catalysed by prolyl 4-hydroxylase and prolyl 3-hydroxylase. The former reacts on Pro with the minimum sequence X-Pro-Gly and the latter appears to require a Pro-4-Hyp-Gly sequence (Kivirikko and Myllyla, 1982). These enzymes require cofactors such as Fe^{2+}, 2-oxoglutarate, O_2 and ascorbate for activity (Prockop, 1976). Active 4-prolyl-hydroxylase in vertebrates is a tetrameric protein composed of 2 α and 2 β subunits (Kivirikko *et al.*, 1989). This enzyme plays a crucial role in collagen synthesis since the 4-hydroxyproline residues are essential for the folding of the procollagen α chains into triple helical molecules (i.e. the β-subunit is identical to the enzyme and chaperone protein, PDI, see below) and to stabilise the triple helical conformation by providing hydrogen bonds and water bridges (Prockop and Kivirikko, 1995). Recently, an isoform of the α subunit, α (II), was cloned and characterised in mouse and human (Annunen *et al.*, 1997; Kivirikko and Pihlajaniemi, 1998). Thus, the previously known α subunit is designated as α (I) and the tetrameric enzyme, $[\alpha(I)]_2\beta_2$, as type I enzyme, and its isoform, $[\alpha(II)]_2\beta_2$, type II enzyme. The expression pattern of these two isoenzymes in several human connective tissues appears to be spatially and temporally different from each other, i.e. type I enzyme is expressed especially by cells of mesenchymal origin and in developing/malignant tissues while type II in more differentiated cells (Nissi *et al.*, 2001). Their specificities with respect to specific collagen types, the presence of additional isoforms (Myllyharju and Kivirikko, 2001), and their respective biological function need to be elucidated.

The hydroxylation of Lys is another important modification since it is a critical determinant for covalent intermolecular cross-linking (see below) and Hyl is the only glycosylation site (O-glycosylation) in a collagen molecule. This modification is catalysed by lysyl hydroxylases (LHs) and for their activities, like prolyl hydroxylases, Fe^{2+}, 2-oxoglutarate, O_2 and ascorbate are required as cofactors (Kivirikko and Myllyla, 1987). The Lys hydroxylation occurs in the X-Lys-Gly sequence within the triple helical domain of the collagen α chains, however, it also occurs in the N- and C-telopeptide domains where the Gly of X-Lys-Gly sequence is replaced by either a serine (in the N-telopeptide of an α1 chain) or an alanine residue (in the C-telopeptide of an α1 chain). This sequence difference together with the difference in the extent of Lys hydroxylation between helical and telopeptide domains in collagen α chains led investigators to speculate the presence of isoforms of LHs that are specific to telopeptidyl Lys residues. For instance, in hypertrophic tendon there is a significant increase in the extent of Lys hydroxylation in the telopeptides of type I collagen while that in the helical domain is unchanged (Gerriets *et al.*, 1993). Furthermore, a highly purified LH fails to hydroxylate the Lys residues in the telopeptide domains (Royce and Barnes, 1985). Recently, two other isoenzymes with LH activity have been characterised and designated as LH2 and LH3 (the originally described one being LH1) (Valtavaara *et al.*, 1997; 1998). In addition, an alternatively spliced form of LH2, LH2*alt*, has been reported (Yeowell and Walker, 1999). All three enzymes appear to be inactive against Lys in the triple helical structure, therefore, these modifications must occur before the formation of the triple helix. The genes for LH1, 2 and 3 have been localised to human chromosomes 1, 3 and 7, respectively. The direct evidence of the importance of LH is demonstrated in patients with Ehlers-Danlos syndrome type VI whose clinical phenotypes are caused by a deficiency of LH activity due to the mutations in LH1 gene (Yeowell, in press). A study of patients with Bruck Syndrome, a rare autosomal recessive disorder characterised by fragile bones and congenital joint fractures, indicated the presence of bone specific telopeptide LH on chromosome 17p12 (Bank *et al.*, 1999), thus different from those of LH1, 2 and 3. Our recent study on the expression pattern of LH1, 2 and 3 genes during osteoblastic cell differentiation, suggested that LH2 expression may be associated with Lys hydroxylation of the telopeptide domains of the type I collagen molecule (Uzawa *et al.*, 1999). The tissue specific pattern of Lys hydroxylation/cross-linking (see below) of type I collagen is likely due, in part, to differential expression of these or other as yet unknown LH isoform genes by the cells in the respective tissues (Yeowell, in press). In human type I collagen, there are 38 residues of Lys in an α1 chain (36 in the helical, 1 in the C- and 1 in the N-telopeptide domains) and 31 in an α2 chain (30 in the helical, 1 in the N-telopeptide and none in the C-telopeptide domains) (Sequence accession #P02452, P02464 and P08123). Within the triple helical domain, 23 Lys residues are located in the X-Lys-Gly sequence in an α1 chain and 21 in an α2 chain (Sequence accession #P02452, P02464 and P08123). The extent of Lys hydroxylation of collagen is much more variable than that of Pro hydroxylation. It greatly varies not only from one genetic type to another but also within the same genetic type I collagen in different tissues in the same organism (Miller, 1984). It also significantly varies depending on the physiological conditions (Shiiba, in press; Uzawa *et al.*, 1998; Yamamoto and Yamauchi, 1999; Yamauchi and Shiiba, 2002).

For *O*-linked glycosylation of Hyl, two specific enzymes, hydroxylysyl galactosyl transferase and galactosylhydroxylysyl glucosyltransferase are involved. The former catalyses an addition of a galactose to Hyl (Hyl-gal) and the latter glucose to the galactosyl-Hyl residue (Hyl-gal-glc). Recently, LH 3 has been demonstrated to have galactosylhydroxylysyl glucosyltransferase activity (Heikkinen *et al.*, 2000). Thus, it is a multifunctional enzyme possessing both LH and galactosylhydroxylysyl glucosyltransferase activities. The role of Hyl glycosides is not fully understood, however, several potential functions have been proposed. Since these carbohydrates are located at the surface of collagen triple helix and point outward from the backbone of the collagen molecule, it may control the lateral growth of collagen fibril by placing steric hindrance (i.e. gal-glc can shield 3 or 4 amino acid residues) (Yang *et al.*, 1993). A more direct evidence for this potential role was demonstrated by a study comparing the fibrillogenesis using collagen species that are genetically identical but differ markedly in the extent of hydroxylation/glycosylation (Notbohm *et al.*, 1999). In this study, over hydroxylated/glycosylated collagen formed significantly thinner fibrils in comparison with that of under-hydroxylated/glycosylated collagen. Those carbohydrates also may regulate the rate of collagen degradation by adding their highly hydrophilic moiety around the enzymatic cleavage site, thus increasing its resistance to enzymatic digestion (Yang *et al.*, 1993). It has been also proposed that they may regulate cross-link maturation (Yamauchi *et al.*, 1982; 1986b). It is interesting to note that the Hyl residues that are involved in cross-linking of type I collagen (see below) are also often glycosylation sites.

The glycosylation of an asparagyl (Asn) residue in the C-propeptide also occurs at the Asn-X-Ser/Thr consensus sequence for N-linked oligosaccharide addition by oligosaccharyl transferase complex (OTC). A series of processing events, including both trimmings and additions, results in a wide variety of high mannose-containing oligosaccharide structure found in this domain. The N-linked oligosaccharide processing begins in the ER and appears to continue in the Golgi complex. This high-mannose N-linked oligosaccharide complex does not appear to be involved in collagen assembly, secretion and extracellular processing since the absence of this complex does not affect these events (Lamande and Bateman, 1995).

3.2.1. Procollagen formation

Three individual procollagen α chains are associated through the C-propeptides and folded into a procollagen molecule. For the initial chain recognition and association, a cluster of hydrophobic amino acid sequences that is conserved in all fibril-forming collagens appears to be critical. Once the chain association and alignment takes place, the folded three procollagen α chains are stabilised by the formation of interchain disulfide bonds which are catalysed by protein disulfide isomerase (PDI), β-subunit of prolyl-4-hydroxylase (Kivirikko and Pihlajaniemi, 1998). Another important factor in this step is rate-limiting of the peptide folding by the *cis-trans* isomerisation of prolyl peptide bonds in the α chains catalysed by peptidyl-prolyl *cis-trans* isomerase (PPI) (Lang *et al.*, 1987; Steinmann *et al.*, 1991). Due to this specific direction of folding from the C- to the N-terminus, point mutations of Gly residues that occur closer to the C-terminus of the triple helical domain generally cause more severe forms of osteogenesis imperfecta while there are exceptions (Bateman *et al.*, 1996). Several chaperone molecules such as Bip (78 kDa ER resident binding protein), Hsp 47 (ER-resident stress inducible 47 kDa glycoprotein) and likely others (e.g. members of ER resident glucose-regulated proteins such as GRP94) (Lee, 2001) are also involved in the correct folding, assembly and maturation of procollagen (Chessler and Byers, 1993; Nagata, 1996). Of a number of ER-resident chaperones, Hsp47 is unique in terms of its substrate specificity since it exclusively binds to procollagens and collagens. Although it binds to the nascent procollagen α chains in the ER and controls the mode of chain folding (Nagata and Hosokawa, 1996), it appears to have the highest affinity for the triple helical form of procollagens and collagens (Koide *et al.*, 2000; Tasab *et al.*, 2000). The precise roles of Hsp 47 in the collagen biosynthesis are not completely understood as yet. However, it has recently been shown that targeted disruption of the Hsp47 gene causes severe deficiency in the mature, processed form of type I collagen and fibril structures in mesenchymal tissues, disrupted basement membrane and ruptured blood vessels leading to embryonic lethality (Nagai *et al.*, 2000). Furthermore, the conformation of procollagen triple helices appears to be defective. This study clearly indicates that Hsp-47, a collagen-specific molecular chaperone, is essential for collagen maturation and for normal connective tissue development. Another recent study also indicates that Hsp47 is essential for prevention of procollagen aggregation in the ER (Thomson and Ananthanarayanan, 2000).

3.2.2. Secretion and extracellular processing

The properly folded procollagen molecules pass through Glogi and are packaged into secretory vacuoles forming a large laterally-aligned aggregate which finally moves to the surface of the cell, and is secreted (except for transmembrane collagens) into the extracellular matrix space via secretory vesicles (Bateman *et al.*, 1996). Then, in the extracellular matrix, both N- and C-propeptides are cleaved off by specific enzymes, a procollagen-N-proteinase and a procollagen-C-proteinase, respectively. Both enzymes are Zn^{2+}-dependent metalloendopeptidases. This process (removal of the bulky globular domains at both N-and C-terminal ends) must take place in order for the collagen molecules to self-assemble into fibrils under physiological conditions. Procollagen-N-proteinase cleaves the N-propeptides of both type I and II procollagens and possibly some other procollagens, however, for the cleavage of the N-propeptide of procollagen type III another isoenzyme may be required (Prockop *et al.*, 1998).

The procollagen-C-proteinase, identical to bone morphogenetic protein 1 (BMP1), cleaves the C-propeptides of procollagen types I, II, III and possibly V (Kessler *et al.*, 2001). It is also known to cleave prolaminin 5 gamma 2 chain (Amano *et al.*, 2000), prolysyl oxidase (Panchenko *et al.*, 1996; Uzel *et al.*, 2001), chordin (Mullins, 1998) and probiglycan (Scott *et al.*, 2000). Thus, this enzyme appears to be involved not only in procollagen processing but also in overall extracellular matrix organisation and animal development. The activity of the procollagen C-proteinase/BMP-1 during the C-terminal processing of fibrillar procollagens is enhanced by a 55 kDa glycoprotein, procollagen C-proteinase enhancer (PCPE) (Hulmes *et al.*, 1997). This protein specifically binds to the C-propeptides of type I procollagen and this interaction may be important for the correct conformation of the cleavage site and/or to enhance the interaction between the procollagen C-proteinase/BMP-1 and the procollagen substrate (Kessler and Adar, 1989). The amino-terminal 36 kDa portion of this protein appears to be the most conserved domain among different species and may be essential for its activity (Takahara *et al.*, 1994).

Both N- and C-propeptides released by the enzymes described above are thought to play a role in modulating procollagen synthesis by a process of feedback inhibition. The former specifically binds to fibroblast cell membrane and is internalized by endocytosis (Schlumberger *et al.*, 1988). It has also been shown that a metallothionein-inducible α1(I) N-propeptide minigene inhibits type I collagen synthesis in fibroblasts indicating its modulatory role in procollagen synthesis in intact cells (Fouser *et al.*, 1991). The latter may control procollagen synthesis at the transcriptional (Wu *et al.*, 1991) or post-transcriptional level (Aycock *et al.*, 1986).

3.2.3. Self-assembly into a fibril and its stabilisation by covalent cross-linking

The processed collagen molecules then spontaneously self-assemble into the fibrils. This is mainly an entropy-driven process through the loss of solvent molecules from the surface of the molecules resulting in molecular assembly and is organised through clusters of charge and hydrophobicity of the triple helical portion of the neighbouring molecules (Kadler *et al.*, 1996). This implies that all information necessary to initiate the fibrillogenesis is intrinsic to the collagen molecules themselves. Several factors are then involved in the regulation of the fibril growth including:

1. The presence of the C- and N-telopeptides since the loss of either one of those has profound effects on collagen fibril growth. Previous studies indicate that each telopeptide has a specific role in fibrillogenesis (see a review by Veis and George, 1994).
2. The extent of post-translational modifications, i.e. Lys hydroxylation/glycosylation (Notbohm *et al.*, 1999; Yang *et al.*, 1993).
3. Presence and/or quantity of minor fibrillar collagens such as type V collagen with the N-terminal domain unprocessed (see above) (Birk, 2001).
4. Presence of collagen-binding small leucine-rich proteoglycans (Iozzo, 1999; Keene *et al.*, 2000).
5. FACIT collagens (Olsen, 1995).

Type I collagen molecules appear to be packed together into fibrils in a very specific manner axially and azimuthally such that the molecules are longitudinally staggered with respect to one another by specific multiples of ~ 67 nm and have specific lateral orientations with respect to one another (Mechanic *et al.*, 1987; Veis and George, 1994; Yamauchi *et al.*, 1986b). Although type I collagen, a product of two single genes encoding pro α1 and α2, is the most ubiquitous protein in the body, its physicochemical properties differ significantly from tissue to tissue because of the tissue-specific post-translational modifications. In addition, there is more than one packing modality of type I collagen fibrils (Brodsky *et al.*, 1980; Yamauchi and Katz, 1993). These tissue-specific features of type I collagen (molecular packing and post-translational modifications, in particular, cross-linking chemistries) are most likely important for the tissue's physiological functions.

The intermolecular cross-linking is the final post-translational modification in collagen biosynthesis and is crucial in providing the connective tissue matrices with tensile strength and viscoelasticity (Nimni and Harkness, 1988).

4. Collagen Cross-Linking

4.1. Lysyl oxidase

The process of cross-linking is initiated by the oxidative deamination of ε-amino groups on Lys and Hyl residues located in the C- and N-telopeptides to the respective aldehydes, 5-amino-5-carboxypentanal (α-amino adipic acid-δ-semialdehyde, allysine or Lys^{ald}) and 2-hydroxy-5-amino-5-carboxypentanal (δ-hydroxy-α-amino adipic acid-δ-semialdehyde, hydroxyallysine or Hyl^{ald}), respectively, through the action of an enzyme, lysyl oxidase (Protein-lysine 6-oxidase) (Fig. 4). Lysyl oxidase (LO) is a copper-dependent amine oxidase which is secreted as the 50-kDa N-glycosylated prolysyl oxidase to the extracellular space where it is processed by procollagen C-proteinase/BMP1 (see above). The processed functional enzyme, lysyl oxidase, is a 32 kDa protein and contains two cofactors, Cu(II) and a peptide-linked carbonyl residue at its

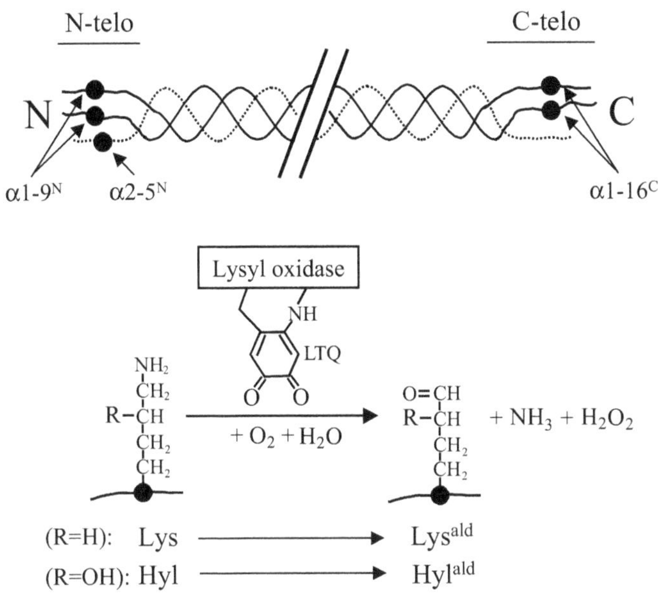

Fig. 4. Initiation of collagen cross-linking by lysyl oxidase. There are five lysyl/hydroxylysyl residues which are potentially subject to enzymatic conversion to aldehyde, two in the C- and three in the N-telopeptides. $\alpha 1\text{-}16^C$: 16th residue from the N-terminus of the C-telopeptide of an α1 chain, $\alpha 1\text{-}9^N$: 9th residue from the N-terminus of an α1 chain, $\alpha 2\text{-}5^N$: 5th residue from the N-terminus of an α2 chain. Solid line represents an α1 chain, and dotted line an α2 chain. See details in the text.

catalytic site. The latter is a lysine tyrosylquinone (LTQ) derived from covalent bonding between a post-translationally modified tyrosine and a lysine residue (Kagan, 2000; Smith-Mungo and Kagan, 1998) Apparently, for the initiation of oxidative deamination of telopeptidyl Lys/Hyl of collagen, one of the C = O groups of LTQ has to bind to the Lys/Hyl amino group.

In addition to the well known catalytic activity for conversion of the telopeptidyl Lys/Hyl residues to the respective aldehydes, recently, a number of novel biological functions have been attributed to this enzyme. This includes tumour suppression (Hamalainen *et al.*, 1995; Kenyon *et al.*, 1991), senescence (Sharma *et al.*, 1997), developmental regulation (Di Donato *et al.*, 1997) and cell growth control and chemotaxis (Li *et al.*, 2000) which involve additional substrates that reside extracellularly and intracellularly. Recently, it has been shown that the mature extracellular LO can be taken up by intact cells and rapidly enter into the cell nucleus (Nellaiappan *et al.*, 2000) suggesting its intracellular and intranuclear oxidative function. These multiple functions suggest the presence of a family of several LOs and, in fact, at present, four isoforms have been characterised (Maki and Kivirikko, 2001). The precise functions of these new lysyl oxidase family members are yet to be elucidated.

An early study indicated that LO did not bind to nor was very reactive to solutions of a collagen molecule or denatured collagen. It was highly active to the macromolecular fibrils of collagen (staggered native collagen fibril) (Siegel, 1974), indicating its binding site is on growing collagen fibrils. A study using pepsin or chymotrypsin treated native collagen as a substrate indicated that it may bind to the triple helical portion of the collagen molecule in a fibrillar array (Cronlund *et al.*, 1985). Most likely, the binding sites are at both opposite ends of the triple helical regions of the molecules and thus can oxidize Lys and Hyl located in both C- and N-termini in adjacent staggered collagen molecules. Selective inhibition of the LO activity by lathyrogens such as β-aminopropionitrile causes severe connective tissue disorders due to the lack of cross-linking (the condition called lathyrism). The disorders include the loss of tensile strength, spontaneous aortic rupture, skin fragility, abnormal mineralisation and bone deformation (Kagan, 2000; Tanzer, 1976).

4.2. Cross-linking pathway (Fig. 5)

It is hard to overemphasize the biological importance of covalent intermolecular cross-linking of collagen as exemplified by lathyrism (see above). Thus, it is not surprising that altered cross-linking is associated with many connective tissue diseases (Bank *et al.*, 1999; Brinckmann *et al.*, 2001; Brinckmann *et al.*, 1999; Masse *et al.*, 1995; Ono and Yamauchi, 1992; Rodeo *et al.*, 1998; Tamura and Ishikawa, 2001; Uzawa *et al.*, 1998; Wojtowicz *et al.*, 1997; Yamauchi and Mechanic, 1988). Once the aldehydes (either Lys^{ald} or Hyl^{ald}) are

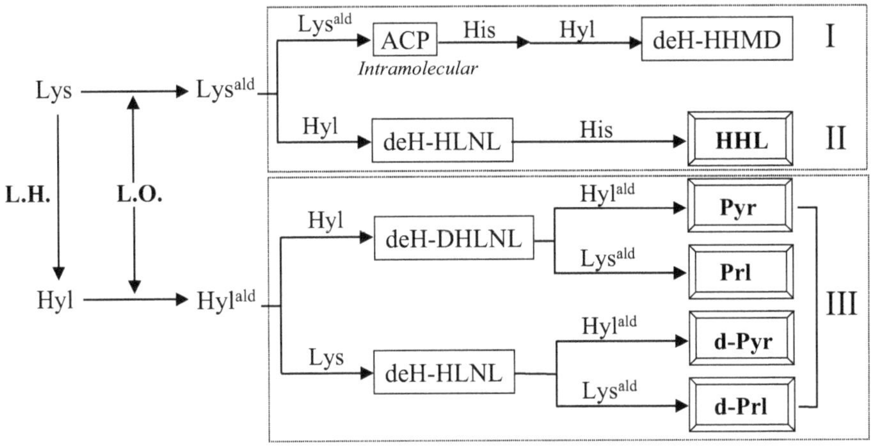

Fig. 5. Major cross-linking pathways of collagen. Predominant in I: soft tissues, II: skin and cornea, III: skeletal tissues L.H.: Lysyl hydroxylase, L.O.: Lysyl oxidase, ACP: Aldol condensation product (Intramolecular cross-link), deH: dehydro HLNL: hydroxylysinonorleucine, DHLNL: dihydroxylysinonorleucine, HHMD: histidino-hydroxymerodesmosine HHL: histidinohydroxylysinonorleucine, Pyr: pyridinoline, d-: deoxy, Prl: pyrrole, ▭ Reducible cross-links, ▭ Non-reducible cross-links.

formed by the action of lysyl oxidase in the C- and the N-telopeptide domains, they undergo a series of condensation reactions involving another aldehyde in the same molecule and/or the juxtaposed Lys, Hyl and histidine (His) residues on the neighbouring molecules. The re-actions result in the formation of covalent intra- and intermolecular cross-links (Eyre *et al.*, 1984; Tanzer, 1976; Yamauchi and Mechanic, 1988). As described below, the cross-linking chemistry/pattern varies from tissue to tissue rather than particular collagen types since a number of tissue-specific factors govern the chemistries. Those factors include the state of hydroxylation of Lys residues that are involved in cross-linking (Eyre *et al.*, 1984; Yamauchi and Mechanic, 1988), the maturation/turnover rate (Bailey and Shimokomaki, 1971; Eyre *et al.*, 1988; Moriguchi and Fujimoto, 1978; Yamauchi *et al.*, 1988b), the details of molecular packing structure (Katz and David, 1992; Mechanic *et al.*, 1987; Yamauchi *et al.*, 1986b), the physical force, e.g. the presence and growth of mineral in the fibrils (Eyre, 1981; Otsubo, 1992, Yamauchi, 1995), possibly other factors such as the extent of glycosylation of Hyl residues that are involved in cross-linking (Yamauchi *et al.*, 1982; 1986b). and the binding of a small leucine-rich proteoglycan (Keene *et al.*, 2000). Among those, the state of hydroxylation of Lys is the most critical factor to determine the subsequent cross-linking pathways.

In the past, the chemistry of lysyl oxidase-mediated collagen cross-linking has been extensively reviewed by several authors (Bailey, 2001; Eyre *et al.*, 1984; Tanzer, 1976; Yamauchi, 1995; Yamauchi and Mechanic, 1988). The non-enzymatic cross-linking of collagen has also been reviewed (Reiser *et al.*, 1992; Reiser, 1998). Therefore, in this section, basic biochemistry of LO-mediated cross-linking and some new developments in this field will be described.

4.2.1. Lys^{ald}-pathway

There are two types of cross-linking reactions in the Lys^{ald} pathway: one leads to the formation of a tetravalent cross-link, dehydro-histidinohydroxymerodesmosine (deH-HHMD) via intramolecular aldol condensation product (ACP), and another the formation of a trivalent stable cross-link, histidinohydroxylysinonorleucine (HHL) via the iminium cross-link, dehydro-hydroxylysinonorleucine (deH-HLNL).

4.2.1.1. *Lys^{ald} –pathway leading to a histidine-based tetravalent reducible cross-link via aldol (Fig. 5.I)*

Lys^{ald} can condense with another Lys^{ald} to form an aldol condensation product (ACP, α,β-unsaturated aldol) which occurs as an intramolecular cross-link located in the N-telopeptide domain of the molecule (Kang *et al.*, 1969; Tanzer *et al.*, 1973). Then Michael addition of N3 imidazole of His to the β-carbon of the unsaturated α,β unsaturated bond of ACP produces aldolhistidine (AH) (Tanzer *et al.*, 1973). The latter then condenses with ε-amino group of Hyl forming an iminium bond to produce dehydro-histidinohydroxymerodesmosine (deH-HHMD) (Tanzer *et al.*, 1973)

(Fig. 5). Although this tetravalent cross-link, deH-HHMD, was proposed to be a base catalysed artifact by reduction (Robins and Bailey, 1973), this issue has been resolved by strong evidence that the native form of deH-HHMD occurs in the collagen fibrils (Bernstein and Mechanic, 1980) and it is clearly one of the major cross-links in soft connective tissue collagen (Tanzer, 1976; Yamamoto and Yamauchi, 1999; Yamauchi *et al.*, 1986b). It is not clear whether or not this complex cross-link is the final product of this pathway or not. We have just recently isolated and partially characterised the HHMD-cross-linked peptides from enzymatic digests of the peripheral layer of turkey leg tendon (unpublished). This portion of the tendon never gets mineralised while the central portion does upon maturation of the animal. The deH-HHMD cross-link is abundant in this never-mineralised portion of tendon and minimal in the mineralised portion. The results indicated that the preferred molecular locus of this cross-link was: two Lys^{ald} residues located in the N-telopeptides and both Hyl and His residues from $\alpha1$ and $\alpha2$ chains located in the helical domain near the C-termini. Several other molecular loci are also possible. Interestingly, the Hyl residue that is involved in deH-HHMD was not glycosylated while that involved in the divalent iminium cross-link, deH-HLNL, at the same location was significantly glycosylated. These results suggest the stereospecific packing structure of the type I collagen fibrils and a potential role of carbohydrate moieties of Hyl in cross-link formation that was suggested by early studies (Yamauchi *et al.*, 1982; 1986b).

4.2.1.2. *Lys^{ald} –pathway leading to a stable histidine-based trifunctional cross-link (Fig. 5.II)*

Lys^{ald} located in the C- and N-telopeptide domains can also cross-link to the juxtaposed ε-amino group of helical Hyl or Lys on the neighbouring molecules to form iminium intermolecular cross-links, dehydro-hydroxylysinonorleucine (deH-HLNL) and dehydro-lysinonorleucine (deH-LNL), respectively. However, since many of these helical Lys residues are hydroxylated, the former (deH-HLNL) is generally the major product. DeH-HLNL involved in C-telopeptide domain of the molecule can then form a stable, non-reducible cross-link, histidinohydroxylysinonorleucine (HHL) by involving a helical His residue on another molecule (Mechanic *et al.*, 1987).

The HHL cross-linking compound was isolated from an acid hydrolysate of mature bovine skin collagen and the structure was determined (Yamauchi *et al.*, 1987) The data indicated that imidazole C-2 of His is linked to C-6 of norleucine (ε-deaminated Lys residue) which in turn is linked to the C-6 amino group of Hyl (Fig. 6). Quantitative analysis demonstrated that the concentration of HHL in skin collagen rapidly increased from birth through maturational stage and a gradual but steady increase with aging in both cow and humans. This indicates that HHL is an age related cross-link. In 89 year-old human skin, the content of HHL reached almost one residue per type I collagen molecule (Yamauchi *et al.*, 1988b).

The HHL cross-link is abundant in skin and cornea collagen (Yamauchi *et al.*, 1996) but almost absent from skeletal tissues such as dentin, bone, ligament and Achilles tendon collagen (Yamauchi *et al.*, 1988c) where pyridinium cross-links (see below) are dominant. This specific tissue distribution is interesting in the

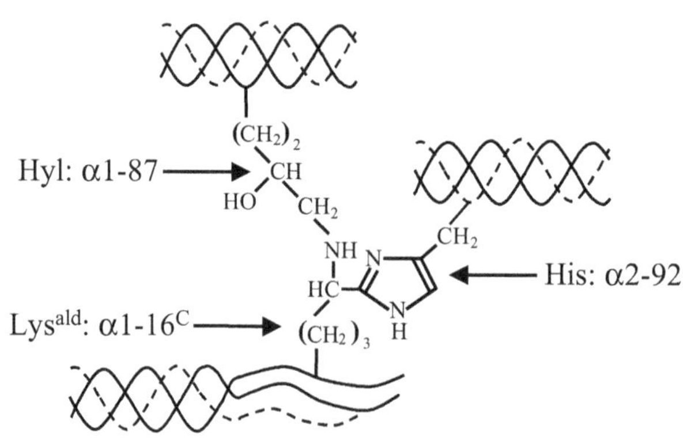

Fig. 6. Structure and molecular location of histidinohydroxylysinonorleucine (HHL) (Yamauchi *et al.*, 1982, 1987; Mechanic *et al.*, 1987). α1-87: 87th residue from the N-terminus of the triple helical domain of an α1 chain (solid line). α2-92: 92nd residue from the N-terminus of the triple helical domain of an α2 chain (dotted line). α1-16^{C}: 16th residue from the N-terminus of the C-telopeptide of an α1 chain (solid line).

light of the susceptibility of these cross-links against UV light, i.e. HHL is stable against UV light while pyridinoline is not (Sakura *et al.*, 1982; Yamauchi *et al.*, 1991). The three-chain peptide cross-linked by HHL was isolated and the molecular locus of this cross-link was determined as Lys^{ald} at α1-16^{C}, Hyl at α1-87 and His at α2-92 (Mechanic *et al.*, 1987) (Fig. 6). This, together with deH-HHMD (see above), is unique because it ties one helix to another. This molecular locus indicates that the packing structure of skin collagen fibril is different from those of skeletal tissues and fits an early observation of a shorter periodicity of skin collagen fibril in comparison with others (Brodsky *et al.*, 1980; Mechanic *et al.*, 1988). Thus, there is more than one modality of collagen packing structure (Yamauchi and Katz, 1993). It has been shown that the concentration of this cross-link is altered in various skin pathologies like amyotrophic lateral sclerosis (Ono and Yamauchi, 1992), systemic sclerosis (Ishikawa *et al.*, 1998), lipodermatosclerosis (Brinckmann *et al.*, 2001), morphea (Tamura and Ishikawa, 2001) and irradiated skin (Sassi *et al.*, 2001).

4.2.2. Hyl^{ald} pathway

When Hyl^{ald} in the C- or the N-telopeptides pairs with helical Hyl, an iminium cross-link, dehydro-dihydroxylysinonorleucine (deH-DHLNL) is formed and when paired with helical Lys, dehydro-hydroxylysinonorleucine (deH-HLNL). The unreduced compound is more stable than other iminium cross-links because of the formation of the stable keto form by a spontaneous Amadori rearrangement (Robins and Bailey, 1973). However, whether this rearrangement is complete in situ in various skeletal tissues is still not fully determined. There are two types of mature, non-reducible cross-links derived from this pathway: one pyridinium and another pyrrole type (Fig. 7).

4.2.2.1. *Pyridinium cross-links: Pyridinoline/deoxypyridinoline*

A non-reducible, fluorescent stable cross-linking compound was first isolated from an acid hydrolysate of bovine tendon collagen

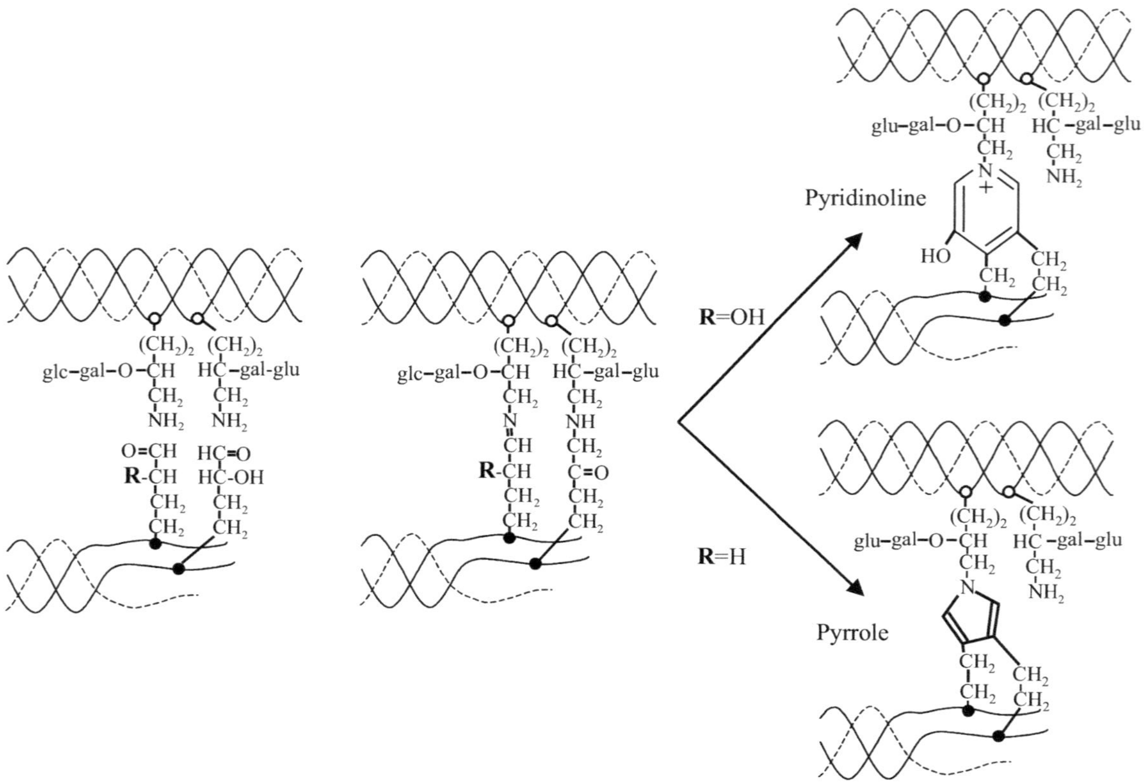

Fig. 7. Formation of trivalent collagen cross-links (upper: pyridinoline, lower: pyrrole) in skeletal tissues. Solid line: α1 chain, Dotted line: α2 chain.

and characterised by Fujimoto and his colleagues (Fujimoto, 1977; Fujimoto *et al.*, 1978). This naturally fluorescent 3-hydroxypyridinium compound was named "pyridinoline" (Pyr). Although the natural occurrence of this cross-link was first disputed (Elsden *et al.*, 1980), overwhelming data published in the next few years strongly supported "Pyr" as a natural collagen cross-link (Eyre and Oguchi, 1980; Fujimoto, 1980; Kuboki *et al.*, 1981; Ogawa *et al.*, 1982; Robins, 1983; Robins and Duncan, 1983).

Pyr is present in a wide variety of collagenous tissues in differing amounts. It is likely a maturational product of deH-DHLNL by the condensation of their two keto-amines with elimination of a Hyl (Eyre and Oguchi, 1980), by the condensation of a keto-amine and a Hyl^{ald} (Robins and Duncan, 1983) or by the condensation of a deH-DHLNL and its keto-amine (Yamauchi and Mechanic, 1988). Whether Pyr ties three collagen molecules (Eyre *et al.*, 1984) or only two molecules (Robins and Duncan, 1983) is not certain at this point. It has been demonstrated that in cartilage Pyr ties not only type II collagen molecules but also type II and IX collagens (thus, between a fibrillar collagen and a FACIT collagen) (Eyre *et al.*, 1987; van der Rest and Mayne, 1988).

As a minor form, a lysyl analog of Pyr, deoxypyridinoline (d-Pyr) (or lysyl pyridinoline), was also identified (Ogawa *et al.*, 1982). This compound is derived from two Hyl^{ald} and one helical Lys and is relatively abundant in mineralised tissues such as dentin and bone (Eyre *et al.*, 1984) although many other skeletal tissues do contain this cross-link to a lesser extent (Gineyts *et al.*, 2000). Interestingly, in the mineralising turkey tendon, d-Pyr is formed almost exclusively in the mineralising portion of tendon and its formation coincides with the onset of mineralisation (Yamauchi and Katz, 1993).

For the last 15 years, there has been considerable interest in measurements of urinary Pyr and d-Pyr as biomarkers of bone resorption. Both Pyr and d-Pyr are excreted to urine in free (~ 40%) as well as peptide-bound (~ 60%) forms. Although Pyr is present in various connective tissues other than bone (cartilage, ligament, tendon, bone, dentin, muscle, etc.), the urinary excretion from these tissues is much lower than that from bone because of their slow turnover rate. Bone is a dynamic tissue undergoing remodelling throughout life. During the bone matrix resorption, the Pyr and d-Pyr cross-links/their peptides are released into circulation system, cleared by the kidneys and excreted in urine. Probably a significant portion of the peptide-bound Pyr/d-Pyr is cleaved to the free form in the kidney (Colwell and Eastell, 1996). Due to the structural stability, the Pyr/d-Pyr cross-links are not metabolised within the body. The Pyr and d-Pyr measurements in urine seem to have several major advantages over the conventional urinary Hyp measurement in terms of specificity and sensitivity. First, they are derived only from mature collagen fibrils not from the intracellular degradation or newly synthesised collagen. Second, they do not seem to be significantly influenced by diet. Third, d-Pyr is relatively bone/mineralised tissue collagen specific (Eyre, 1992; Kraenzlin and Seibel, 1999; Fontana and Delmas, 2000). Currently, there are two major techniques to measure the urinary Pyr and d-Pyr. The first involves acid-hydrolysis of urine followed by purification and measurements by a fluorescence flow monitor connected to HPLC. Since the acid hydrolysis releases all peptide-bound cross-links to the free forms, this method measures the total Pyr/d-Pyr (i.e. both peptide-bound and free forms) contents. This well-established, accurate and standard method is, however, labour intensive and time-consuming, thus not practical for clinical use. The second method is an immunoassay using antibodies that specifically recognise the Pyr or d-Pyr cross-linking amino acid or their associated peptides (peptide sequence of either the C- or the N-telopeptide). These assays are much more practical when compared to the HPLC method, due to the ease and speed of the measurements. Several immunoassays for free Pyr and d-Pyr (Gomez *et al.*, 1996; Robins *et al.*, 1994; Seyedin *et al.*, 1993), N-telopeptide (Clemens *et al.*, 1997; Hanson *et al.*, 1992), C-telopeptides (Bonde *et al.*, 1994; Risteli *et al.*, 1993) have been developed and demonstrated, to a various degree, their usefulness as indices of bone resorption. These assays have been used as diagnostic tools not only for metabolic bone diseases (osteoporosis, Paget's disease, etc.) directly but also for many other diseases that involve bone metabolism such as breast cancer, prostate cancer, multiple myeloma (see a review by Fontana and Delmas, 2000) and liver fibrosis (Stone, 2000).

4.2.2.2. *Pyrrolic cross-link: Pyrrole/deoxypyrrole*

As described above, when the Hyl^{ald}-derived bifunctional cross-link is combined with another Hyl^{ald} or Hyl^{ald}-derived bifunctional cross-link, Pyr (when the helical donor is Hyl) or d-Pyr (when helical donor is Lys) would be formed (see Fig. 7). An N-substituted pyrrolic cross-link which gives a pink colour when reacted with Ehrlich's reagent, p-dimethylaminobenzaldehyde, was proposed to be another collagen cross-link (Scott *et al.*, 1981). However, due to the inherent in-stability of this cross-link, its isolation and characterisation have been severely hampered. In a later study, by characterising the coupled p-phenol-azo-pyrrole cross-linked peptides isolated from bovine tendon collagen, Kuyper and co-workers proposed that pyrrole is located at the similar loci as Pyr but concentrated in the N-telopeptide region involving a Hyl^{ald}-derived bifunctional ketoamine cross-link and Lys^{ald}-containing component (Kuypers *et al.*, 1992). Hanson and Eyre, by isolating and characterising pyrrole-containing cross-linked peptides from a bacterial collagenase digest of demineralised human bone collagen, confirmed that this cross-link was concentrated at the N-telopeptide region ($\alpha 1^N$ X $\alpha 2^N$ X $\alpha 2$ -helix) and proposed that it may be formed by a condensation between Hyl^{ald}-derived bifunctional ketoamine cross-link and Lys^{ald}-derived iminium cross-link (Hanson and Eyre, 1996). Several studies suggested some correlation between this cross-link and thermal/mechanical properties of collagen (Horgan *et al.*, 1990; Knott and Bailey, 2000). A recent study confirmed the structure of this cross-link and its predominant molecular locus at the N-telopeptide region involving mainly $\alpha 1^N$ X $\alpha 1^N$ X $\alpha 1/\alpha 2$ -helix (Brady and Robins, 2001) which is slightly different from the report by Hanson and Eyre (1996). Recently, we have found that pyrrole is present in dentin collagen with similar quantities as Pyr and that the major molecular loci were the same as those in bone collagen (Srisawasdi, 2000). The precise molecular loci, biological function,

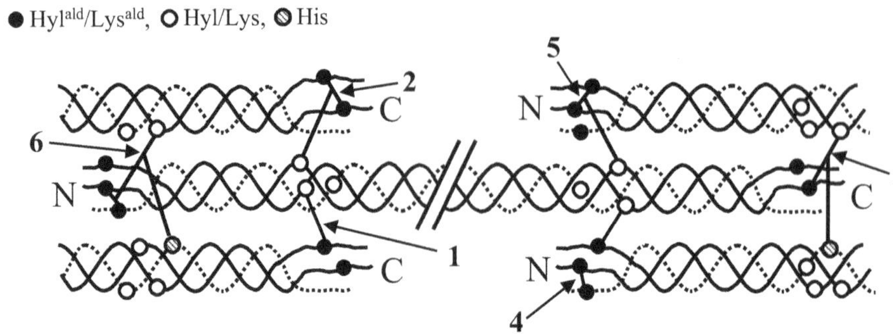

Fig. 8. Preferential molecular loci of cross-links in a type I collagen fibril. C-telopeptidyl aldehyde derived cross-links,

1. Intermolecular divalent cross-links: deH-DHLNL (Hyl^{ald} X Hyl) or deH-HLNL (Hyl^{ald} X Lys or Lys^{ald} X Hyl).
2. Intermolecular trivalent cross-link: Pyrr (Hyl^{ald} X Hyl^{ald} X Hyl). It is also formed at the N-telopeptide region.
3. Intermolecular trivalent cross-link: HHL (Lys^{ald} X Hyl X His).

N-telopeptidyl aldehyde derived cross-links,

4. Intramolecular aldol (Lys^{ald} X Lys^{ald}).
5, Intermolecular trivalent cross-links: Pyrrole (Hyl^{ald} X Lys^{ald} X Hyl), d-Pyr (Hyl^{ald} X Hyl^{ald} X Lys). They can also be formed at the N-telopeptide region.
6. Intermolecular tetravalent cross-link: deH-HHMD: aldol (Lys^{ald} X Lys^{ald}) X His X Hyl

● Lys^{ald}/Hyl^{ald} ○ Hyl/Lys ◎ His
Solid line: α1 chain, Dotted line: α2 chain

tissue distribution/specificity of this cross-link need to be elucidated by employing a more consistent/reproducible assay.

Preferential molecular loci of the type I collagen cross-links described are illustrated in Fig. 8.

4.3. Other cross-links

The potential bifunctional nonreducible stable cross-links, lysinoalanine and histidinoalanine, were isolated from an acid hydrolysate of calcified tissues (Fujimoto *et al.*, 1981; 1982) and the authors speculated that these cross-links were derived from collagen in calcitied tissues. However, studies by Kuboki and his colleagues indicate that these cross-links are rather associated with phosphoproteins (Kuboki *et al.*, 1984; 1991). The dehydroalanine residues, generated by β-elimination that occurs to the phosphoserine residues in phosphoproteins, can condense with imidazole N-3 of His to form histidinoalanine or with ε-amino group of Lys to form lysinoalanine. The biological significance and the precise involvement of collagen for these cross-links are not clear at this point.

5. Other Post-Translational Modifications

It has been reported that the L-aspartic acid residue (αL) located in the C-telopeptide region of type I collagen can be modified, with aging, to an isomerised (βL), a racemized (αD) and an isomerised/racemized (βD) form (Fledelius *et al.*, 1997). The relative rate of these isoforms in bone matrix with age is shown to be βL > βD > αD (Cloos and Fledelius, 2000). Therefore, the normal form of Asp (αL) seems to represent the newly synthesised bone matrix and its isoforms βL, βD and αD old to very old matrices. Increased ratios of αL to βL and αL to αD in urine of patients with Paget's disease support this hypothesis (Garnero *et al.*, 1997; 1998). These modifications in collagen also appear to occur in soft tissues including dermis, arteries, skeletal muscle, heart, tendon, ligament, etc. (Gineyts *et al.*, 2000). Based on the rate of these modifications in collagen, the authors speculate that the rate of collagen turnover of most soft tissues examined is higher than that of bone. Likely the occurrence of such isomerisation is influenced by the tissue-specific microenvironment of the telopeptide region, e.g. the peptide conformation, the nature of juxtaposed amino acids and their modifications (e.g. glycosylation) on the neighbouring molecules and the molecular packing configuration that may be influenced by other matrix molecules (Maroudas *et al.*, 1991). Further careful studies are warranted to correlate these modifications with biological aging.

There are many other important subjects to be covered in collagen biochemistry, e.g. collagen-cell interaction, collagen degradation and its inhibition, roles of collagen in mineralisation, collagen-proteoglycan interaction, collagen as a biomaterial, collagen glycation (non-enzymatic glycosylation) and its relation to diabetes and aging, etc. The readers may refer to a number of recent review articles in these respective fields.

6. Acknowledgements

This is supported in prt by grants from NIH-NIDCR (DE10489) and NASA (NAG 2-1188). The author would like to thank Ms. G. Chandler and Dr. S. Pornprasertsuk for their technical/editorial assistance.

7. References

AMANO, S. *et al.* (2000). Bone morphogenetic protein 1 is an extracellular processing enzyme of the laminin 5 gamma 2 chain, *J. Biol. Chem.* **275**, 22728–22735.

ANNUNEN, P., HELAAKOSKI, T., MYLLYHARJU, J., VEIJOLA, J., PIHLAJANIEMI, T. and KIVIRIKKO, K.I. (1997). Cloning of the human prolyl 4-hydroxylase alpha subunit isoform alpha(II) and characterisation of the type II enzyme tetramer. The alpha(I) and alpha(II) subunits do not form a mixed alpha(I)alpha(II)beta2 tetramer, *J. Biol. Chem.* **272**, 17342–17348.

AREIDA, S.K., REINHARDT, D.P., MULLER, P.K., FIETZEK, P.P., KOWITZ, J., MARINKOVICH, M.P. and NOTBOHM, H. (2001). Properties of the collagen type XVII ectodomain. Evidence for n- to c-terminal triple helix folding, *J. Biol. Chem.* **276**, 1594–1601.

AYCOCK, R.S., RAGHOW, R., STRICKLIN, G.P., SEYER, J.M. and KANG, A.H. (1986). Post-transcriptional inhibition of collagen and fibronectin synthesis by a synthetic homolog of a portion of the carboxyl-terminal propeptide of human type I collagen, *J. Biol. Chem.* **261**, 14355–14360.

BAILEY, A.J. (2001). Molecular mechanisms of ageing in connective tissues, *Mech. Ageing Dev.* **122**, 735–755.

BAILEY, A.J. and SHIMOKOMAKI, M. (1971). Age-related changes in the reducible cross-links of collagen, *FEBS Lett.* **16**, 86–88.

BANK, R.A., ROBINS, S.P., WIJMENGA, C., BRESLAU-SIDERIUS, L.J., BARDOEL, A.F., VAN DER SLUIJS, H.A., PRUIJS, H.E. and TEKOPPELE, J.M. (1999). Defective collagen crosslinking in bone, but not in ligament or cartilage, in Bruck syndrome: Indications for a bone-specific telopeptide lysyl hydroxylase on chromosome 17, *Proc. Natl. Acad. Sci. USA* **96**, 1054–1058.

BATEMAN, J.F., LAMANDE, S.R. and RAMSHAW, J.A.M. (1996). Collagen superfamily. In: *Extracellular Matrix*, W.D. COMPER, ed., Harwood Academic Publisher, Amsterdam, Netherland, pp. 22–67.

BERNSTEIN, P.H. and MECHANIC, G.L. (1980). A natural histidine-based imminium cross-link in collagen and its location, *J. Biol. Chem.* **255**, 10414–10422.

BIRK, D.E. (2001). Type V collagen: Heterotypic type I/V collagen interactions in the regulation of fibril assembly, *Micron.* **32**, 223–237.

BIRK, D.E., FITCH, J.M., BABIARZ, J.P., DOANE, K.J. and LINSENMAYER, T.F. (1990). Collagen fibrillogenesis *in vitro*: Interaction of types I and V collagen regulates fibril diameter, *J. Cell. Sci.* **95**, 649–657.

BONDE, M., QVIST, P., FLEDELIUS, C., RIIS, B.J. and CHRISTIANSEN, C. (1994). Immunoassay for quantifying type I collagen degradation products in urine evaluated, *Clin. Chem.* **40**, 2022–2025.

BOUTAUD, A., BORZA, D.B., BONDAR, O., GUNWAR, S., NETZER, K.O., SINGH, N., NINOMIYA, Y., SADO, Y., NOELKEN, M.E. and HUDSON, B.G. (2000). Type IV collagen of the glomerular basement membrane. Evidence that the chain specificity of network assembly is encoded by the noncollagenous NC1 domains, *J. Biol. Chem.* **275**, 30716–30724.

BRADY, J.D. and ROBINS, S.P. (2001). Structural characterisation of pyrrolic cross-links in collagen using a biotinylated Ehrlich's reagent, *J. Biol. Chem.* **276**, 18812–18818.

Brewton, R.G., Wright, D.W. and Mayne, R. (1991). Structural and functional comparison of type IX collagen-proteoglycan from chicken cartilage and vitreous humor, *J. Biol. Chem.* **266**, 4752–4757.

BRINCKMANN, J. *et al.* (2001). Different pattern of collagen cross-links in two sclerotic skin diseases: Lipodermatosclerosis and circumscribed scleroderma, *J. Invest. Dermatol.* **117**, 269–273.

BRINCKMANN, J., NOTBOHM, H., TRONNIER, M., ACIL, Y., FIETZEK, P.P., SCHMELLER, W., MULLER, P.K. and BATGE, B. (1999). Overhydroxylation of lysyl residues is the initial step for altered collagen cross-links and fibril architecture in fibrotic skin, *J. Invest. Dermatol.* **113**, 617–621.

BRODSKY, B., EIKENBERRY, E.F. and CASSIDY, K. (1980). An unusual collagen periodicity in skin, *Biochim. Biophys. Acta* **621**, 162–166.

BRUCKNER-TUDERMAN, L., HOPFNER, B. and HAMMAMI-HAUASLI, N. (1999). Biology of anchoring fibrils: Lessons from dystrophic epidermolysis bullosa, *Matrix Biol.* **18**, 43–54.

BYERS, P.H. (2000). Collagens: Building blocks at the end of the development line, *Clin. Genet.* **58**, 270–279.

BYERS, P.H. (2000). Osteogenesis imperfecta: Perspectives and opportunities, *Curr. Opin. Pediatr.* **12**, 603–609.

CHAVANAS, S., GACHE, Y., TADINI, G., PULKKINEN, L., UITTO, J., ORTONNE, J.P. and MENEGUZZI, G. (1997). A homozygous in-frame deletion in the collagenous domain of bullous pemphigoid antigen BP180 (type XVII collagen) causes generalized atrophic benign epidermolysis bullosa, *J. Invest. Dermatol.* **109**, 74–78.

CHERNOUSOV, M.A., ROTHBLUM, K., TYLER, W.A., STAHL, R.C. and CAREY, D.J. (2000). Schwann cells synthesise type V collagen that contains a novel alpha 4 chain. Molecular cloning, biochemical characterisation, and high affinity heparin binding of alpha 4(V) collagen, *J. Biol. Chem.* **275**, 28208–28215.

CHESSLER, S.D. and BYERS, P.H. (1993). BiP binds type I procollagen pro alpha chains with mutations in the carboxyl-terminal propeptide synthesised by cells from patients with osteogenesis imperfecta, *J. Biol. Chem.* **268**, 18226–18233.

ClemeNS, J.D., HERRICK, M.V., SINGER, F.R. and EYRE, D.R. (1997). Evidence that serum NTx (collagen-type I N-telopeptides) can act as an immunochemical marker of bone resorption, *Clin. Chem.* **43**, 2058–2063.

CLOOS, P.A. and FLEDELIUS, C. (2000). Collagen fragments in urine derived from bone resorption are highly racemized and isomerised: A biological clock of protein aging with clinical potential, *Biochem. J.* **345**(Pt. 3), 473–480.

COLWELL, A. and EASTELL, R. (1996). The renal clearance of free and conjugated pyridinium cross-links of collagen, *J. Bone Miner. Res.* **11**, 1976–1980.

CRONLUND, A.L., SMITH, B.D. and KAGAN, H.M. (1985). Binding of lysyl oxidase to fibrils of type I collagen, *Connect. Tissue Res.* **14**, 109–119.

DI DONATO, A., LACAL, J.C., DI DUCA, M., GIAMPUZZI, M., GHIGGERI, G. and GUSMANO, R. (1997). Micro-injection of recombinant lysyl oxidase blocks oncogenic p21-Ha-Ras and progesterone effects on Xenopus laevis oocyte maturation, *FEBS Lett.* **419**, 63–68.

EHNIS, T., DIETERICH, W., BAUER, M., KRESSE, H. and SCHUPPAN, D. (1997). Localisation of a binding site for the proteoglycan decorin on collagen XIV (undulin), *J. Biol. Chem.* **272**, 20414–20419.

ELSDEN, D.F., LIGHT, N.D. and BAILEY, A.J. (1980). An investigation of pyridinoline, and putative collagen cross-link, *Biochem. J.* **185**, 531–534.

EYRE, D. (1992). New biomarkers of bone resorption, *J. Clin. Endocrinol. Metab.* **74**, 470A–470C.

EYRE, D.R. (1981). Crosslink maturation in bone collagen, *Dev. Biol.* **22**, 51–55.

EYRE, D.R., APON, S., WU, J.J., ERICSSON, L.H. and WALSH, K.A. (1987). Collagen type IX: Evidence for covalent linkages to type II collagen in cartilage, *FEBS Lett.* **220**, 337–341.

EYRE, D.R., DICKSON, I.R. and VAN NESS, K. (1988). Collagen cross-linking in human bone and articular cartilage. Age-related changes in the content of mature hydroxypyridinium residues, *Biochem. J.* **252**, 495–500.

EYRE, D.R. and OGUCHI, H. (1980). The hydroxypyridinium crosslinks of skeletal collagens: Their measurement, properties and a proposed pathway of formation, *Biochem. Biophys. Res. Commun.* **92**, 403–410.

EYRE, D.R., PAZ, M.A. and GALLOP, P.M. (1984). Cross-linking in collagen and elastin, *Ann. Rev. Biochem.* **53**, 717–748.

FELBOR, U., DREIER, L., BRYANT, R.A., PLOEGH, H.L., OLSEN, B.R. and MOTHES, W. (2000). Secreted cathepsin L generates endostatin from collagen XVIII, *EMBO J.* **19**, 1187–1194.

FLEDELIUS, C., JOHNSEN, A.H., CLOOS, P.A., BONDE, M. and QVIST, P. (1997). Characterisation of urinary degradation pro-ducts derived from type I collagen. Identification of a beta-isomerised Asp-Gly sequence within the C-terminal telopeptide (alpha1) region, *J. Biol. Chem.* **272**, 9755–9763.

FONTANA, A. and DELMAS, P.D. (2000). Markers of bone turnover in bone metastases, *Cancer* **88**, 2952–2960.

FORLINO, A. and MARINI, J.C. (2000). Osteogenesis imperfecta: Prospects for molecular therapeutics, *Mol. Genet. Metab.* **71**, 225–232.

FOUSER, L., SAGE, E.H., CLARK, J. and BORNSTEIN, P. (1991). Feedback regulation of collagen gene expression: A Trojan horse approach, *Proc. Natl. Acad. Sci. USA* **88**, 10158–10162.

FUJIMOTO, D. (1977). Isolation and characterisation of a fluorescent material in bovine achilles tendon collagen, *Biochem. Biophys. Res. Commun.* **76**, 1124–1129.

FUJIMOTO, D. (1980). Evidence for natural existence of pyridinoline crosslink in collagen, *Biochem. Biophys. Res. Commun.* **93**, 948–953.

FUJIMOTO, D., HIRAMA, M. and IWASHITA, T. (1981). Occurrence of lysinoalanine in calcified tissue collagen, *Biochem. Biophys. Res. Commun.* **103**, 1378–1383.

FUJIMOTO, D., HIRAMA, M. and IWASHITA, T. (1982). Histidinoalanine, a new crosslinking amino aicd, in calcified tissue collagen, *Biochem. Biophys. Res. Commun.* **104**, 1102–1106.

FUJIMOTO, D., MORIGUCHI, T., ISHIDA, T. and HAYASHI, H. (1978). The structure of pyridinoline, a collagen crosslink, *Biochem. Biophys. Res. Commun.* **84**, 52–57.

GARNERO, P., FLEDELIUS, C., GINEYTS, E., SERRE, C.M., VIGNOT, E. and DELMAS, P.D. (1997). Decreased beta-isomerisation of the C-terminal telopeptide of type I collagen alpha 1 chain in Paget's disease of bone, *J. Bone Miner. Res.* **12**, 1407–1415.

GARNERO, P., GINEYTS, E., SCHAFFER, A.V., SEAMAN, J. and DELMAS, P.D. (1998). Measurement of urinary excretion of nonisomerised and beta-isomerised forms of type I collagen breakdown products to monitor the effects of the bisphosphonate zoledronate in Paget's disease, *Arthritis. Rheum.* **41**, 354–360.

GAROFALO, S., METSARANTA, M., ELLARD, J., SMITH, C., HORTON, W., VUORIO, E. and DE CROMBRUGGHE, B. (1993). Assembly of cartilage collagen fibrils is disrupted by overexpression of normal type II collagen in transgenic mice, *Proc. Natl. Acad. Sci. USA* **90**, 3825–3829.

GATALICA, B., PULKKINEN, L., LI, K., KUOKKANEN, K., RYYNANEN, M., MCGRATH, J.A. and UITTO, J. (1997). Cloning of the human type XVII collagen gene (COL17A1), and detection of novel mutations in generalized atrophic benign epidermolysis bullosa, *Am. J. Hum. Genet.* **60**, 352–365.

GERRIETS, J.E., CURWIN, S.L. and LAST, J.A. (1993). Tendon hypertrophy is associated with increased hydroxylation of nonhelical lysine residues at two specific cross-linking sites in type I collagen, *J. Biol. Chem.* **268**, 25553–25560.

GINEYTS, E., CLOOS, P.A., BOREL, O., GRIMAUD, L., DELMAS, P.D. and GARNERO, P. (2000). Racemization and isomerisation of type I collagen C-telopeptides in human bone and soft tissues: Assessment of tissue turnover, *Biochem. J.* **345(**Pt. 3), 481–485.

GOMEZ, B., JR., ARDAKANI, S., EVANS, B.J., MERRELL, L.D., JENKINS, D.K. and KUNG, V.T. (1996). Monoclonal antibody assay for free urinary pyridinium cross-links, *Clin. Chem.* **42**, 1168–1175.

HAGG, P., REHN, M., HUHTALA, P., VAISANEN, T., TAMMINEN, M. and PIHLAJANIEMI, T. (1998). Type XIII collagen is identified as a plasma membrane protein, *J. Biol. Chem.* **273**, 15590–15597.

HAGG, P.M., HAGG, P.O., PELTONEN, S., AUTIO-HARMAINEN, H. and PIHLAJANIEMI, T. (1997). Location of type XV collagen in human tissues and its accumulation in the interstitial matrix of the fibrotic kidney, *Am. J. Pathol.* **150**, 2075–2086.

HALFTER, W., DONG, S., SCHURER, B. and COLE, G.J. (1998). Collagen XVIII is a basement membrane heparan sulfate proteoglycan, *J. Biol. Chem.* **273**, 25404–25412.

HAMALAINEN, E.R., KEMPPAINEN, R., KUIVANIEMI, H., TROMP, G., VAHERI, A., PIHLAJANIEMI, T. and KIVIRIKKO, K.I. (1995). Quantitative polymerase chain reaction of lysyl oxidase mRNA in malignantly transformed human cell lines demonstrates that their low lysyl oxidase activity is due to low quantities of its mRNA and low levels of transcription of the respective gene, *J. Biol. Chem.* **270**, 21590–21593.

HANSON, D.A. and EYRE, D.R. (1996). Molecular site specificity of pyridinoline and pyrrole cross-links in type I collagen of human bone, *J. Biol. Chem.* **271**, 26508–26516.

HANSON, D.A., WEIS, M.A., BOLLEN, A.M., MASLAN, S.L., SINGER, F.R. and EYRE, D.R. (1992). A specific immunoassay for monitoring human bone resorption: Quantitation of type I collagen cross-linked N-telopeptides in urine, *J. Bone Miner. Res.* **7**, 1251–1258.

HATA, R., KURATA, S. and SHINKAI, H. (1988). Existence of malfunctioning pro alpha2(I) collagen genes in a patient with a pro alpha 2(I)-chain-defective variant of Ehlers-Danlos syndrome, *Eur. J. Biochem.* **174**, 231–237.

HEIKKINEN, J., RISTELI, M., WANG, C., LATVALA, J., ROSSI, M., VALTAVAARA, M. and MYLLYLA, R. (2000). Lysyl hydroxylase 3 is a multifunctional protein possessing collagen glucosyltransferase activity, *J. Biol. Chem.* **275**, 36158–36163.

HORGAN, D.J., KING, N.L., KURTH, L.B. and KUYPERS, R. (1990). Collagen crosslinks and their relationship to the thermal properties of calf tendons, *Arch. Biochem. Biophys.* **281**, 21–26.

HULMES, D.J., MOULD, A.P. and KESSLER, E. (1997). The CUB domains of procollagen C-proteinase enhancer control collagen assembly solely by their effect on procollagen C-proteinase/bone morphogenetic protein-1, *Matrix Biol.* **16**, 41–45.

IOZZO, R.V. (1999). The biology of the small leucine-rich proteoglycans. Functional network of interactive proteins, *J. Biol. Chem.* **274**, 18843–18846.

ISHIKAWA, O., KONDO, A. and MIYACHI, Y. (1998). Mature type of skin collagen crosslink, histidinohydroxylysinonorleucine, is significantly increased in the skin of systemic sclerosis patients, *Arthritis. Rheum.* **41**, 376–377.

KADLER, K.E., HOLMES, D.F., TROTTER, J.A. and CHAPMAN, J.A. (1996). Collagen fibril formation, *Biochem. J.* **316**, 1–11.

KAGAN, H.M. (2000). Intra- and extracellular enzymes of collagen biosynthesis as biological and chemical targets in the control of fibrosis, *Acta Trop.* **77**, 147–152.

KANG, A.H., FARIS, B. and FRANZBLAU, C. (1969). Intramolecular cross-link of chick skin collagen, *Biochem. Biophys. Res. Commun.* **36**, 345–349.

KATZ, E.P. and DAVID, C.W. (1992). Unique side-chain conformation encoding for chirality and azimuthal orientation in the molecular packing of skin collagen, *J. Mol. Biol.* **228**, 963–969.

KEENE, D.R., LUNSTRUM, G.P., MORRIS, N.P., STODDARD, D.W. and BURGESON, R.E. (1991). Two type XII-like collagens localise to the surface of banded collagen fibrils, *J. Cell. Biol.* **113**, 971–978.

KEENE, D.R., SAN ANTONIO, J.D., MAYNE, R., MCQUILLAN, D.J., SARRIS, G., SANTORO, S.A. and IOZZO, R.V. (2000). Decorin binds near the C terminus of type I collagen, *J. Biol. Chem.* **275**, 21801–21804.

KENYON, K., CONTENTE, S., TRACKMAN, P.C., TANG, J., KAGAN, H.M. and FRIEDMAN, R.M. (1991). Lysyl oxidase and rrg messenger RNA, *Science* **253**, 802.

KESSLER, E. and ADAR, R. (1989). Type I procollagen C-proteinase from mouse fibroblasts. Purification and demonstration of a 55-kDa enhancer glycoprotein, *Eur. J. Biochem.* **186**, 115–121.

KESSLER, E., FICHARD, A., CHANUT-DELALANDE, H., BRUSEL, M. and RUGGIERO, F. (2001). Bone morphogenetic protein-1

(BMP-1) mediates C-terminal processing of procollagen V homotrimer, *J. Biol. Chem.* **276**, 27051–27057.

KIELTY, C.M. and SHUTTLEWORTH, C.A. (1997). Microfibrillar elements of the dermal matrix, *Microsc. Res. Tech.* **38**, 413–427.

KIVIRIKKO, K.I. and MYLLYLA, R. (1982). Posttranslational enzymes in the biosynthesis of collagen: Intracellular enzymes, *Methods Enzymol.* **82**, 245–304.

KIVIRIKKO, K.I. and MYLLYLA, R. (1987). Recent developments in posttranslational modification: Intracellular processing, *Methods Enzymol.* **144**, 96–114.

KIVIRIKKO, K.I., MYLLYLA, R. and PIHLAJANIEMI, T. (1989). Protein hydroxylation: Prolyl 4-hydroxylase, an enzyme with four cosubstrates and a multifunctional subunit, *FASEB J.* **3**, 1609–1617.

KIVIRIKKO, K.I. and PIHLAJANIEMI, T. (1998). Collagen hydroxylases and the protein disulfide isomerase subunit of prolyl 4-hydroxylases, *Adv. Enzymol. Relat. Areas Mol. Biol.* **72**, 325–3298.

KNOTT, L. and BAILEY, A.J. (2000). Collagen cross-links in mineralised tissues: A role for the pyrrole cross-link. In: *The Chemistry and Biology of Mineralised Tissue*, M. GOLDBERG, A. BOSKEY and C. ROBINSON, eds., Am. Acad. Orthop. Surg., pp. 125–128.

KOCH, M., FOLEY, J.E., HAHN, R., ZHOU, P., BURGESON, R.E., GERECKE, D.R. and GORDON, M.K. (2001). Alpha 1(Xx) collagen, a new member of the collagen subfamily, fibril-associated collagens with interrupted triple helices, *J. Biol. Chem.* **276**, 23120–23126.

KOIDE, T., ASO, A., YORIHUZI, T. and NAGATA, K. (2000). Conformational requirements of collagenous peptides for recognition by the chaperone protein HSP47, *J. Biol. Chem.* **275**, 27957–27963.

KRAENZLIN, M.E. and SEIBEL, M.J. (1999). Measurement of biochemical markers of bone resorption. In: *Dynamics of Bone and Cartilage Metabolism*, M.J. SEIBEL, S.P. ROBINS and J.P. BILEZIKIAN, eds., Academic Press, San Diego, pp. 411–426.

KUBOKI, Y., FUJISAWA, R., TSUZAKI, M., LIU, C.F. and SASAKI, S. (1984). Presence of lysinoalanine and histidinoalanine in bovine dentin phosphoprotein, *Calcif. Tissue Int.* **36**, 126–128.

KUBOKI, Y., TAKITA, H., FUJISAWA, R., YAMAGUCHI, H., YAMADA, H., TAZAKI, M., OHNUMA, Y., MIZUNO, M. and TSUZAKI, M. (1991). Histidinoalanine-containing phosphoprotein in bone, *Calcif. Tissue Int.* **48**, 196–203.

KUBOKI, Y., TSUZAKI, M., SASAKI, S., LIU, C.F. and MECHANIC, G.L. (1981). Location of the intermolecular cross-links in bovine dentin collagen, solubilization with trypsin and isolation of cross-link peptides containing dihydroxylysinonorleucine and pyridinoline, *Biochem. Biophys. Res. Commun.* **102**, 119–126.

KUYPERS, R., TYLER, M., KURTH, L.B., JENKINS, I.D. and HORGAN, D.J. (1992). Identification of the loci of the collagen-associated Ehrlich chromogen in type I collagen confirms its role as a trivalent cross-link, *Biochem. J.* **283**, 129–136.

LAMANDE, S.R. and BATEMAN, J.F. (1995). The type I collagen pro alpha 1(I) COOH-terminal propeptide N-linked oligosaccharide. Functional analysis by site-directed mutagenesis, *J. Biol. Chem.* **270**, 17858–17865.

LANG, K., SCHMID, F.X. and FISCHER, G. (1987). Catalysis of protein folding by prolyl isomerase, *Nature* **329**, 268–270.

LEE, A.S. (2001). The glucose-regulated proteins: Stress induction and clinical applications, *Trends Biochem. Sci.* **26**, 504–510.

LI, D., CLARK, C.C. and MYERS, J.C. (2000). Basement membrane zone type XV collagen is a disulfide-bonded chondroitin sulfate proteoglycan in human tissues and cultured cells, *J. Biol. Chem.* **275**, 22339–22347.

LI, K., TAMAI, K., TAN, E.M. and UITTO, J. (1993). Cloning of type XVII collagen. Complementary and genomic DNA sequences of mouse 180-kilodalton bullous pemphigoid antigen (BPAG2) predict an interrupted collagenous domain, a transmembrane segment, and unusual features in the 5′-end of the gene and the 3′-untranslated region of the mRNA, *J. Biol. Chem.* **268**, 8825–8834.

LI, W., LIU, G., CHOU, I.N. and KAGAN, H.M. (2000). Hydrogen peroxide-mediated, lysyl oxidase-dependent chemotaxis of vascular smooth muscle cells, *J. Cell. Biochem.* **78**, 550–557.

LI, Y. *et al.* (1995). A fibrillar collagen gene, Col11a1, is essential for skeletal morphogenesis, *Cell* **80**, 423–430.

MAKI, J.M. and KIVIRIKKO, K.I. (2001). Cloning and characterisation of a fourth human lysyl oxidase isoenzyme, *Biochem. J.* **355**, 381–387.

MARNEROS, A.G. and OLSEN, B.R. (2001). The role of collagen-derived proteolytic fragments in angiogenesis, *Matrix Biol.* **20**, 337–345.

MAROUDAS, A., WACHTEL, E., GRUSHKO, G., KATZ, E.P. and WEINBERG, P. (1991). The effect of osmotic and mechanical pressures on water partitioning in articular cartilage, *Biochim. Biophys. Acta* **1073**, 285–294.

MASSE, P.G., YAMAUCHI, M., MAHUREN, J.D., COBURN, S.P., MUNIZ, O.E. and HOWELL, D.S. (1995). Connective tissue integrity is lost in vitamin B-6-deficient chicks, *J. Nutr.* **125**, 26–34.

MECHANIC, G.L., KATZ, E.P., HENMI, M., NOYES, C. and YAMAUCHI, M. (1987). Locus of a histidine-based, stable trifunctional, helix to helix collagen cross-link: Stereospecific collagen structure of type I skin fibrils, *Biochemistry* **26**, 3500–3509.

MILLER, E.J. (1984). Chemistry of the collagens and their distribution. In: *Extracellular Matrix Biochemistry*, K.A. PIEZ and

A.H. REDDI, eds., Elsevier Science Publishing Co., New York, pp. 41–81.

MORIGUCHI, T. and FUJIMOTO, D. (1978). Age-related changes in the content of the collagen crosslink, pyridinoline, *J. Biochem. (Tokyo)* **84**, 933–935.

MULLINS, M.C. (1998). Holy tolloido: Tolloid cleaves SOG/chordin to free DPP/BMPs, *Trends Genet.* **14**, 127–129.

MURAGAKI, Y., TIMMONS, S., GRIFFITH, C.M., OH, S.P., FADEL, B., QUERTERMOUS, T. and OLSEN, B.R. (1995). Mouse Col18a1 is expressed in a tissue-specific manner as three alternative variants and is localised in basement membrane zones, *Proc. Natl. Acad. Sci. USA* **92**, 8763–8767.

MYLLYHARJU, J. and KIVIRIKKO, K.I. (2001). Collagens and collagen-related diseases, *Ann. Med.* **33**, 7–21.

NAGAI, N., HOSOKAWA, M., ITOHARA, S., ADACHI, E., MATSUSHITA, T., HOSOKAWA, N. and NAGATA, K. (2000). Embryonic lethality of molecular chaperone hsp47 knockout mice is associated with defects in collagen biosynthesis, *J. Cell. Biol.* **150**, 1499–1506.

NAGATA, K. (1996). Hsp47: A collagen-specific molecular chaperone, *Trends Biochem. Sci.* **21**, 22–26.

NAGATA, K. and HOSOKAWA, N. (1996). Regulation and function of collagen-specific molecular chaperone, HSP47, *Cell. Struct. Funct.* **21**, 425–430.

NELLAIAPPAN, K., RISITANO, A., LIU, G., NICKLAS, G. and KAGAN, H.M. (2000). Fully processed lysyl oxidase catalyst translocates from the extracellular space into nuclei of aortic smooth-muscle cells, *J. Cell. Biochem.* **79**, 576–582.

NIMNI, M.E. and HARKNESS, R.D. (1988). Molecular structures and functions of collagen. In: *Collagen*, M.E. NIMNI, ed., CRC Press, Boca Raton, Florida, pp. 1–77.

NISSI, R., AUTIO-HARMAINEN, H., MARTTILA, P., SORMUNEN, R. and KIVIRIKKO, K.I. (2001). Prolyl 4-hydroxylase isoenzymes I and II have different expression patterns in several human tissues, *J. Histochem. Cytochem.* **49**, 1143–1153.

NOTBOHM, H., NOKELAINEN, M., MYLLYHARJU, J., FIETZEK, P.P., MULLER, P.K. and KIVIRIKKO, K.I. (1999). Recombinant human type II collagens with low and high levels of hydroxylysine and its glycosylated forms show marked differences in fibrillogenesis *in vitro*, *J. Biol. Chem.* **274**, 8988–8992.

OGAWA, T., ONO, T., TSUDA, M. and KAWANISHI, Y. (1982). A novel fluor in insoluble collagen: A crosslinking moiety in collagen molecule, *Biochem. Biophys. Res. Commun.* **107**, 1252–1257.

OH, S.P., WARMAN, M.L., SELDIN, M.F., CHENG, S.D., KNOLL, J.H., TIMMONS, S. and OLSEN, B.R. (1994). Cloning of cDNA and genomic DNA encoding human type XVIII collagen and localisation of the alpha 1(XVIII) collagen gene to mouse chromosome 10 and human chromosome 21, *Genomics* **19**, 494–499.

OLSEN, B.R. (1995). New insights into the function of collagens from genetic analysis, *Curr. Opin. Cell. Biol.* **7**, 720–727.

OLSEN, B.R., WINTERHALTER, K.H. and GORDON, M.K. (1995). FACIT collagens and their biological roles, *Trends Glycosci. Glycotechnol.* **7**, 115–127.

ONO, S. and YAMAUCHI, M. (1992). Collagen cross-linking of skin in patients with amyotrophic lateral sclerosis, *Ann. Neurol.* **31**, 305–310.

O'REILLY, M.S., BOEHM, T., SHING, Y., FUKAI, N., VASIOS, G., LANE, W.S., FLYNN, E., BIRKHEAD, J.R., OLSEN, B.R. and FOLKMAN, J. (1997). Endostatin: An endogenous inhibitor of angiogenesis and tumour growth, *Cell* **88**, 277–285.

OTSUBO, K., KATZ, E.P., MECHANIC, G.L. and YAMAUCHI, M. (1992). Cross-linking connectivity in bone collagen fibrils: The

COOH-terminal locus of free aldehyde, *Biochemistry* **31**, 396–402.

PANCHENKO, M.V., STETLER-STEVENSON, W.G., TRUBETSKOY, O.V., GACHERU, S.N. and KAGAN, H.M. (1996). Metalloproteinase activity secreted by fibrogenic cells in the processing of prolysyl oxidase. Potential role of procollagen C-proteinase, *J. Biol. Chem.* **271**, 7113–7119.

PELTONEN, S., REHN, M. and PIHLAJANIEMI, T. (1997). Alternative splicing of mouse alpha1(XIII) collagen RNAs results in at least 17 different transcripts, predicting alpha1(XIII) collagen chains with length varying between 651 and 710 amino acid residues, *DNA Cell. Biol.* **16**, 227–234.

PIHLAJANIEMI, T. and REHN, M. (1995). Two new collagen subgroups: Membrane-associated collagens and types XV and XVII, *Prog. Nucleic Acid. Res. Mol. Biol.* **50**, 225–262.

PROCKOP, D.J. (1976). Intracellular steps in the biosynthesis of collagen. In: *Biochemistry of Collagen*, G.N. RAMACHANDRAN and A.H. REDDI, eds., Plenum Press, New York, pp. 163–273.

PROCKOP, D.J. and KIVIRIKKO, K.I. (1995). Collagens: Molecular biology, diseases, and potentials for therapy, *Ann. Rev. Biochem.* **64**, 403–434.

PROCKOP, D.J., SIERON, A.L. and LI, S.W. (1998). Procollagen N-proteinase and procollagen C-proteinase. Two unusual metalloproteinases that are essential for procollagen processing probably have important roles in development and cell signaling, *Matrix Biol.* **16**, 399–408.

REISER, K., MCCORMICK, R.J. and RUCKER, R.B. (1992). Enzymatic and nonenzymatic cross-linking of collagen and elastin, *FASEB J.* **6**, 2439–2449.

REISER, K.M. (1998). Nonenzymatic glycation of collagen in aging and diabetes, *Proc. Soc. Exp. Biol. Med.* **218**, 23–37.

RICARD-BLUM, S. (2000). *Unconventional Collagens: Types VI, VII, VIII, IX, X, XII, XIV, XVI, and XIX*. M. VAN DER REST, B. DUBLET and S. RICARD-BLUM, eds. Oxford Univ. Press.

RISTELI, J., ELOMAA, I., NIEMI, S., NOVAMO, A. and RISTELI, L. (1993). Radioimmunoassay for the pyridinoline cross-linked carboxy-terminal telopeptide of type I collagen: A new serum marker of bone collagen degradation, *Clin. Chem.* **39**, 635–640.

ROBINS, S.P. (1983). Cross-linking of collagen. Isolation, structural characterisation and glycosylation of pyridinoline, *Biochem. J.* **215**, 167–173.

ROBINS, S.P. and BAILEY, A.J. (1973). The chemistry of the collagen cross-links. The characterisation of fraction C, a possible artifact produced during the reduction of collagen fibres with borohydride, *Biochem. J.* **135**, 657–665.

ROBINS, S.P. and BAILEY, A.J. (1973). Relative stabilities of the intermediate reducible crosslinks present in collagen fibres, *FEBS Lett.* **33**, 167–174.

ROBINS, S.P. and DUNCAN, A. (1983). Cross-linking of collagen. Location of pyridinoline in bovine articular cartilage at two sites of the molecule, *Biochem. J.* **215**, 175–182.

ROBINS, S.P., WOITGE, H., HESLEY, R., JU, J., SEYEDIN, S. and SEIBEL, M.J. (1994). Direct, enzyme-linked immunoassay for urinary deoxypyridinoline as a specific marker for measuring bone resorption, *J. Bone Miner. Res.* **9**, 1643–1649.

RODEO, S.A., SUZUKI, K., YAMAUCHI, M., BHARGAVA, M. and WARREN, R.F. (1998). Analysis of collagen and elastic fibers in shoulder capsule in patients with shoulder instability, *Am. J. Sports Med.* **26**, 634–643.

ROYCE, P.M. and BARNES, M.J. (1985). Failure of highly purified lysyl hydroxylase to hydroxylate lysyl residues in the non-helical regions of collagen, *Biochem. J.* **230**, 475–480.

SAARELA, J., REHN, M., OIKARINEN, A., AUTIO-HARMAINEN, H. and PIHLAJANIEMI, T. (1998). The short and long forms of type XVIII collagen show clear tissue specificities in their expression and location in basement membrane zones in humans, *Am. J. Pathol.* **153**, 611–626.

SAKURA, S., FUJIMOTO, D., SAKAMOTO, K., MIZUNO, A. and MOTEGI, K. (1982). Photolysis of pyridinoline, a cross-linking amino acid of collagen, by ultraviolet light, *Can. J. Biochem.* **60**, 525–529.

SANDELL, L.J. and BOYD, C.D. (1990). Conserved and Divergent sequence and Functional Elements within Collagen Genes. In: *Extracellular Matrix Genes*, L.J. SANDELL and C.D. BOYD, eds., Academic Press, San Diego, pp. 1–56.

SASAKI, T., LARSSON, H., TISI, D., CLAESSON-WELSH, L., HOHENESTER, E. and TIMPL, R. (2000). Endostatins derived from collagens XV and XVIII differ in structural and binding properties, tissue distribution and anti-angiogenic activity, *J. Mol. Biol.* **301**, 1179–1190.

SASSI, M., JUKKOLA, A., RIEKKI, R., HOYHTYA, M., RISTELI, L., OIKARINEN, A. and RISTELI, J. (2001). Type I collagen turnover and cross-linking are increased in irradiated skin of breast cancer patients, *Radiother. Oncol.* **58**, 317–323.

SCHLUMBERGER, W., THIE, M., VOLMER, H., RAUTERBERG, J. and ROBENEK, H. (1988). Binding and uptake of Col 1(I), a peptide capable of inhibiting collagen synthesis in fibroblasts, *Eur. J. Cell. Biol.* **46**, 244–252.

SCHNIEKE, A., HARBERS, K. and JAENISCH, R. (1983). Embryonic lethal mutation in mice induced by retrovirus insertion into the alpha 1(I) collagen gene, *Nature* **304**, 315–320.

SCOTT, I.C., IMAMURA, Y., PAPPANO, W.N., TROEDEL, J.M., RECKLIES, A.D., ROUGHLEY, P.J. and GREENSPAN, D.S. (2000). Bone morphogenetic protein-1 processes probiglycan, *J. Biol. Chem.* **275**, 30504–30511.

SCOTT, J.E., HUGHES, E.W. and SHUTTLEWORTH, A. (1981). A collagen-associated Ehrlich chromogen: A pyrrolic cross-link? *Biosci. Rep.* **1**, 611–618.

SEYEDIN, S.M., KUNG, V.T., DANILOFF, Y.N., HESLEY, R.P., GOMEZ, B., NIELSEN, L.A., ROSEN, H.N. and ZUK, R.F. (1993). Immunoassay for urinary pyridinoline: The new marker of bone resorption, *J. Bone Miner. Res.* **8**, 635–641.

SHARMA, R., KRAMER, J.A. and KRAWETZ, S.A. (1997). Lysyl oxidase, cellular senescence and tumour suppression, *Biosci. Rep.* **17**, 409–414.

SHIIBA, M., ARNAUD, S.B.,TANZAWA, H., UZAWA, K. and YAMAUCHI, M. (In press). Alterations of bone collagen matrix in weight-bearing bones during skeletal unloading, *Connect. Tissue Res.*

SIEGEL, R.C. (1974). Biosynthesis of collagen crosslinks: Increased activity of purified lysyl oxidase with reconstituted collagen fibrils, *Proc. Natl. Acad. Sci. USA* **71**, 4826–4830.

SMITH-MUNGO, L.I. and KAGAN, H.M. (1998). Lysyl oxidase: Properties, regulation and multiple functions in biology, *Matrix Biol.* **16**, 387–398.

SRISAWASDI, S., VERDELIS, K, and YAMAUCHI, M. (2000). Pyrrole cross-link in dentin collagen, *J. Dent. Res.* **79**, 250.

STEINMANN, B., BRUCKNER, P. and SUPERTI-FURGA, A. (1991). Cyclosporin A slows collagen triple-helix formation *in vivo*: Indirect evidence for a physiologic role of peptidyl-prolyl *cis-trans*-isomerase, *J. Biol. Chem.* **266**, 1299–1303.

STONE, P.J. (2000). Potential use of collagen and elastin degradation markers for monitoring liver fibrosis in schistosomiasis, *Acta Trop.* **77**, 97–99.

TAKAHARA, K., KESSLER, E., BINIAMINOV, L., BRUSEL, M., EDDY, R.L., JANI-SAIT, S., SHOWS, T.B. and GREENSPAN, D.S.

(1994). Type I procollagen COOH-terminal proteinase enhancer protein: Identification, primary structure, and chromosomal localisation of the cognate human gene (PCOLCE), *J. Biol. Chem.* **269**, 26280–26285.

TAMURA, M. and ISHIKAWA, O. (2001). An increase of mature type skin collagen cross-link, histidinohydroxylysinonorleucine, in the sclerotic skin of morphea, *J. Dermatol. Sci.* **25**, 83–86.

TANZER, M.L. (1976). Cross-linking. In: *Biochemistry of Collagen*, G.N. RAMACHANDRAN and A.H. REDDI, eds., Plenum Press, New York, pp. 137–162.

TANZER, M.L., HOUSLEY, T., BERUBE, L., FAIRWEATHER, R., FRANZBLAU, C. and GALLOP, P.M. (1973). Structure of two histidine-containing crosslinks from collagen, *J. Biol. Chem.* **248**, 393–402.

TASAB, M., BATTEN, M.R. and BULLEID, N.J. (2000). Hsp47: A molecular chaperone that interacts with and stabilises correctly-folded procollagen, *EMBO J.* **19**, 2204–2011.

THOMSON, C.A. and ANANTHANARAYANAN, V.S. (2000). Structure-function studies on hsp47: pH-dependent inhibition of collagen fibril formation *in vitro, Biochem. J.* **349(**Pt. 3), 877–883.

TIMPL, R. and CHU, M.L. (1997). Microfibrillar collagen type VI. In: *Extracellular Matrix Assembly and Structure*. P.D. YURCHENCO, D.E. BIRK and R.P. MECHAM, eds., Academic Press, San Diego, pp. 207–242.

TRUEB, B. and ODERMATT, B.F. (2000). Loss of type VI collagen in experimental and most spontaneous human fibrosarcomas, *Int. J. Cancer* **86**, 331–336.

TUCKWELL, D. (200). Identification and analysis of collagen alpha 1 (XXI), a novel member of the FACIT collagen family, *Matrix Biol.* **21**, 63–66.

UZAWA, K., GRZESIK, W.J., NISHIURA, T., KUZNETSOV, S.A., ROBEY, P.G., BRENNER, D.A. and YAMAUCHI, M. (1999). Differential expression of human lysyl hydroxylase genes, lysine hydroxylation, and cross-linking of type I collagen during osteoblastic differentiation *in vitro*, *J. Bone Miner. Res.* **14**, 1272–1280.

UZAWA, K., MARSHALL, M.K., KATZ, E.P., TANZAWA, H., YEOWELL, H.N. and YAMAUCHI, M. (1998). Altered post-translational modifications of collagen in keloid, *Biochem. Biophys. Res. Commun.* **249**, 652–655.

UZEL, M.I., SCOTT, I.C., BABAKHANLOU-CHASE, H., PALAMAKUMBURA, A.H., PAPPANO, W.N., HONG, H.H., GREENSPAN, D.S. and TRACKMAN, P.C. (2001). Multiple bone morphogenetic protein 1-related mammalian metalloproteinases process pro-lysyl oxidase at the correct physiological site and control lysyl oxidase activation in mouse embryo fibroblast cultures, *J. Biol. Chem.* **276**, 22537–22543.

VALTAVAARA, M., PAPPONEN, H., PIRTTILA, A.M., HILTUNEN, K., HELANDER, H. and MYLLYLA, R. (1997). Cloning and characterisation of a novel human lysyl hydroxylase isoform highly expressed in pancreas and muscle, *J. Biol. Chem.* **272**, 6831–6834.

VALTAVAARA, M., SZPIRER, C., SZPIRER, J. and MYLLYLA, R. (1998). Primary structure, tissue distribution, and chromosomal localisation of a novel isoform of lysyl hydroxylase (lysyl hydroxylase 3), *J. Biol. Chem.* **273**, 12881–12886.

VAN DER REST, M. and MAYNE, R. (1988). Type IX collagen proteoglycan from cartilage is covalently cross-linked to type II collagen, *J. Biol. Chem.* **263**, 1615–1618.

VEIS, A. and GEORGE, A. (1994). Fundamentals of interstitial collagen self-assembly. In: *Extracellular Matrix Assembly and Structure*, P.D. YURCHENCO, D.E. BIRK and R.P. MECHAM, eds., Academic Press, San Diego, pp. 15–45.

VUORIO, E. and DE CROMBRUGGHE, B. (1990). The family of collagen genes, *Ann. Rev. Biochem.* **59**, 837–872.

WALCHLI, C., KOCH, M., CHIQUET, M., ODERMATT, B.F. and TRUEB, B. (1994). Tissue-specific expression of the fibril-associated collagens XII and XIV, *J. Cell. Sci.* **107**, 669–681.

WOJTOWICZ, A., DZIEDZIC-GOCLAWSKA, A., KAMINSKI, A., STACHOWICZ, W., WOJTOWICZ, K., MARKS, S.C., Jr. and YAMAUCHI, M. (1997). Alteration of mineral crystallinity and collagen cross-linking of bones in osteopetrotic toothless (tl/tl) rats and their improvement after treatment with colony stimulating factor-1, *Bone* **20**, 127–132.

WU, C.H., WALTON, C.M. and WU, G.Y. (1991). Propeptide-mediated regulation of procollagen synthesis in IMR-90 human lung fibroblast cell cultures. Evidence for transcriptional control, *J. Biol. Chem.* **266**, 2983–2987.

YAMAMOTO, K. and YAMAUCHI, M. (1999). Characterisation of dermal type I collagen of C3H mouse at different stages of the hair cycle, *Br. J. Dermatol.* **141**, 667–675.

YAMAUCHI, M. (1995). Collagen: The major matrix molecule in mineralised tissues. In: *Calcium and Phosphorus in Health and Disease*, J.J.B. ANDERSON and S.C. GARNER, eds., CRC Press, Boca Raton, Florida, pp. 127–145.

YAMAUCHI, M., CHANDLER, G.S., TANZAWA, H. and KATZ, E.P. (1996). Cross-linking and the molecular packing of corneal collagen, *Biochem. Biophys. Res. Commun.* **219**, 311–315.

YAMAUCHI, M. and KATZ, E.P. (1993). The post-translational chemistry and molecular packing of mineralising tendon collagens, *Connect. Tissue. Res.* 29, 81-98.

YAMAUCHI, M., KATZ, E.P. and MECHANIC, G.L. (1986b). Intermolecular cross-linking and stereospecific molecular packing in type I collagen fibrils of the periodontal ligament, *Biochemistry* **25**, 4907–4913.

YAMAUCHI, M., KUBOKI, Y., SASAKI, S. and MECHANIC, G.L. (1986a). New pepsin-solubilized low molecular weight collagenous component possibly unique to periodontal ligament, *Biochemistry* **25**, 1997–2002.

YAMAUCHI, M., LONDON, R.E., GUENAT, C., HASHIMOTO, F. and MECHANIC, G.L. (1987). Structure and formation of a stable histidine-based trifunctional cross-link in skin collagen, *J. Biol. Chem.* **262**, 11428–11434.

YAMAUCHI, M. and MECHANIC, G.L. (1988). Cross-linking of collagen. In: *Collagen*, M.E. NIMNI, ed., CRC Press, Boca Raton, Florida, pp. 157–172.

YAMAUCHI, M., NOYES, C., KUBOKI, Y. and MECHANIC, G.L. (1982). Collagen structural microheterogeneity and a possible role for glycosylated hydroxylysine in type I collagen, *Proc. Natl. Acad. Sci. USA* **79**, 7684–7688.

YAMAUCHI, M., PRISAYANH, P., HAQUE, Z. and WOODLEY, D.T. (1991). Collagen cross-linking in sun-exposed and unexposed sites of aged human skin, *J. Invest. Dermatol.* **97**, 938–941.

YAMAUCHI, M. and SHIIBA, M. (2002). Lysine hydroxylation and cross-linking of collagen. In: *Post-translational Modifications of Proteins: Tools for Functional Proteomics*, C. KANNICHT, ed. Human Press, New Jersey, *Meth. Mol. Biol.* **194**, 277–290.

YAMAUCHI, M., WOODLEY, D.T. and MECHANIC, G.L. (1988b). Aging and cross-linking of skin collagen, *Biochem. Biophys. Res. Commun.* **152**, 898–903.

YANG, C.L., RUI, H., MOSLER, S., NOTBOHM, H., SAWARYN, A. and MULLER, P.K. (1993). Collagen II from articular cartilage and annulus fibrosus. Structural and functional implication of tissue specific posttranslational modifications of collagen molecules, *Eur. J. Biochem.* **213**, 1297–1302.

YEOWELL, H.N. (In press). Lysyl hydroxylases. In: *Encyclopedia of Molecular Medicine*, T. CREIGHTON, ed.

YEOWELL, H.N. and WALKER, L.C. (1999). Tissue specificity of a new splice form of the human lysyl hydroxylase 2 gene, *Matrix Biol.* **18**, 179–187.

YURCHENCO, P.D. and O'REAR, J.J. (1994). Basal lamina assembly, *Curr. Opin. Cell. Biol.* **6**, 674–681.4

SECTION VI:

FACTORS INFLUENCING GLAND TRANSPLANTATION

Advances in Tissue Banking Vol. 6

26 AUTOTRANSPLANTATION OF COMBINED ADRENAL GLANDS IN RATS

S.I. ISMAILOV, M.M. RASHITOV
and L.B. NUGMANOVA

Department of Endocrine Surgery
Institute of Endocrinology
Ministry of Health of Uzbekistan

1. Introduction

It is known that the treatment of such diseases as a Multiple Endocrine Neoplasia (MEN) type 2 with bilateral pheochromocytomas, ectopic ACTH-dependent Cushing's Disease (CD) requires performing of bilateral adrenalectomy (BA). However, one of threatening complications of this treatment is the inevitable development of chronic adrenal insufficiency, development of Nelson's syndrome in periods from 2–4 years until 25–30 years after operation, that forces the patient lifelong accepting of hormonal replacement therapy (HRT). (Commisarenko, 1988; Kemink-SA *et al.*, 1999; Klempa *et al.*, 1980; Till, 1997; Oki 1998).

One of perspective direction in the treatment of these patients is the transplantation of the adrenals after surgery (Artishevskiy, 1976; Okamoto, 1992; Kaukinen, 1999).

Research on the efficiency of autotransplantation of adrenals after BA in 6 patients suffering by CD, Erdogan *et al.* (1994) reported good functional activity of a graft. And in one patient the graft completely substituted the function of adrenals such that there was no necessity of carrying out HRT. However, completely contradictory data have been found in number of authors, who under the same conditions did not receive such encouraging results (Okamoto, 1992; Lucon, 1994; Till, 1997).

Because of such contradicting opinions about the autotransplantation of adrenal glands in clinical and experimental conditions, more detailed investigation of features of this complex process is necessary.

The purpose of present work was to study functional ability of native and cryopreserved adrenal autografts in bilateral adrenalectomised rats in short and postponed periods after transplantation.

2. Materials and Methods

In experiments white male rats were used weighing 120–150 g, keeping in vivariums with an inspected temperature schedule, standard diet and light mode from 8 AM till 20 PM. After operation all rats received saline solution (0.9% NaCl) as a drink according to the generally accepted method. Rats were assigned according to Table 1.

The amount of autograft consisted of one whole adrenal gland in Groups 3, 4 and two whole adrenal glands in Group 5.

Adrenalectomy was performed by lumbar access under introperitoneal injection of callipsol in dose of 6 Mk/kg. Adrenals were transplanted in the large omentum, by virtue of its rich blood supply.

The blood collection for hormonal investigation was carried out under the collipcol injection from heart cavities after thoracotomy. Peritoneal cavity and retroperitoneal space of slaughtered rats were inspected for definition of oddments of adrenal glands or additional adrenal tissue. Additional adrenal tissue in various degrees of development were found in 7 rats in Groups 2, 3, 5 in paranephral space. They were eliminated from further investigation.

Native and cryopreserved adrenal glands were utilised in grafting. Transplantation of native tissue of adrenals was carried immediately after its removal.

Table 1. Schedule of series and periods of experiment observation.

Series of experiments \ Periods of observations	Months after surgery			
	1	3	6	12
Controls	10 rats	10	10	10
Bilateral adrenalectomy (BA)	12	–	–	–
BA with immediate autotransplantation of one native adrenal gland	10	9	9	8
BA with postponed (24 hours) autotransplantation of one cryopreserved adrenal gland	9	9	9	7
BA with autotransplantation of both adrenal glands (native + cryopreserved)	10	9	9	8
Total	51	37	37	33

Tissue for cryopreservation was prepared in the operation room after removing of adrenals in rats. The whole gland was incubated in cryoprotective solution (12.5% of dimexidum in the Henks solution) for 20 min and then placed in special plastic containers. Capsulation of containers was reached by soldering. For simplification, a double-staged cryopreservation freezing agent, providing program cooling of samples, was utilised. One-stage-cooling utilised a speed 8–10°C/min up to –70°C; Two-stage-cooling used speed 80°C/min up to –196°C in liquid nitrogen. Defrosting of cryopreserved tissue was carried out in the operation room right before transplantation. The container was left in a water bath at temperature 37–38°C for 5 min.

2.1. Estimation of efficiency of a transplantation

A functional activity of adrenal autograft was estimated by the level of hormones — ACTH and corticosterone in a blood plasma,

1, 3, 6 and 12 months after a transplantation. Viability of graft was demonstrated by morphological and morphometric methods of investigation.

2.2. Statistics

The analysis of obtained data was carried out on the personal computer IBM on the "Statgraphics" programme (USA). The reliability of differences was determined according to a *t*-student criterion.

3. Results

BA in rats induced a clinical manifestations such as lowering of appetite, hypodynamia expressed dyspnea and lowering of weight. The indicated tags of a hypocorticoidism in the absence of adrenal autotransplantation, resulted to an acute adrenal insufficiency and lethal outcome, where 100% mortality of animals of the BA group was found during 14–21 days after the operation, and the larger number belonged to 4–7th days (Table 2).

The content of corticosterone in blood after BA at the 3rd day is reduced by 71%, and ACTH raised by 173% in relation to control values (Table 3). From the following table the data shifts are even more pronounced the 5th day after operation.

Mortality indices in rats of Groups 3, 4 and 5 sharply decreased. The survival rate after operations rose to 40–55%.

Table 2. Changes of mortality among rats exposed BA without autotransplantation.

Mortality date	Day of operation	1–3 days after surgery	4–7 days after surgery	7–14 days after surgery	14–21 days after surgery	Total
Amount of dead animals	0	5	19	6	3	33

Table 3. Plasma corticosterone and ACTH levels in rats undergone BA on the first week.

Values	Controls Intact rats (n = 10)	Sham operated rats (n = 10)	3rd day after surgery (n = 8)	5th day after surgery (n = 7)
Corticosterone (M ± m) nmol/l	81.8 ± 1.48	81.3 ± 1.95 $P > 0.05$	24.42 ± 0.98 $P < 0.05$	14.62 ± 0.82 $P < 0.05$
ACTH (M ± m) pg/ml	15.3 ± 0.65	15.5 ± 0.5 $P > 0.05$	41.63 ± 0.88 $P < 0.05$	54.0 ± 1.5 $P < 0.05$

The results of hormonal investigations of corticosterone and ACTH in serum of rats with different kinds of autotransplantation are reflected in Tables 4 and 5.

The observations have shown, that in 1 month after autotransplantation of one native adrenal gland, the concentration of corticosterone in serum is higher by 264% relatively to the BA group, but remains below control values at 38%. The ACTH concentration is reduced to 26% comparable to the BA group. Nevertheless they are higher than control values on 166%.

At 3, 6 and 12 months the state of animals is essentially not changed. The signs of a hypocorticoidism were found at the end of the experimental period.

During autotransplantation of one cryopreserved adrenal gland, the appearance of animals, the values of ACTH, and corticosterone were close to those in the group which had autotransplantation of one native adrenal gland (see Tables 4, 5).

The results of combined autotransplantation of adrenals (5th group) differ from those in the previous series. One month after autotransplantation, the animals became more dynamic, their appetite improved, and they reacted adequately to an external stimulus.

During this period of the research, the concentration of corticosterone in the blood increased to 75.8 ± 1.4 nmol/l ($P_1 < 0.05$) and the ACTH decreased to 23.7 ± 1.4 pg/ml, whereas control

Table 4. Changes of corticosterone (nmol/l) value in blood of BA rats with different kinds of autotransplantation.

Periods of observations / Series of experiments	Months after surgery: 1 (M ± m)	3 (M ± m)	6 (M ± m)	12 (M ± m)
Control group 1 (n = 10)	81.8 ± 1.48	82 ± 1.73	81 ± 1.85	82.1 ± 1.64
Bilateral adrenalectomy (BA) (n = 12)	14.62 ± 0.82‡	–	–	–
BA with immediate autotransplantation of one native adrenal gland	51 ± 1.1†† n = 10	39.8 ± 1.07†† n = 9	33.6 ± 1.05†† n = 9	30.75 ± 1.35†† n = 8
BA with postponed (24 hours) autotransplantation of one cryopreserved adrenal gland	38.25 ± 1.7†† n = 9	54.1 ± 1.43†† n = 9	60.5 ± 1.64†† n = 9	45.85 ± 1.76†† n = 7
BA with autotransplantation of both adrenal glands (native + cryopreserved)	75.8 ± 1.4†† n = 10 $P_1 < 0.05$	63.5 ± 4.2†† n = 9 $P_1 < 0.05$	71.8 ± 1.82†† n = 9 $P_1 < 0.05$	74.5 ± 2†† n = 8 $P_1 < 0.05$

*Not significant relative to control values ($P > 0.05$).
†Significant relative to control values ($P < 0.05$).
‡Corticosterone value on 5th day of BA.
P_1 — significance of Group 5's values relative to Groups 3, 4.

Table 5. Changes of ACTH (pg/ml) value in blood of BA rats with different kinds of autotransplantation.

Series of experiments \ Periods of observations	Months after surgery			
	1 (M ± m)	3 (M ± m)	6 (M ± m)	12 (M ± m)
Control group 1 (n = 10)	15.3 ± 0.65	15.1 ± 0.58	15.5 ± 0.54	15.6 ± 0.63
Bilateral adrenalectomy (BA) (n = 12)	54.0 ± 1.5‡	–	–	–
BA with immediate autotransplantation of one native adrenal gland	40.2 ± 0.64†† n = 10	36.3 ± 0.78†† n = 9	39.4 ± 0.58†† n = 9	38.1 ± 1.11†† n = 8
BA with postponed (24 hours) autotransplantation of one cryopreserved adrenal gland	33.7 ± 0.92†† n = 9	29.6 ± 1.17†† n = 9	31 ± 0.75†† n = 9	37.1 ± 1.7†† n = 7
BA with autotransplantation of both adrenal glands (native + cryopreserved)	23.7 ± 1.4†† n = 10 $P_1 < 0.05$	25.7 ± 1.3†† n = 9 $P_1 < 0.05$	28.7 ± 1.42†† n = 9 $P_1 < 0.05$	21.6 ± 0.73†† n = 8 $P_1 < 0.05$

*Not significant relative to control values ($P > 0.05$).
†Significant relative to control values ($P < 0.05$).
‡ACTH value on 5th day of BA.
P_1 — significance of Group 5's values relative to Groups 3, 4.

values were 81.8 ± 1.48 nmol/l and 15.3 ± 0.65 pg/ml ($P_2 < 0.05$), accordingly.

Three months after a combined transplantation, the adrenals of rats became even more active. The appetite of the animals improved and they began to add weight. Over this period the concentration of corticosterone and ACTH in the serum was 63.5 ± 4.2 nmol/l and 25.7 ± 1.3 pg/ml, respectively ($P_1 < 0.05$; $P_2 < 0.05$).

Six months after operation obvious features of hypocorticoidism were not found. The level of corticosterone increased up to 71.8 ± 1.82 nmol/l (407% above from BA values), and the ACTH dropped to 28.7 ± 1.42 pg/ml (49% below from BA values). However, these values did not reach control values and significantly differed from before (when corticosterone was below on –13%, ACTH was higher on 86%).

By the end of year, no signs of severe adrenal insufficiency were observed. The remarkable rise of a corticosterone and lowering ACTH in blood on 428% and 62% (74.5 ± 2 nmol/l and 21.6 ± 0.73 pg/ml accordingly) is revealed in comparison with BA animals, for who levels of control values did not reach ($P < 0.05$).

4. Discussion

It is known, that endocrine organs can be replaced without surgical restoration of vascular links, that is the so called "free" grafting (Klempa, 1980; Abobakr, 1994; Islam, 1995). The nutritional supply of such grafts is achieved by simple contact with the plasmatic environment of the recipient, by diffusion and osmosis. Worse fed central sites can lead to necrosis. The medullary substance of adrenals degenerates rapidly after transplantation, while the cortical part, undergoes initial degenerative changes, but soon begins to regenerate from the cells of subcapsular layer. Regeneration of the adrenal cortex finishes by the end of the month after autotransplantation (Spanal, 1993; Vendeira, 1992; Kaukinen, 1999; Leong, 1999). Our previous investigations confirmed the data of these authors. The morphological study of autografts removed from rats (native and cryopreserved) has shown the absence of differences in the square of cells and the size of nuclei of the various zones of the adrenal cortex, which testifies to the safety of grafted tissue of the adrenal cortex in the recipient. Thus, it is markedly higher density of cells in cryopreserved adrenal graft which is the best indication of safety of a tissue. The morphologic features of native and cryopreserved adrenal autograft 1 month after transplantation almost correspond to intact glands. Only the medullar zone is completely destroyed. The restoring of morphologic frame of adrenal autograft after BA is due to underlying capsule cells.

After detailed investigation of adrenal tissue transplantation Spanal *et al.* (1993) concluded that following transplantation in the adult rats after BA, there is an increase in survival rate up to 30% (6 of 20 rats survived). In our series, survival rate after BA irrespective of sort of a transplantation was even more, i.e. 40–55%, in comparison with BA rats where the mortality was 100%.

Okamoto *et al.* (1992) studied the influence of the "quantitative" and "quality" factors on the outcome of adrenal autotransplantation. They transplanted 1/4, 1 and 2 adrenal glands after BA in rats. The results showed that after grafting of 1/4 of a adrenal gland provided normal production of corticosterone in blood, which corresponded to those in the intact group. But the response of production after stress was at a very low level. The transplantation of two adrenals did not make rats better able to withstand stress than at transplantation of one adrenal. Our data on a level of corticosterone somewhat contradicted to Okamoto *et al.*'s (1992). In the rats underwent BA with autotrans-plantation of only one (either native or cryopreserved adrenal gland) expressed signs of hypocorticoidism were retained. However, the combined auto-transplantation of two adrenals rats did not differ significantly from the intact ones.

Thus, from our results, it is possible to confium the safety of the main frames of adrenal cortex after "free" autografting. The most effective results were obtained with a series of combined adrenal autotransplantation. The most important factor in the success of autotransplantation is the amount of replaced tissue, i.e. the more adrenal tissue replaced, the more pronounced clinical effect. It was also shown that in the process of cryopreservation, the adrenal cortex is well preserved and can be utilised for the treatment of hypocorticoid states.

5. Conclusions

From our results, the following conclusions are decluced:

1. Using of a model of a hypocorticoidism in white male rats, created by BA, it has been shown that after adrenal autotransplantation (native or cryopreserved or combined), the survival rate of animals reaches 40–55%, whereas without it, the lethal outcome is 100%.
2. Autotransplantation of one adrenal gland (native or cryopreserved) causes an improvement of clinical states of the animals, increases the corticosterone content in blood by 3 times and lowers ACTH by 1.5 times relative to the values of BA animals.
3. Combined autotransplantation of two adrenals results in the best clinical effect. The tendency to normalisation of corticosterone rises by 5 times and ACTH was lowered by 2 times relative to values of BA animals.

6. References

ABOBAKR, A.M. (1994). Free and microencapsulated adrenal cortical cells produce similar cortisol responses when stimulated by ACTH: An *in vitro* study, *Int. J. Artif. Organs* **17**(3), 171–174.

ARTISHEVSKIY, A.A. (1976). Results of cryopreserved fetal adrenal gland transplantation in experiment. Transplantation of organs and tissues, *Publ. Medicina Rostov-na-Donu*, 317.

COMMISARENKO, I.V. (1988). Transplantation of adrenals. Reveiw, *J. Vrachebnoe delo* **11**, 29–32.

ERDOGAN, G., KOLOGLU, S. and KAMEL, N. (1994). Adrenal autotransplantation after total adrenalectomy: Delayed determined function, *Endocr. J.* **41**(1), 45–48.

ISLAM, A., HENRIKSSON, B. and MOHAMMED, A. (1995). Behavioural deficits in adult rats following long-term adrenalectomy, *Neurosci. Lett.* **94**(1–2), 49–52.

KAUKINEN, K., COILLIN, P. and MYKKANEN, A.H. (1999). Celiac disease and autoimmune endocrinologic disorders. *Dig. Dis. Sci.* **44**(7), 1428–1433.

KEMINK, S.A. (1999). Progression of a Nelson's adenoma to pituitary carcinoma: A case report and review of the literature, *J. Endocrinol.* **22**(1), 70–75.

KLEMPA, I. and SCHWEDES, U. (1980). Adrenal cortex transplantation. Animal experimental models for the human Cushing and Conn syndromes — Clinical aspects of adrenal cortex transplantation, *Steinau* **51**(10), 634–643.

LEONG, K.S. *et al.* (1999). Clinical presentation of thyroid dysfunction and Addison's disease in young adults with type 1 diabetes, *Postgrad. Med. J.* **75**(886), 467–470.

LUCON, A.M., MENDONCA, B.B. and DOMENICE, S. (1993). Adrenal autografts following bilateral adrenalectomy, *J. Urol.* **149**(5), 977–979.

OKAMOTO, T. (1992). Experimental study on adrenal autografts in rats to preserve normal adrenocortical function after bilateral adrenalectomy, *Eur. Surg. Res.* **24**(2), 112–118.

OKI, Y., IWABUCHI, M. and MASUZAWA, M. (1998). Distribution and concentration of urocortin, and effect of adrenalectomy on its content in rat hypothalamus, *Life Sci.* **62**(9), 807–812.

SARRIA, R. (1994). Analysis of the pituitary-thyroid axis in bilaterally adrenalectomized or adrenal transplanted rats, *Donate Oliver F. Anat. Histol. Embryol.* **23**(3), 257–268.

SPANAL, M. (1993). Morphological and physiological findings of fetal adrenal grafts in rats who underwent bilateral staged adrenalectomy. *Eur. J. Pediatr. Surg.* **3**(6), 339–342.

TILL, H., KELLNER, S. and BOHM, R. (1997). Fetal adrenal transplantation: Success of a laparoscopic technique in rats, *J. Pediatr. Surg.* **32**(10), 1455–1457.

TILL, H., KELLNER, S. and STRASSBURGER, C.J. (1998). Fetal adrenal transplants respond to ACTH and prevent addisonian crisis in adrenalectomized rats, *Pediatr. Surg. Int.* **13**(4), 271–273.

VENDEIRA, P. and MAGALHAES, M.M. (1992). Autotransplantation of the adrenal cortex: A morphological and autoradiographic study, *Anat. Rec.* **232**(2), 262–272.

VENDEIRA, P. and NEVES, D. (1996). Modulation of autotransplanted adrenal gland by endothelin-1: A morphological and biochemical study, *Anat. Rec.* **246**(1), 98–106.

Advances in Tissue Banking Vol. 6

27 ACTIVITY OF PITUITARY-THYROID AXIS IN DEPENDS ON SEVERITY OF HYPOCORTICOIDISM IN RATS

S.I. ISMAILOV, M.M. RASHITOV
and L.B. NUGMANOVA

Department of Endocrine Surgery
Institute of Endocrinology
Ministry of Health of Uzbekistan

1. Introduction

It is generally accepted that a close functional inter-relation between adrenal and thyroid glands exists. It is considered that these glands are part of a complex functional inter-regulatory neuroendocrine system providing hormonal homeostasis of the organ (Aversa, 1998; Gerhard, 1999; Kaukinen, 1999; Leong, 1999).

The study of the inter-relationship of adrenals and thyroid gland (T) assists in our understanding of the pathogenesis of diseases connected with infection and to select a rational therapy (Addison, 1855; Henkin, 1970; Gozariu, 1988). Most papers on this problem are based on clinical observation. However, clinical observation of the endocrine organ's function cannot always give an objective answer to this interested question. The age, sex, duration of disease, emotional excitability of the patient, the influence of the various factors of the external environment can remarkably influence the end result. Morphological investigation of the clinical conditions can almost never be carried out, unless the patient has undergone an operation. Moreover, when surgical treatment for adrenal disease is concerned, no one will remove the thyroid to investigate it. For all of these reasons, rather contrary views have been expressed about the nature of the inter-relationship of glands (Fang, 1981; Islam, 1995; Sarria 1994; Shibata 1999).

Thus, the purpose of the present work is to achieve a dynamic study of the functional indices of the pituitary-thyroid axis in animals, depending on severity of hypocorticoidism.

2. Materials and Methods

White male rats weighing 120–150 g were used and kept in vivariums with an controlled temperature schedule, with standard diet and light mode for 8 mornings and 20 evenings. After operation all rats received saline solution (0.9% NaCl) as a drink according to the generally accepted method. Rats were assigned according to Table 1.

The amount of autograft consisted of one whole adrenal gland in Group 3 and two whole adrenal glands in Group 4.

Table 1. Schedule of series and periods of experiment observation.

Periods of observations / Series of experiments	Months after surgery			
	1	3	6	12
Controls	10 rats	10	10	10
Bilateral adrenalectomy (BA)	12	–	–	–
• BA with immediate autotransplantation of one native adrenal gland	10	9	9	8
• BA with postponed (24 hours) autotransplantation of one cryopreserved adrenal gland	9	9	9	7
BA with autotransplantation of both adrenal glands (native + cryopreserved)	10	9	9	8
Total	51	37	37	33

Adrenalectomy was performed by lumbar access under intro-peritoneal injection of callipsol with a dose of 6 Mk/kg. Adrenals were transplanted in the large omentum because of its rich blood supply.

The blood collection for hormonal investigation was carried out by collipcol injection from heart cavities after thoracotomy. The peritoneal cavity and retroperitoneal space of the sacrificed rats were inspected for definition of oddments of adrenal glands or additional adrenal tissue. Additional adrenal tissue, at various degrees of development were found in 7 rats in Groups 2–4 in the paranephral space. They were eliminated from the further investigation.

Native and cryopreserved adrenal glands were utilised in grafting. Transplantation of native tissue of adrenals was carried out immediately after its removal.

Tissue for cryopreservation was prepared in the operation room after removal of adrenals in rats. Whole gland was incubated in cryoprotective solution (12.5% of dimexidum in the Henks solution) for 20 min, and then placed in special plastic containers. Capsulation of containers was achieved by soldering. A double-staged cryo-preservation freezing agent, providing program cooling of samples, was utilised. First-stage-cooling with speed 8–10°C/min up to −70°C; Second-stage-cooling with speed 80°C/min up to −196°C in liquid nitrogen. Defrosting of cryopreserved tissue was carried out in the operation room immediately before transplantation. The container was kept in a water bath at temperature 37–38°C for 5 min.

2.1. Hormonal assay

TSH, T3 and T4 of blood were investigated for evaluation of functional activity of pituitary-thyroid axis (PTA). Changes were confirmed by morphometric analysis of thyroid gland and pituitary tissues. A degree of severity of hypocorticoidism in rats was defined by clinical manifestation of the symptoms, estimating ACTH and corticosterone levels in blood after 1, 3, 6 and 12 months autotrans-plantation of adrenal glands. Hormones were determined by RIA kits supplied by the Immunothec company.

2.2. Statistics

The analysis of obtained data was carried out on the personal computer IBM on the "Statgraphics" programme (USA). The reliability of differences was determined according to a *t*-student criterion.

3. Results

In our early work on the creation of a model of hypocorticoidism (H) and the evaluation of the efficiency of the various forms of autotransplantation, it was found that bilateral adrenalectomy in rats without consequent adrenal autotransplantation caused severe expressed hypocorticoidism which reached its peak moment on the 5–6th day of operation, resulting in an inevitable lethal outcome. Autotransplantation of one adrenal gland (either native or cryopreserved) after total adrenalectomy markedly restored corticosterone synthesis and secretion, sufficient for maintenance of life. However, the latter was at a very low level, which was the reason for saving hypocorticoidism in animals to the end of the experimental period. A combined autotransplantation of both adrenals after BA resulted in the most positive effect, i.e. corticosterone and ACTH values were close to normal level. Clinical signs of H were insignificant.

On this basis, for evaluation of functional features of PTA in the animals with experimental H, experimental series were grouped according to the degree of H severity: Group 1 — control (intact, sham operated), Group 2 — animals with critical or decompensated degree of H (BA), Group 3 — animals with severe degree of H (BA with autotransplantation of one adrenal gland either native or cryopreserved), Group 4 — animals with a mild degree H (BA with combined autotransplantation of two adrenals).

3.1. Control group of experimental animals

This group consists of intact and sham operated rats. Corticosterone and ACTH values among these series did not differ from each other

Table 2. Changes of TSH (mM/l) value in blood of BA rats with different kinds of autotransplantation.

Periods of observations / Series of experiments	Months after surgery			
	1 (M ± m)	3 (M ± m)	6 (M ± m)	12 (M ± m)
Control 1 ($n = 10$)	2.19 ± 0.023	2.22 ± 0.029	2.34 ± 0.03	2.41 ± 0.027
Bilateral adrenalectomy (BA) ($n = 7$)	3.3[‡] ± 0.08[††]	–	–	–
• BA with immediate autotransplantation of one native adrenal gland	3.44 ± 0.06[††] $n = 10$	3.28 ± 0.035[††] $n = 9$	3.55 ± 0.05[††] $n = 9$	3.7 ± 0.026[††] $n = 8$
• BA with postponed (24 hours) autotransplantation of one cryopreserved adrenal gland	3.37 ± 0.031[††] $n = 8$	3.45 ± 0.032[††] $n = 8$	3.55 ± 0.06[††] $n = 8$	3.7 ± 0.1[††] $n = 7$
BA with autotransplantation of both adrenal glands (native + cryopreserved)	3.16 ± 0.04[††] $n = 9$ $P_1 < 0.05$	2.9 ± 0.08[††] $n = 9$ $P_1 < 0.05$	2.9 ± 0.08[††] $n = 9$ $P_1 < 0.05$	3.15 ± 0.042[††] $n = 8$ $P_1 < 0.05$

*Not significant relative to control values ($P > 0.05$).
[†]Significant relative to control values ($P < 0.05$).
[‡]TSH value on 5th day of BA.
P_1 — significance of 4th group values relative to 3 other groups.

Table 3. Changes of T3 (nmol/l) value in blood of BA rats with different kinds of autotransplantation.

Periods of observations / Series of experiments	Months after surgery			
	1 (M ± m)	3 (M ± m)	6 (M ± m)	12 (M ± m)
Control 1 (n = 10)	0.76 ± 0.056	0.69 ± 0.067	0.71 ± 0.06	0.63 ± 0.068
Bilateral adrenalectomy (BA) (n = 7)	$1.41^{\ddagger} \pm 0.06^{\dagger\dagger}$	–	–	–
• BA with immediate autotransplantation of one native adrenal gland	$1.81 \pm 0.09^{\dagger\dagger}$ $n = 10$	$1.67 \pm 0.1^{\dagger\dagger}$ $n = 9$	$1.83 \pm 0.097^{\dagger\dagger}$ $n = 9$	$2.68 \pm 0.063^{\dagger\dagger}$ $n = 8$
• BA with postponed (24 hours) autotransplantation of one cryopreserved adrenal gland	$1.6 \pm 0.046^{\dagger\dagger}$ $n = 8$	$2 \pm 0.068^{\dagger\dagger}$ $n = 8$	$1.98 \pm 0.095^{\dagger\dagger}$ $n = 8$	$2.21 \pm 0.1^{\dagger\dagger}$ $n = 7$
BA with autotransplantation of both adrenal glands (native + cryopreserved)	$1.15 \pm 0.09^{\dagger\dagger}$ $n = 9$	$1.4 \pm 0.07^{\dagger\dagger}$ $n = 9$	$1.21 \pm 0.1^{\dagger\dagger}$ $n = 8$	$1.18 \pm 0.044^{\dagger\dagger}$ $n = 8$

*Not significant relative to control values ($P > 0.05$).

†Significant relative to control values ($P < 0.05$).

‡T3 value on 5th day of BA.

P_1 — significance of 4th group values relative to 3 other groups.

Table 4. Changes of T4 (nmol/l) value in blood of BA rats with different kinds of autotransplantation.

Periods of observations / Series of experiments	Months after surgery			
	1 (M ± m)	3 (M ± m)	6 (M ± m)	12 (M ± m)
Control 1 (n = 10)	49.4 ± 1.1	50.9 ± 0.7	48.9 ± 1.0	49 ± 1.2
Bilateral adrenalectomy (BA) (n = 7)	94.4‡ ± 1.5††	–	–	–
• BA with immediate autotransplantation of one native adrenal gland	96.2 ± 1.0†† n = 10	89 ± 1.6†† n = 9	95 ± 0.6†† n = 9	94 ± 0.9†† n = 8
• BA with postponed (24 hours) autotransplantation of one cryopreserved adrenal gland	93 ± 1.2†† n = 8	86 ± 1.0†† n = 8	89 ± 1.0†† n = 8	100.2 ± 1.4†† n = 7
BA with autotransplantation of both adrenal glands (native + cryopreserved)	79.6 ± 1.1†† n = 9	75.4 ± 0.9†† n = 9	80.6 ± 1.4†† n = 9	71 ± 1.2†† n = 8

*Not significant relative to control values ($P > 0.05$).
†Significant relative to control values ($P < 0.05$).
‡T4 value on 5th day of BA.
P_1 — significance of 4th group values relative to 3 other groups.

(81.8 ± 1.48 nmol/l and 15.3 ± 0.65 pg/ml, respectively). The data of TSH, T3 and T4 in blood after statistical processing showed that there was no significant difference either, so they were taken as control indices. The results of hormonal assay of TSH, T3 and T4 are given in Tables 2–4, accordingly.

3.2. Activity PTA in rats with critical or decompensated degree of H

The term of critical degree of H refers to a group of animals exposed to BA without consequent adrenal autotransplantation. In our previous work, we found that the animals in the given series of experiment died during average 4–7 days after the initial operation. The corticosterone and ACTH assay reveal a critical level of 14.62 ± 0.82 nmol/l and 54.0 ± 1.5 pg/ml, respectively.

It was shown that on the 5th day after operation TSH concentration in blood significantly increases (up to 3.3 ± 0.08 mM/l vs. 2.19 ± 0.023 mm/π, $P < 0.05$), which causes superactivity of T. The concentration T3, T4 rises up to 1.41 ± 0.06 nmol/l and 94.4 ± 1.5 nmol/l accordingly against control values 0.76 ± 0.056 nmol/l and 49.4 ± 1.1 nmol/l (Tables 2–4).

Thus, the data indicates that BA results in PTA activation, which is positively expressed even over a weekend.

3.3. Activity PTA in rats with severe degree of H

According to clinical features and hormonal assay in this series, some secretion of glucocorticoids is noticed which increases the viability of the rats. Nevertheless, signs of hypocorticoidism were retained (delicacy, hypodynamia, loss of weight etc.), despite adrenal cortex functional activity indices increasing in relation to the second group which had critical adrenal insufficiency ($P < 0.05$). Nevertheless these were far from control values (corticosterone — 45.85 ± 1.76 nmol/l, ACTH — 33.7 ± 0.92 pg/ml).

On this basis, we identified those animals with severe H and attempted to evaluate PTA over the study period.

Table 2 shows that in 1 month after operation of this group, TSH level rises to 54.5% from the control value. Therefore, the concentrations of T3 and T4 increased 123% and 69%, accordingly. In longer periods TSH, T3 and T4 concentrations tend to rise over a year by 68%, 160%, 84% accordingly ($P < 0.05$).

3.4. Activity PTA in rats with mild degree of H

The results show the remarkable improvement in clinical signs of adrenal insufficiency levels of hormones (corticosterone — 71.8 ± 1.85 nmol/l, ACTH — 23.7 ± 1.4 pg/ml). However, in spite of the improvement of the conditions of rats in this series, glucocorticoids and ACTH did not reach normal values and mild degree of adrenal insufficiency continued.

For this group, after one month of surgery, a lower activation of PTA occurred. Tables 2–4 show that TSH rises by 44% and T3, T4 by 59% and 61% in relation to control respectively. However, in comparison with the group of rats with severe degree of H TSH T3 and T4 concentration was noticeably reduced by 18%, 21% and 18%, respectively.

Analysis of dynamics of hormones over one year showed that there was no statistically significant variation in the short and longer periods (Tables 2–4).

4. Discussion

Over the past 40 years, researchers have made numerous attempts to investigate the nature of adrenal-thyroid glands inter-relationships. A great deal of attention was given to the way one gland reacts after a rise or lowering of activity of another gland (Malendowicz, 1975; Shibata 1999; Takeda, 1999).

Henkin *et al.* (1970) on the basis of their research and the literature information conclude that there is antagonism between a thyroid gland and adrenals, but the possibility of some synergy is not excluded. Malendowicz (1975) specifies a depression of T function and weight after injection of corticosterone and desoxicortisone-acetat. Other authors report opposite results (Troncone, 1999;

Valueva, 1976; Fang *et al.*, 1991) and even others deny any kind of action of steroids on T status.

Valueva (1976) shows that the mutual relation of adrenals and T does not have an opposing effect. However, in conclusion, she concedes that there is no clear quantitative and qualitative change in the biosynthesis of thyroid hormones under influence of corticosteroids, which require extensive experimental research.

We have studied the dependence of PTA activity on various functional activities of the adrenals. Four groups of rats were selected, with clinically and hormonally chronic adrenal insufficiency with various degree of severity.

The hormonal assay of TSH, T3 and T4 in plasma showed that there is an observable correlative between functional activity of adrenals and thyroid gland in hypocorticoidism.

There was a marked activation of PTA in rats with severe H (3 groups). TSH, T3 and T4 concentration increased by 54%, 123%, 69%, respectively in relation to control values. The activation continued to the end of the experimental period.

Statistical processing of obtained data for the 4th group of rats with a mild degree of H showed that activation of PTA also is marked until the end of experiment. TSH, T3 and T4 concentrations rise from the control to 44%, 59%, 61%, respectively. However, compared to rats with severe degree of H, these values were significantly lower at 18%, 21.5%, 18%, respectively ($P < 0.05$).

The morphological observations confirmed the increased functional activity of thyrotroph cells of pituitary and thyroid gland follicular cell over the series.

It is necessary to emphasise that in evaluating these values in groups with the compensated adrenal insufficiency, we detected a dependency of TSH, T3 and T4 concentrations at a corticosterone level ($r = -0.84$). As more H is expressed, there is greater activation of PTA with a higher content of TSH and thyroid hormones found.

However, when the dynamic changes of TSH and thyroid hormones were measured in rats of Group 2 with critical or decompensated adrenal insufficiency, such a correlation link was absent. Here, the expressed H did not yield greater activation of PTA as expected before. TSH, T3 and T4 concentration were not statistically significant varied from the group with severe H, although these rats demonstrated a noticeable reduced level of corticosterone.

Thus, it can be concluded that there is some antagonism between adrenals and T which yield to correlation analysis up to a definite level.

The rise of TSH secretion and therefore, T3, T4 in H with all outgoing clinical effects is probably a compensatory response of the organ to developing chronic adrenal insufficiency and originally compensates it. However, because of more penetrating violation with consequent reduction in chronic adrenal insufficiency, the compensatory effects of PTA are probably lost due to deep metabolic changes. It has given rise to a vicious circle. For this reason, asynchronisation of pituitary-thyroid-adrenal system indices occurs.

5. Conclusions

1. Chronic adrenal insufficiency results PTA activation, which results in a rise of TSH, T3 and T4 concentrations in blood.
2. The degree of PTA activity depends inversely on the severity of any compensated chronic adrenal insufficiency. That is greater adrenal insufficiency causes more functional activity of PTA.
3. In uncompensated adrenal insufficiency, the correlation dependence of adrenal and thyroid glands functional activities is lost and has a asynchronous character.

6. References

ADDISON, T. (1855). *On the Constitutional and Local Effects of Disease of the Suprarenal Capsules*. Highley, London.

AVERSA, A., PALLESCHI, S. and CRUCCU, G. (1998). Rapid decline of fertility in a case of adrenoleukodystrophy, *Hum. Reprod.* **13**(9), 2474–2479.

FANG, V.S. *et al.* (1981). Adrenal influence on pituitary secretion of thyrotropin and prolactin in rats, *J. Endocrinol.* **108**(4), 1545–1551.

GERHARD, I., FRICK, A. and MONGA, B. (1999). Pentachlorophenol exposure in women with gynecological and endocrine dysfunction, *Environ. Res.* **80**(4), 383–388.

GOZARIU, L., ORBAI, P. and BARABAS E. (1988). Changes in calcitonin secretion and in serum magnesium following adrenalectomy in rats, *J. Endocrinol.* **26**(1), 21–26.

HENKIN, V.L. (1970). Vliaynie adrenalectomii i posleduyushey peresadki nadpochechnikov na funksiu shitovidnoy jelezi v eksperimente, *J. Klinicheskaya Chirurgia.* **9**, 13–18.

ISLAM, A., HENRIKSSON, B. and MOHAMMED, A. (1995). Behavioural deficits in adult rats following long-term adrenalectomy, *Neurosci. Lett.* **194**(1–2), 49–52.

KAUKINEN, K., COLLIN, P. and MYKKANEN, A.H. (1999). Celiac disease and autoimmune endocrinologic disorders, *Dig. Dis. Sci.* **44**(7), 1428–1433.

LEONG, K.S. (1999). Clinical presentation of thyroid dysfunction and Addison's disease in young adults with type 1 diabetes, *Postgrad. Med. J.* **75**(886), 467–470.

MALENDOWICZ, L.K. and FILIPIAK, B. (1975). The effects of adrenalectomy and hydrocortisone replacement on the thyriod of the adult male rat I. Morphometrical data and histochemistry of some oxidative enzymes, *Endokrinologie* **64**(2), 223–231.

SARRIA, R. and LOSADA, J. (1994). Analysis of the pituitary-thyroid axis in bilaterally adrenalectomized or adrenal transplanted rats, *J. Anat. Histol. Embryol.* **23**(3), 257–268.

SHIBATA, T., OEDA, T. and SAITO, Y. (1999). Severe hyponatremia caused by hypothalamic adrenal insufficiency, *Int. Med.* **38**(5), 42–32.

TAKEDA, R. (1999). Schmidt's syndrome: autoimmune polyglandular disease of the adrenal and thyroid glands, *Isr. Med. Assoc. J.* **1**(4), 285–286.

TRONCONE, R. (1999). Unusual association of thyroiditis, Addison's disease, ovarian failure and celiac disease in a young woman, *J. Endocrinol. Invest.* **22**(5), 390–394.

VALUEVA, G.V. (1976). Age and features of the functional activity of the thyroid gland in experimental hypocorticism, *Probl. Endokrinol. (Mosk.)* **22**(1), 59–64.

INDEX

www.ingramcontent.com/pod-product-compliance
Lightning Source LLC
La Vergne TN
LVHW051057210926
848486LV00002B/140
* 9 7 8 9 8 1 2 3 8 0 8 8 3 *